FIFTY YEARS WITH SCIENCE

BY THE SAME AUTHOR

Science in Modern Society
The Social Relations of Science

Francis Bacon: The First Statesman of Science
Founders of British Science
Scientists of the Industrial Revolution
British Scientists of the Nineteenth Century
British Scientists of the Twentieth Century
Statesmen of Science
Scientific Types

Famous American Men of Science
Soviet Science
Discoveries and Inventions of the Twentieth Century
A Short History of Science
Science Unfolds the Future
The Sciences of Energy
Nuclear Energy in Industry
Radioastronomy and Radar
Science in Liberated Europe
Industry and Education in Soviet Russia
Science in Soviet Russia
An Outline of the Universe
The Progress of Science
The Science of Petroleum
The Young Man's Guide to Civil Engineering
Electricity
The A.B.C. of Chemistry
The Story of Agriculture
Six Great Scientists
Six Great Inventors
Six Great Engineers
Six Great Doctors
Six Great Astronomers
Osiris and the Atom
Short Stories in Science
Science for You
Science and Life

FIFTY YEARS WITH SCIENCE

J. G. Crowther

BARRIE & JENKINS
LONDON

© 1970 by J. G. Crowther

*First published 1970
by Barrie & Jenkins
(Barrie Books Ltd.)
2 Clement's Inn, London WC2*

SBN 248 65220 6

*Printed in Great Britain
by Ebenezer Baylis and Son Limited
The Trinity Press, Worcester, and London*

CONTENTS

		PREFACE	7
1	1918	Assisting in Research	9
2	1919–23	Teaching Mathematics and Science	19
3	1924–26	Technical and Scientific Publishing	28
4	1927–28	Inventing Scientific Journalism	41
5	1929	Science in Soviet Russia	52
6	1930	Berlin	60
7	1930	Producing the Engineers for the First Five Year Plan	70
8	1931	Social Crisis and Scientific Inspiration	76
9	1932	The Neutron	89
10	1932	The Top of the Wave	101
11	1933	The Wave Rolls On, Above the Crises	115
12	1933	German Science and Culture in Exile	123
13	1934	Politics Change Scientific Circumstances	130
14	1935	Realism in Science and Politics	147
15	1936	Scientific Perspectives Modified by Historical Conditions	158
16	1937	Harvard and Paris	171
17	1938	Near the End of an Epoch	191
18	1939–40	Frustration	204
19	1941–42	Science from the Administrative Aspect	223
20	1941–42	Individual Work and Institutional Influence	236
21	1942–46	Social Needs of Scientists	240
22	1944–46	Reviving Scientific Relations	261
23	1947	International Science, Science Writing, and Poland	271
24	1947	Homage to Rutherford, and Conference in Mexico	284

25	1947	Conservation of Resources: New York and Washington	294
26	1948	Science in British Occupied Germany	300
27	1948	A Redistribution of Effort	307
28	1949	London, Paris, Milan and Moscow	313
29	1950–54	Foreshadowings	318
30	1955–57	The Post-War Epoch Ended	321
31	1958–59	Planned Scientific Writing	326
32	1960–68	Completing the Account	333
		Appendix (British Council in 1945)	337
		INDEX	339

PREFACE

I HAVE BEEN with science, though not in it, for fifty years. My first adult connection with it was as an assistant in research during the First World War. After this, I was engaged in teaching for five years, and then, in 1924, in the publishing of technical and scientific books. I was drawn, quite unexpectedly, into writing on science, and in 1928 was appointed the first Scientific Correspondent of the *Manchester Guardian*. By 1930 scientific writing had become my main occupation.

As a scientific journalist I visited Germany during the Weimar Republic, the U.S.S.R. at the beginning of the First Five Year Plan and on other occasions; the United States before and after the Second World War, France during the Popular Front, the Second World War, and after; and many other countries. I reported the scientific developments that I saw, and became acquainted with leading scientists, and also Bauhaus architects and artists.

The rise of Nazism and the Second World War made these connections more important, for they had provided knowledge of the forthcoming scientific and technological power of the U.S.S.R., and enabled me to help in welcoming scientists and others who had had had to leave Germany, and after the outbreak of the war, the occupied countries of Europe.

These events, and the personalities and knowledge with which they acquainted me, sharpened my perception of the importance of the social and political relations of science, both in national and international affairs. I began to write on science from that point of view.

The threat and the outbreak of the Second World War caused me to join with many others in attempts to strengthen the national and international relations of science, as defences against reaction and Nazism.

During the Second World War I had the task of developing a Science Department in the British Council. This gave me administrative experience in the national and international fields of science. I took steps that led to the foundation of the Society for Visiting Scientists. I initiated the foundation of the Association of British Science Writers, and was its first Chairman. I was involved in the activities which ultimately led to the creation of UNESCO and the World Federation of Scientific Workers, the latter of which I became Secretary-General.

In this book I have recounted these activities. They presented many of the problems that arise in dealing with science and scientists as factors and actors in social life.

I began to consider more concretely what the social role of scientists should be, and how their social awareness might be extended and deepened. How should they be brought together on the social and political aspects of scientific matters? How could the public understanding of science be increased? What should be the relations between private and state scientific organizations?

I have described how these and related problems confronted me, how I attempted to deal with them, and how I tried in my writings to apply the knowledge I had acquired to the interpretation of the history and development of science, and the role of scientists in society.

I hope that my account may add to the understanding of the problems of science and scientists in modern society, and the kind of difficulties that have to be overcome in solving them.

I am conscious of the great help I have received from many, and I wish to express my gratitude to all, especially to the scientists, in various countries, who have given me information and encouragement, to my literary agent, and to those who have published my writings.

J. G. CROWTHER

CHAPTER ONE

1918
ASSISTING IN RESEARCH

IN DECEMBER 1917, during the First World War, I sat for a scholarship at Trinity College, Cambridge, and was awarded an exhibition in mathematics and physics. The English essay paper had contained a question on the history of science, which I had answered at length. After the examination Dr Montagu Butler, the Master of Trinity, wrote to my teachers at Bradford Grammar School that my papers on mathematics and physics were quite good, and he himself had read my English essay, which was 'very creditable for a science man'. Dr Butler died not long afterwards, and I never met him.

Following the award of the exhibition, I was put in touch with official sources in order to find a scientific job in the war effort. I was taken on by the Anti-Aircraft Experimental Section, Ministry of Munitions Inventions Department, under A. V. Hill. This was engaged in research on anti-aircraft gunnery, making use of an invention by Hill for discovering the speed of air at any height. A group of about a dozen scientists and mathematicians worked on the theoretical and practical problems, with headquarters in H.M.S. *Excellent*, the name of the Royal Navy's Gunnery School on Whale Island in Portsmouth Harbour.

The leader of the group at Portsmouth was R. H. Fowler, who later became Rutherford's son-in-law. The group included E. A. Milne and D. R. Hartree, who became respectively an eminent cosmologist and the leading British authority on computers and numerical analysis, the geometers H. W. Richmond and T. L. Wren, and the applied mathematician W. R. Dean. The scientific characteristics of these subsequently eminent men were already evident and in operation.

I joined this group in 1918, at the age of eighteen. The Navy was accommodating itself to the new phenomenon of independent scientists in its midst. The work at H.M.S. *Excellent* was a forerunner of the way in which science was to be used in the Second World War; it contained elements of what was later to be known as operational research.

The atmosphere of a naval establishment was strange to a

grammar school boy who had hitherto never been outside Yorkshire and Lancashire, except to sit for scholarships and look for a job. The first thing that struck me was the superiority of the food, which in Yorkshire had become pretty poor in the later stages of the war. Then the large and complicated structure of naval life became gradually unfolded to me. The social chasm between the upper and lower decks was immense and absolute. I only came across a few ratings, usually through some menial duty. Ratings used to be sent for to carry messages. One would march into the mess and stand to attention while the officer, surrounded by fellow officers drinking aperitifs and engaged in informal conversation, would gaze at or above the man's forehead and give him the message. The man thereupon made a roundabout turn, and marched off like an automaton. On one occasion, the automaton returned, and announced that he had forgotten the message. He was sternly reprimanded in an awkward general awareness of the ridiculousness of the situation.

The gap between executive and engineering officers was equally clear-cut, though not so wide. The relations between them appeared to be civil but distant. I had an impression that the engineer officers preferred to eat by themselves if they could, while executive officers appeared to regard conversation as an essential part of meals.

However, conversation at breakfast was taboo to all officers. Each used to come into the mess singly, and sit down in isolation. An orderly came up with a wire frame, placed it in front of him, with some morning newspaper open upon it. If the frame, and breakfast, did not arrive at sufficient speed, a voice of thunder sometimes roared: 'Gather round, gather round.' Orderlies then rushed forward from various directions.

I was delighted with the copious supplies of butter and grilled plaice, which by 1918 had become almost unknown in the West Riding. Occasionally, a ship's band was borrowed to play in the gallery during dinner; I found table music very pleasant. The officers dressed for dinner on Sundays, though we scientists, in civilian clothes or a variety of uniforms, were excused. As guests in the establishment we dined at the staff table. When the gong sounded for dinner the officers on the staff and the officers under instruction formed two files, and went through separate doors into the mess. As the junior person attached to the staff, I used to walk in at the tail of the staff file.

H.M.S. *Excellent*, though an island, had various parts designated as if it were a ship. It had a quarter-deck, which consisted of a

Assisting in Research

rectangular gravel-covered expanse. Some of our group were Army officers who had been badly wounded in the war, and being mathematicians had been sent to work with us. They felt that they had already made a fair contribution through active service, and took their new duties rather informally. One of them actually walked across the quarter-deck without wearing his Sam Browne belt; there was a slight row about this. I did not discover until shortly before I left Portsmouth that my cloth cap, known as a miners' cap and a standard form of headgear in Yorkshire, had been unpopular. At the end of the course, a mass photograph of the staff and officers under instruction, in all about three hundred people, was taken. I accidentally overheard an officer imply that of course the picture had been spoiled by my cap and deportment.

H.M.S. *Excellent* had a large and good library. In it I found fine editions of the works of Edgar Allan Poe and Dostoyevsky, in particular. Officers were not often seen in the library, which I frequented a good deal, but I believe they used it more than they cared to show. On the night of Armistice Day in 1918, most of the sailors engaged in wild celebrations in Portsmouth. H.M.S. *Excellent* was almost deserted, and I went to the library, which, as so often, was empty. On this occasion a young officer walked in, and we had a long and interesting conversation. He raised the question of our future careers. He was thinking of retiring early and following a civilian occupation. I suddenly saw him not as a military man, but a quiet reflective civilian in uniform. He asked me what I intended to do. He supposed I would pursue the usual scientific career. I answered in entirely conventional terms. After graduating at Cambridge, I would become a scientist or engineer.

When I arrived at Portsmouth, I reported to R. H. Fowler. He was then twenty-nine years old, very much a Wykhamist and a pure mathematics don. He had been one of G. H. Hardy's pupils, and immersed in Hardy's approach to mathematics. He pursued abstract problems with relentless vigour and energy. At that time, his powers of physical thinking had not been developed; he depended on others for physical ideas, and applied mathematics to them with immense energy. I remember listening to a discussion on a problem concerning the calculation of the trajectories of anti-aircraft shells. The trajectory is considerably affected by the degree of cant of the shell immediately after it has left the muzzle of the gun. The shell does not proceed absolutely straight ahead, because neither the gun barrel, the shell, nor the explosive forces are perfectly symmetrical in shape and action. Consequently, the shell leaves the gun slightly sideways, which causes it to perform a corkscrew motion in its trajectory. The degree of this departure from the straight symmetrical position affects the range of the shell.

Fowler was searching for some way of estimating this initial cant of the shell, so that the mathematical apparatus could then be let loose on working out its exact trajectory, including the effect of the cant. He gave the impression of having absolutely no idea of how to attack this problem. In the discussion, someone said that A. V. Hill was wanted, who had a physical imagination and experimental insight, and could guess what the initial cant would be. The difference between Fowler's powerful abstract mind and Hill's inventiveness of physical ideas and devices was striking.

Fowler subsequently was more and more involved in applied mathematics. Soon after the war, he came within the orbit of Rutherford, whose daughter he married. He became the leader of the teaching of theoretical physics at Cambridge. His greatest pupil was P. A. M. Dirac. Fowler's physical imagination developed, and enabled him to contribute great creative work on the conception and theory of super-dense matter in stars, which explained the nature of the very small but very heavy White Dwarf stars.

At Portsmouth, Fowler assigned me to work under E. A. Milne, who came from Hymers College at Hull. It was thought that one Yorkshire grammar school boy might understand another, and the thought proved quite well founded. Milne was then twenty-two, and at the height of his intellectual powers. Before he came to Portsmouth he had made a tremendous impression at Cambridge, where he had not yet graduated. Some people said that his powers were such that he might prove to be another Isaac Newton.

Fowler was an extremely able and energetic man, but Milne appeared to me to be about three times as energetic and able. He then had normal health, and exhibited a continuous intense intellectual activity almost unparalleled in my experience. My task was to carry out the arithmetical computing of the solutions of equations describing the trajectories of anti-aircraft shells. The equations were worked out by Fowler, Milne and others. J. E. Littlewood came in as a consultant, though he was not a member of our group.

At week-ends Milne sometimes invited me to go for a walk. On one occasion we went on to Portsdown, the hill outside Portsmouth. We walked for miles, while Milne recited to me long passages of argument from Whitehead and Russell's *Principia Mathematica*. Milne's early and passionate interest in mathematical logic was impressed on me, and subsequently made it easier to understand his peculiarly logical attitude to cosmological theory, which seemed to many people unrealistic, even though ingenious. In his cosmology he came to talk about the moment when the universe came into existence as an entity about which precise mathematical calculations could be made. When people too incautiously disagreed with him about his later cosmological ideas, he was apt to become neurotically angry.

Milne was not like that when I first knew him at Portsmouth, though his temper was short. He was impatient with opposition, and did not conceal his belief that people who disagreed with him were foolish, but after a little while he recovered his composure. It seemed to me that the severe illness he had, about the time he went to Manchester as professor of applied mathematics soon after the war, left his nervous constitution permanently impaired. He stayed with S. Chapman at Manchester while he was convalescing, and I noticed that he was profoundly changed since I knew him at Portsmouth. He showed signs of a nervous argumentativeness, which Chapman dealt with kindly but firmly. This was different from the robust impatience I had known. The mixture of logic and theology which appeared in Milne's later cosmological work seemed to be connected with his early interest in mathematical logic and his illness.

In my last meeting with J. H. C. Whitehead, not long before he died, our conversation came round to E. A. Milne. Whitehead told me a story, no doubt apocryphal, but not psychologically uncharacteristic of Milne in his later years, though not when I knew him at Portsmouth. He said that when the Second World War started, Milne offered his services to the War Office, and received some cyclostyled document informing him that his services would be called upon, if necessary. Milne, in view of his services in the First World War, and his eminence since, was infuriated by what he regarded as a discourteous reply. He used his connections to have his disapproval brought to the attention of the higher ranks of the War Office. He thereupon received an invitation, signed by a brigadier-general, to call at the Office. Milne arrived, fulminating with criticism. He told the brigadier, who listened quietly to his harangue, that the War Office ought to know that the war would be a scientific one. In that case, was the way they had treated him the way to make the best use of science and eminent scientists?

The brigadier waited until Milne had run out of his first breath, and then asked: 'Did you win the Adams Prize in your year?' 'No,' replied Milne angrily, 'but what has that got to do with it?' 'I did,' said the brigadier.

The member of the Portsmouth group who was most at home with the work we were doing was D. R. Hartree. He was twenty-one, and had not yet been academically very successful. His genius was for computing and numerical analysis, which were not then much esteemed at Cambridge. At Portsmouth this was just what was wanted for calculating shell trajectories. He revelled in computing, and said so. He wrote a treatise on the calculation of trajectories which became the textbook for those who worked in this field. This was before he completed his course at Cambridge. After he had already made this important contribution, he did not do very well in the Cambridge tripos, one of the classical instances of the fallibility of examinations.

Hartree's father, W. Hartree, was a coach in mechanics at Cambridge, and his mother a well-known Cambridge personality, who became mayor of the town. W. Hartree had been sent to Hill to work in his section, and Hill did not at first quite know what to make of him. W. Hartree had a passion for order and arithmetic, which was useful, if not creative, in computing. After the war, Hill invited him to assist in his physiological researches. He had an insatiable appetite for tedious routine observations and arithmetical calculations, which was invaluable in assisting in Hill's experimental observations on heat production in muscle, and in working out the mass of results.

D. R. Hartree's talents were creative, and of quite a different degree. The father and his eldest son did not seem to understand each other very well. W. Hartree used to appear in the work-room on the stroke of nine every morning, or even before. His son had not the same sense of order, and was sometimes late. I remember his coming in at about ten past nine on one morning, his mind probably full of original reflections on new methods and objects of computation, to be suddenly carpeted before the whole office by his father. The situation was comic, for D. R. Hartree was really the most important man at Portsmouth, so far as the theory of gunnery was concerned. W. Hartree's attitude seemed to be in conflict with his principles, for he and his wife had sent their son to Bedales School, which was then considered educationally rather advanced. No doubt D. R. Hartree's informality had been strengthened there, or at least not diminished.

D. R. Hartree's work on the computing of shell trajectories at Portsmouth was the basis of his subsequent work on the theory and construction of computers, and on numerical analysis. He applied his technique to the calculation of the exact dimensions of atoms, which became one of the foundations of atomic physics. His younger brother Colin was also at H.M.S. *Excellent*. He was a charming young man, who most unfortunately died prematurely.

A. V. Hill was stationed in London in the Ministry of Munitions Inventions Department, and came down to Portsmouth only occasionally. He was then thirty-two, an athletic figure, with hair already handsomely greying. Hill's inventive genius was most striking. During his visits he always poured out ideas and suggestions for new devices. His first reaction to any situation was either to ignore it, or suggest something new to cope with it. Anything that caught his attention immediately released new ideas in his mind. He had difficulty in attending to anything that did not interest him and set his creative mind in action.

Once, when asked to write a book, he said he never would, because if he did, he would have to read what other people had said about the subject. He said that when he began to read other people's papers, he had difficulty in finishing them; either he could not go on because they seemed boring, or if they were good they immediately raised in his mind the suggestion of new experiments. He felt then compelled to rush off to his laboratory to try these new experiments. Consequently, he felt he would not be able to do justice to other people's work, which would be necessary in writing a book on any subject.

At Portsmouth Hill's colleagues sometimes felt that he was not sufficiently receptive to their ideas. But, as someone remarked; 'You have to admit that he has more ideas than all the rest of us put together.'

A number of visitors came to H.M.S. *Excellent* to study the work in progress. One of these was the civil servant L. C. Bolton, who came down to write a report on our work for his department. Soon after the war, Bolton sent an article in to the competition organized by the *Scientific American*, for the best popular exposition of the Theory of Relativity, which had become of great public interest after the solar eclipse observations of 1919. These indicated that the sun's mass bent rays of light passing near it, as Einstein had forecast. Months after sending his essay in, and having almost forgotten about it, he returned to his home in Bedford one evening, after one of those strenuous days in Whitehall. He saw there was a letter from America for him, and assuming it was some advertisement, put it on one side to open after dinner. Later in the evening he thought he might as well open it, and when he did so, out fell a cheque for $5,000. He had won the *Scientific American*'s prize for the best popular essay on relativity.

Our most amusing visitors were two French officers, a colonel of artillery who was a wine-grower in civilian life, and Haag, the distinguished mathematician. Haag did not say much, but the colonel had ideas of sociability which were novel in the naval establishment. Many of the naval officers had dressed for dinner on the occasion when the French deputation was entertained. Everything began with formality, but gradually became easier under the colonel's guidance. He introduced the game of moulding his bread into small pellets, which he shot with his forefinger at his hosts. Presently, they took up the game, and the mess became littered with pellets.

One day two tall American officers appeared at H.M.S. *Excellent*. The authorities had not sent advance information on their names. They were placed in Richmond's charge, who brought them to our office, and began to explain to them in the simplest language what we were doing. Experience had taught him not to assume that visiting officers knew any mathematics, so he began by avoiding the use of calculus. The two Americans listened politely, and occasionally made quite sensible comments. Presently Richmond said in his gentle voice: 'Perhaps you know the calculus?' The taller of them, a fair, middle-aged man of agreeable manner, smiled slightly and said, 'Yes, we know the calculus.'

Richmond sighed with relief and said that they would now be able to get on. He went into things a little more deeply, and the Americans' comments became still more intelligent. After a while, Richmond looked up and said: 'Perhaps you are mathematicians?' The taller, fair American smiled quizzically and replied, 'My name is Veblen.' I was seated about four yards away, and I can still see Richmond jump in his chair, murmuring a series of inarticulate 'Oh's'. He and Oswald Veblen worked in the same field of mathematics, and he had spent half an hour trying to explain the solution of differential equations to him without mentioning the calculus. Oswald Veblen, whom I called to see in America twenty years later, became the director of the Princeton school which gave refuge to Einstein.

The main problem with which we were concerned at Portsmouth arose from the new conditions set for artillery by aircraft. Hitherto, guns had been used for shooting along the earth or sea. The trajectories of their shells had been comparatively flat, and wide deviations from flatness had not been of great practical importance. With the introduction of aircraft moving in three dimensions instead of two, the accurate calculation of trajectories of shells which had to reach any specified height as well as distance involved new problems. The pressure of the air declined markedly with height, and this involved big changes in resistance which had to be accurately calculated and allowed for.

Hill's height-finder enabled the speed and direction of the wind at any height to be found by measuring those of a puff of smoke from an exploded shell, carried by the wind. It was done by plotting reflections of the puff on a horizontal mirror.

On one occasion, this technique showed that a gale of 120 miles an hour was blowing in the region of 20,000 feet. Commander Maton, who was the link between our section and the naval establishment, remarked that it would be fun if the Zeppelins made a raid that night, for they had adopted the technique of approaching at a great height, and drifting towards England with the engines shut off, so that they could not be heard. Next day we learned that Zeppelins had indeed approached England during the night, and had got into the gale. Several of them were carried into the Mediterranean, and one or more blown into the Atlas Mountains. It seemed that at that time the Germans did not possess an apparatus equivalent to Hill's.

Pilot balloons provided one of the techniques used by the meteorologists for measuring the speed of the wind at various heights. The balloons were designed so that they rose at an approximately constant speed. By taking observations from them at regular intervals with a theodolite, the direction and speed of the air could be determined. This method was useful up to moderate altitudes, but the uncertainty became large at great heights. Richmond and I were sometimes detailed to make observations with these balloons. It was not an onerous task, and I enjoyed Richmond's conversation. He was a distinguished geometer, keenly interested in music, and deeply attached to King's College at Cambridge, of which he was a fellow.

When anti-aircraft gunnery was invented, there were no anti-aircraft guns. Conventional guns, capable of only a limited elevation up to about 45°, were used. The higher angles of elevation were obtained by mounting them on wedges, so that the elevation could be increased up to 90°, that is, the gun became capable of being fired vertically. The guns were operated by crews who were given instructions by those in charge of the posts where the observational equipment was installed. There was an occasion when an officer, in a fit of exasperation, roared down the phone to the crew to give their gun maximum elevation and fire it. He had probably momentarily forgotten that the gun was capable of being elevated to 90°. The crew duly carried out the order; they raised the elevation to the vertical, fired their gun, and then ran to the nearest dug-out. The shell went up and came down many miles, almost vertically, and fell only about two hundred yards from the dug-out.

I was put on the telephone between the observation post and the crew in some trials. It was about my first experience with a telephone, and I was so confused by it that I had to be taken off. It was no doubt as well, considering what could happen.

After the Armistice of 1918, A. V. Hill saw the various members of his team, to discuss their futures. This kindly interest helped to clarify and determine our various plans. I explained that I proposed to go to Cambridge, take my degree, and follow the kind of mathematical or scientific career which would naturally result.

Hill and I agreed that my course appeared to be perfectly clear. I left Portsmouth at the end of 1918, and entered Trinity College, Cambridge, in January 1919.

CHAPTER TWO

1919-23
TEACHING MATHEMATICS AND SCIENCE

WHEN I ARRIVED at Trinity in January 1919, I chose to read pure mathematics. It became clear to me almost at once that I had made a mistake. I tried to change to the study of the history of science, but was told that that was impossible. I was pressed to continue in mathematics, but was unwilling to spend three years on what appeared to me a mistaken course. I decided to leave Cambridge at the end of my first term, and surrender my exhibition and other awards.

Though this crisis excited a variety of troubles, the need to earn a living soon absorbed my attention. This was not so hard in 1919 as one might have expected. Owing to the terrible slaughter in the First World War there was an extreme shortage of science teachers, so I searched the columns of the *Times Educational Supplement*, and applied for jobs as a science teacher. For most of the posts advertised I proved to be the only applicant, so headmasters, desperate for anyone, tactfully passed over my lack of qualifications and experience. My tenuous connection with Trinity was sufficient to give them courage in justifying my employment, and I started by securing some temporary appointments.

My first was in a county secondary school for boys and girls in Cornwall. Among the pupils were L. H. C. Tippett, subsequently a well-known statistician, the historian A. L. Rowse, and R. Ede, the agricultural scientist. Tippett left soon afterwards, and Rowse did not take science, so they were never in my class. Rowse was then about sixteen years old. His exceptional mental energy was conspicuous, and he already showed a mastery of conventional ideas. Ede attended some of my classes. He impressed me by his superior intellectual incisiveness, compared with the hard-working pupils who wrote neat, long and dull answers to questions; Ede handed in two or three lines of scrawls sometimes accompanied with blots, which generally contained more to the point.

About this time my old school friend, Ralph Fox, told me of the summer school organized by the Labour Research Department at Cloughton, near Scarborough. Its secretary was Rose Cohen. The acquaintance I made with her at the school later became of great importance to me. Among the participants in this or succeeding summer schools at Cloughton were Gordon Childe, G. D. H. Cole, Maurice Dobb, J. F. Horrabin, William Paul, Bernard Shaw and C. P. Trevelyan.

Gordon Childe was then known as a political historian, and had not yet achieved eminence as an archaeologist. He had published a well-known book with the title *How Labour Governs*, on the performance of the Labour Party in New South Wales in Australia. Not long after this, Childe, who then had no adequate job and was very hard up, asked me about finding work in the publishing world. Fortunately, his book on *The Dawn of Civilization in Europe*, and other works, attracted attention, and he was appointed professor of prehistoric archaeology in Edinburgh.

Childe had an understanding of technological and scientific factors in the evolution of man which made his view of prehistoric archaeology exceptionally interesting from the scientific point of view. I once asked him what was the social significance of his subject. He said that, for example, it showed that mankind had extraordinary powers of survival. There were moments in the hunting stage of mankind when a single comparatively small incident might have exterminated the human race. The dangers that man had survived were repeated and very great. Frightful though modern dangers might appear, there had probably been many even worse in the prehistoric past, which man had survived. This was a reason why mankind could face the future with optimism and confidence. The modernity and quality of his mind and conceptions were very striking; I was much influenced by them.

The discussions at the Labour Research Department summer schools were very lively, but the entertainments were even more interesting. They were quite elaborate. Shaw read his play *O'Flaherty, V.C.*, giving superb expression to Irish accents and modes of thought, and the attitudes of the Irish and English towards each other. The Coles put in nights of work on the composition of revues, containing references to the foibles of members of the schools.

J. F. Horrabin, the well-known cartoonist, and editor of the *Plebs* magazine, revealed abilities as an actor. This was not the only occasion on which I witnessed the recognition of unsuspected or insufficiently esteemed qualities in this modest man. When the abstract artist L. Moholy-Nagy came to England after the rise of Hitler, I asked him which English artist he particularly esteemed. He pondered, and then surprised me by mentioning Horrabin, whom he considered as outstanding in combining artistic sense with social utility. He regarded his maps as of very special quality. I have heard it said that when H. G. Wells was bringing out his *Outline of History*, he required an illustrator for the work. He suggested that Horrabin should be engaged for a fee of £100. 'What?' he was told. 'A three-figure cheque for an artist? No bloody fear.' 'All right then,' said Wells, 'you are giving me a royalty of ten per cent, give him one per cent.' This was agreed, and Horrabin's first cheque was for over £1,000.

Rose Cohen went to live and work in the U.S.S.R. I wrote to her about possible opportunities there for working in experimental education. In her reply to me, of 18 July 1923, she said that she had not been able to find any opening for one 'coming out here to take his chance' in this field. The most interesting work of the kind she had come across was in the campaign for the liquidation of illiteracy in the Red Army. She said that 'this is a great country, and I don't want to go back to England a bit. By comparison, everything in England seems so decaying, while here everything is real and vital. One feels that one is building up something . . .'

Six years later, while on a visit to Moscow, Ralph Fox took me to call on her, with consequences in the educational field which I will presently describe.

I had become keenly interested in the *Plebs* magazine, and devoted much of my spare time to it, studying its contents and participating in the movement for independent working-class education, of which it was the mouthpiece. My acquaintance with Horrabin reinforced my interest. H. G. Wells had made his celebrated visit to the U.S.S.R., and returned with the view that Lenin was an impractical idealist, 'the dreamer in the Kremlin', and that the new system would not work. Among other evidence, he reported that the electric bell at the entrance of the State Department for Electrification did not work.

I read an account of Wells's return from Russia in a newspaper, which reported that the first place he had gone to after his return was the Foreign Office. I wrote a letter to the *Plebs*, saying that I supposed he had gone to report to his masters. I was astonished and flattered to read a month later a reasoned reply from Wells. He said that he had gone to the Foreign Office to state his views because he thought they might contribute to an improvement of understanding, and he ended his letter with the observation: 'I guess when there is real revolutionary work to do or real order to be established', I would be found, together with certain others, 'in the last, most distant ditch, still disapproving highly of everyone and calling us all bought men.'

I presently secured a permanent post at a junior technical school in Essex. So far, each school I had worked in seemed to me worse than the last. While I was there the school was inspected. I commented on some of the shortcomings to the inspectors, who, I believed, would receive my remarks in confidence, but they reported them to the headmaster. He was understandably infuriated, and said that either I or he would have to go.

About eight years later, when I was concerned with higher technical education in the U.S.S.R. at the beginning of the First Five Year Plan, I was brought into contact with high officials at the Board of Education. One day, on London Bridge railway station, I happened to meet H.M. Chief Inspector of Technical Schools, who had been instructed to be helpful to me about Russian inquiries concerning technical education. He told me that he was on his way to a junior technical school in Essex, from which they had not received certain explanations that they had requested. I did nothing to damp his ardour.

But while criticizing, I had also tried to introduce Sir Percy Nunn's new methods of teaching algebra. In the course of doing this, I had discovered a slip in his presentation of Galileo's treatment of infinite numbers. This led to my becoming acquainted with him. When the row over the inspection arose, I went to see him for advice. It turned out that in his early days he had been a teacher in a private school in Halifax on the opposite side of the road to the the Technical College, where my father was then principal. He advised me to resign, and said he would try to suggest a more suitable job.

Nunn was then the head of the institute which was later incorporated as the Institute of Education in the University of London. He was a mathematician and a noted philosopher. He had published a famous paper with the title 'Are Secondary Qualities independent of Perception?', which had changed A. N. Whitehead's opinion on a fundamental philosophical question. Einstein had recently visited England for the first time, and philosophers, especially those with mathematical training, were particularly anxious to exchange views with him. I asked Nunn whether he had met him, and he told me that he had most keenly desired to do so, but the invitation came at a time when he had to attend an official educational meeting; so, with a pang, he put duty first. I was permanently impressed by Nunn, the creative philosopher who put social duty before intellectual pleasure.

Through Nunn I met R. H. Tawney and Eileen Power. Tawney liked to wear his tattered old sergeant's tunic from the First World

War. For me, it emphasized his Old Rugbeian manner, which I found even more striking than his great gifts of literary and verbal expression. He was very actively working with his old schoolfellow, Archbishop Temple, on behalf of the miners. He had the natural authority and confidence of a member of the ruling class. This gave his scorn for its evil members, such as coalowners, its peculiar force; his denunciations came from the inside, as well as being finely expressed.

Nunn gave me an introduction to Norman McMunn, who ran an anarchist school at Tiptree Hall in Essex. McMunn had about a dozen pupils, several of whom had left orthodox schools for a variety of reasons. Tiptree Hall belonged to the Wilkins's, the Quaker jam manufacturers, who let it to McMunn for a nominal rent. The fees from a dozen pupils could not go far, and I was engaged at a minute salary. I was expected to give about half my time to teaching mathematics, and half assisting in the communal life.

McMunn was a natural teacher, who had made his mark in leading conventional preparatory schools before he launched out on his own. He had published works on new methods of teaching languages before he started his school. He loved children, and loved to be loved by them. His system was based on this bond of affection. I was astonished to observe that a school of a mere dozen boys could present vastly more intellectual variety and activity than a conventional school of four hundred boys.

I then began to realize how very formalized conventional education is, with short periods or 'hours' of work. After McMunn's system, the conventional school seemed like a prison, in which set tasks had to be carried out under set conditions. I saw that no other way of dealing with large numbers of children was economically possible. The universal adoption of McMunn's system would have been impossibly expensive. The character of the conventional education system was determined by economic considerations to a far greater degree than I had imagined.

Nunn's *Modern Algebra* had fired me to attempt new methods of mathematical instruction. I set out with the idea of trying to get rid of the formal and given aspects of the introduction of Euclidian geometry. Euclid makes deductions from perfect points, lines and angles, which do not exist in the environment. How could the beginner's revulsion against these abstractions be mitigated?

I started from an idea in Karl Pearson's *Grammar of Science*, that geometrical conceptions, such as points and lines and angles, are the ultimate product of a process which starts in perceptions of the actual world. The mind abstracts qualities from the perceptions, until pure conceptions only are left. The notion of a point is the result of abstracting colour and the three dimensions of space from solid objects. I constructed a series of irregular solids, painted in mixtures of colours, and hung them in a row high on the wall of the study-room. Beneath this, I hung a parallel row of white tetrahedra, smaller and closer together than the coloured irregular solids; thus in the second row, colour and irregularity of shape had been abstracted. Then a third row of small flat white circles, very close together, was fixed on the wall; this row was beginning to look like a white line made of white points. Below this row, a parallel white line was drawn. Thus the student could see that the fine white line was a late stage in a process of abstraction from the coloured irregularly shaped objects found in the natural world. The aim was to remove the mystery of where the perfect points and lines discussed by Euclid come from.

I reduced the performance of simple operations in complex numbers to a card game which could be played by young children. The aim was to familiarize children with complex numbers and vectors without using these terms. Then, when they came to them later in mathematics lessons, they would find that they already had an idea of what they were. They would not feel that they were confronted with something meaningless, as 'imaginary' numbers usually appear when they are first presented to young students.

I was advised to take the game to a patent agent. His redraft of my idea astonished me. He extended it to cover forms of advertisement and display, which were quite absent from my mind; I was thinking only of instruction. I had used boards with holes at the corners of squares, like the perforated boards now adopted on a vast scale for display of advertisements and goods. I was unable to find a firm interested in manufacturing the game, so the provisional patent lapsed.

I learned from these and other devices for trying to make the beginner's understanding of mathematics easier that the task of revolutionizing the teaching of mathematics is enormous. I began to realize that Euclid's system was the result of a long evolution, and that a new systematization of the presentation of mathematics would require many lifetimes.

While I was with McMunn, I had excellent opportunities for private reading. At the end of the war I had met Dr G. H. Miles, the applied psychologist, who was then in charge of the anti-submarine hydrophone station at Flamborough Head, where my parents had a holiday cottage. He recommended the London

Library to me, so I joined it at an early age. I borrowed many books from it while I was with McMunn.

McMunn was afflicted with asthma, and decided to move his school to Italy, believing the climate would suit him better. I did not want to go to Italy, and began to look for another job. In the meantime, I made a short visit to Germany in 1923. McMunn had a number of visitors to see his school. One of these was C. F. Andrews, a friend of Gandhi. He had just returned from Germany, and spoke with enthusiasm of the people he had met, and I thought that perhaps I might find an interesting job there. He gave me an introduction to a young man in Leipzig. I took an aeroplane flight to Berlin, and then went by train to Leipzig. Air travel in 1923 was not very common, and it was the first time I had been in a plane. I had taken G. E. Moore's *Principia Ethica* with me for a little light reading. Moore introduces his book with a motto from Bishop Butler: 'Everything is what it is, and not another thing.' The passenger beside me, who looked like a Polish intellectual, did not speak English, but evidently read it. He motioned that he would like to look at my book, and when he read the quotation from Butler, he gesticulated ecstatically.

I did not know where to stay in Berlin, and two English officers who had also travelled in the plane advised me to accompany them to the Adlon Hotel. I knew nothing about inflation, but found I could stay there for quite a modest charge, about five shillings a day. I was given a room furnished in red plush and gilt. An intelligent young bell-hop, aged about fourteen, who spoke excellent English, offered to give me a better rate of exchange than the hotel cashier downstairs. I produced a pound note, which he eyed gravely, as a precious object, and asked if he could consult the other bell-hops, as he could not himself change so vast a sum. He went away and came back with several colleagues. The whole group turned out their pockets, putting the contents on the red plush sofa, accumulating what they considered to be an adequate sum in marks.

A little later I went into the restaurant. There were many Americans, dining in great style; for about one dollar one could have the whole orchestra wheeled along the floor until it was beside one's table. Such extravagances were being practised in full.

Presently the two English officers came in, already changed into dinner jackets. They saw me, and came to sit at my table. They were in charge of refugee camps in Asia Minor, and were taking a short leave in the form of flying about Europe. I told them about my experience with the bell-hops, and how ashamed I felt at the excellence of their English, and the badness of my German. One of them said they always spoke English on principle, wherever they were. I asked them what they did if people did not understand. The same officer said: 'We go on speaking English until they *do* understand.'

When I arrived in Germany, the mark was worth about one million to the pound; it was falling at the rate of one million a day. It was difficult to get food, and in ordinary restaurants the food was often bad. Life appeared to be very insecure, and I felt that I had better return to England soon. I forgot my vague idea of a job in Germany.

I took a boat train for the Hook of Holland. When it drew up at one of the stations near the Dutch frontier, a terrified middle-aged man jumped into my compartment. He was fleeing from Hamburg, and was going—anywhere! There had been a *putsch* in Hamburg on that afternoon, and fifty people had been killed in the fighting.

This was my first visit abroad. After my return I continued looking for a job. Dr Nunn gave me an introduction to Dr Jessie White, who had invented apparatus rather on Montessori lines, for teaching arithmetic to children. She engaged me to assist in making the apparatus. She had a small workshop in Great Russell Street, next to the offices of one of my later publishers.

One of my jobs was to look after a stall of Dr White's apparatus at an exhibition at the annual conference of educational associations held at University College, London. Most of the exhibitors were publishers, and among them was the Oxford University Press. As often happens at such exhibitions, the representatives in charge of exhibits spent more time talking to each other than demonstrating to delegates, who came to look at the exhibits only at odd moments.

Shortly after this, the Cambridge University Appointments Board sent me a telegram, informing me that Ralph Peacock, R.A., the noted portrait painter, required a tutor for his son. I went off at once to see Mr Peacock. I arrived at his home in Holland Park before 9 a.m. He was surprised by my call at so early an hour, but saw me, and offered me the job. A few days later he told me that after mine, he had had forty more applications. The tutoring was interesting, and also left me with time for my own reading.

In the meantime, the Oxford University Press had been creating a department for publishing technical books, and wanted a technical representative. Bearing a name that was already known in technical education circles was a useful qualification. The Oxford University Press men who had met me at the educational exhibition remembered me, and I was offered the job. It became permanent, and respectably paid. The Oxford University Press engaged me in 1924, and I have had a connection with them ever since.

This was the end of my career as an educationist. I was glad to escape from teaching, one of the most difficult and ill-rewarded of professions. It seemed to me that one of the most serious defects in the English system had been made permanent by the Burnham committees. They had made salaries dependent on a minimum number of weekly teaching periods which was altogether too long; it left the average teacher with little strength or energy for his own intellectual work. There should be scope for talented teachers to pursue research, by removing teaching chores from their task. In Germany in the nineteenth century, it was possible for secondary school teachers such as Elster and Geitel to do physical research of the first order, parallel with that of C. T. R. Wilson in England. They had only a limited amount of teaching to do, with assistants to help in the routine work. English secondary education would be improved in quality by the introduction of such a system. The Burnham system improved the conditions of salary and tenure, but it did not contribute much to the most necessary reform of all: the inspiration of the teacher by raising the social and intellectual esteem in which he is held.

CHAPTER THREE

1924-26
TECHNICAL AND SCIENTIFIC PUBLISHING

I JOINED THE Oxford University Press in 1924, and started to visit the technical colleges in Britain. I had heard something during my youth of the state of technical education in Britain from my father. He had energetically developed evening class technical education, especially the grouped course system of instruction In this, evening students had to take a balanced group of subjects, not merely one vocational subject. The old type of vocational instruction was converted into an education which broadened the student's mind, as well as improved his skill. The range of science, engineering, crafts and languages taught was striking. It was to be expected that the teaching of engineering and crafts would be good, but one of the most interesting sides was modern languages. The excellence of the teaching in these was due to the earnest interest of young salesmen in the wool trade, who really wanted to learn modern languages in order to sell in the international market.

My father's system attracted visitors from various parts of Britain, and from other countries. I naturally heard a good deal about his struggles to carry out this and other innovations. Years of listening left me with the impression that the chief obstacle to his innovations, and indeed to the development of technical education in England, was lack of social status arising from class prejudice. It was clear that the opposition to technical education could not arise fundamentally from intellectual considerations, but must be rooted in class attitudes. I had absorbed this view almost unconsciously.

Through his work my father was well known in the field, and in the course of years had had many able pupils. Two of the best known were Professor Leonard Bairstow, the chief founder of modern experimental aerodynamics, and Dr Herbert Schofield, the founder of Loughborough College, now Loughborough University.

When I began to visit the technical colleges in Britain I was able to see their condition, and the force of the views on technical education which I had absorbed. Among the many authorities whom I saw, my father's colleagues and old pupils were among the most friendly and informative.

British technical education in 1924 was extremely straitened. It had received an impetus during the First World War, which had, for the first time, impressed on a considerable section of the British ruling classes the vital importance of science and technology to the nation. Germany had almost won the war through her scientific and technological superiority, so far-seeing persons demanded that there should be better scientific and technological education.

During the war the wages of skilled workers were substantially increased, and they devoted a considerable part of them to buying technical and substantial books. For a few years after the war there was a boom in books of this character. It was in these conditions that the enormous demand for H. G. Wells's *Outline of History* arose. This book had a large sale among skilled workers. I heard of an engineering works in a small provincial town, where sixty skilled mechanics had each bought the work in the original edition at a price of several pounds, then a large sum of money.

It was this boom which had caused publishers to give increased attention to technical books. By 1924, however, the post-war industrial boom was over, and recession had begun. The technical colleges in the larger towns carried on, but most of their plans for development were halted. Students were very hard up. In Glasgow, for example, some were attending classes hungry, and they could not buy books; they consequently expected detailed and comprehensive lectures from their teachers. The atmosphere was harsh, but not demoralized.

When I visited the technical schools in the smaller English industrial towns the cultural and social conditions in many of them appeared to be degraded. It was sometimes evident that the principal had given up the hard struggle for live technical education. The basic condition of English technical education was seen there unconcealed; there were discouraged teachers, who held their work in contempt, and unashamedly accepted their low social status. One heard of tyrannical chairmen of education committees, who were proprietors of local businesses and treated the staff of the technical college as if they were their own employees. In the larger technical colleges, there was nearly always some good work going on. Attention was naturally concentrated on it, and away from the large areas where work was dispirited and perfunctory. I now saw my earlier impressions confirmed by observation.

Engineering education was split between the technical colleges and the universities, and in the latter its position had generally not been consolidated. Looked down upon as a technical college product, or not fully accepted in the university, the engineer's status was not equal to his cultural and social importance.

The depressed state of technical education in the second half of the 1920s clearly indicated that the immediate business prospects for the publishing of technical books was limited. The field was dominated by textbooks which had become established long before the First World War; they could now be printed from stereo plates in large editions, and sold at the lowest practicable price. In general, these were the only books the technical student could afford to buy. The introduction of entirely new books competing with these was difficult.

I therefore began to visit the science departments of universities, though this was not originally envisaged in my appointment. Here I found the situation startlingly different. The pure science departments of many universities were full of enterprise, hope and achievement. Scientists were prepared to write books, for which there was a considerable sale. In view of this situation, I did all I could to devote more and more of my time and energy to helping in the science publishing. As I had not been engaged to do this, it was not at first altogether approved.

English publishing business in pure science books was more flourishing than in technical books. Such a situation in an advanced industrial country was symptomatic of a social disorder. I was aware that the general social situation in England was unsatisfactory, but I followed economic pressure rather than philosophical recognition in transferring my attention more and more to pure science. Who could have thought *a priori* that there might be more money in pure science books than in technical books, in the oldest industrial country?

Long afterwards, I realized more clearly that the situation was rather the reverse of that in the nineteenth century, when the sales of technological literature had been comparatively large. British industry had then been rising, while in the twentieth century it was, comparatively, declining.

The extension of my visits to universities widened my acquaintance with scientists. I noticed that while a good British scientist would generally be well known and esteemed, an engineer of the same intellectual calibre was often unknown outside a small technical circle. On the whole, with scientists and engineers of equal calibre, the engineers were the more interesting to talk to. It gradually became impressed on me that Britain was simply failing to make proper use of many exceptionally talented engineers.

Two who impressed me particularly were Professor William

Kerr of the Royal Technical College at Glasgow, and Professor Bernard Hague of Glasgow University. Kerr came from a Scottish mining background, and began to secure higher education through evening classes. He became a draughtsman in a noted Clyde shipbuilding firm. Presently he moved into teaching, and gradually rose to be head of the large department of mechanical engineering at the Royal Technical College, now the University of Strathclyde. Kerr specialized in the difficult problems of vibration in big ships, in which there are so many unknowns that a successful solution depends on engineering intuition as well as command of formal methods of analysis. Owing to Kerr's origin and mode of early education it was a long time before he reached a position of authority, and he was probably then too old for his exceptional talent and critical judgment to have their full creative influence.

Professor Bernard Hague was an Englishman, and another talented engineer who did not reach a senior position until late in his career. He was an authority on electrical circuits, with a command of the appropriate mathematics. He combined with these technical accomplishments an artistic feeling for draughtsmanship, and for music. He was a very good oboist, and his wife a fine violinist.

Not all engineers were prepared to take their situation lying down. In the 1920s J. W. Landon of Cambridge suggested I should see a former student of his, a young assistant lecturer in engineering in Cardiff University College named J. F. Baker. Baker and his colleague Field Foster had a scheme for a book on mathematics for engineering, the publication of which I was able to arrange.

Baker was presently appointed professor of engineering at Bristol, and then at Cambridge. He became a Fellow of the Royal Society, President of the Institution of Civil Engineers, and was knighted. His research on the strength of steel structures is a foundation of modern structural and civil engineering. He showed that large economies could be made in the amount of steel needed to provide a safe structure by taking advantage of the properties of plastic flow in steel. This discovery enabled him to invent the Morrison air-raid shelter.

The technical aspect of this achievement was outstanding, and Baker's success in making statesmen listen to engineering sense was impressive. When Beaverbrook was minister of aircraft production, Baker settled down in his ante-room, and refused to go away without being heard on the precautions necessary to increase the safety in aircraft factories against the effects of bombing from the air.

Baker's engineering department in Cambridge became the largest in the university, and he tirelessly insisted that it should accordingly receive appropriate consideration. No one has done more to raise the intellectual status of engineering in England. Forty years ago, his conception of the equality of engineering with mathematics and physics was rare.

The engineering climates in England and Scotland were rather different. The tradition of James Watt and J. McQuorn Rankine counted for more in Glasgow than that of any English engineers in any English city. Higher engineering education had been founded in Britain by Anderson in Glasgow, whose institution has evolved into the University of Strathclyde. The Glasgow engineers had inherited the Scottish intellectual as well as the engineering tradition. They were devoted to informal philosophical discussion. When I began to visit Glasgow in the 1920s, I was invited to join them in Craig's coffee rooms in the centre of the city. These fine rooms were like a continental café; there was nothing like them in England. They were handsomely furnished and decorated, and comfortable places in which to spend a few hours' interesting discussion.

I was much struck that responsible Scottish engineers were able to take hours off in the middle of the day, and devote them to intellectual and cultural conversation. The circle was joined from time to time by colleagues who had returned to Glasgow from other quarters of the globe. It was not uncommon for a Scottish engineer, who had held a well-paid appointment abroad, to come back to Scotland and take a comparatively minor job. I noticed that in the circle of Scottish engineers whom I met, commercial were rather subordinate to social and cultural considerations. There was a reserve of intelligent and cultivated ability which was not being fully utilized.

After the Second World War the coffee rooms were closed, and the site was taken by a travel agency.

When I found that natural science was then more promising than technology as a medium for publishing, I made all the suggestions I could think of for exploring the possibilities in the science field, which was then almost completely distinct from technology. I suggested that I should go to the meeting of the British Association at Toronto in Canada, in the summer of 1924. It was unusual for junior members of staff to make such proposals in those days. When it was put before Sir Humphrey Milford, the head of that

Technical and Scientific Publishing

part of the Press for which I then worked, he laughed and said: 'Let him go, let him go!'

The Oxford University Press is a large organization with a complex history. The rights in Clarendon's *History of the Great Rebellion* were given to the University, and in memory of this, the University press was named the Clarendon Press. The University's publishing activities are controlled by a committee of dons called Delegates. As the Press's business expanded, it became necessary to set up a distributing organization, with offices in London and other cities in the English-speaking world. The manager of the London office became the head of a large organization, and presently he was given powers to publish books on subjects which the Delegates at Oxford then regarded as outside their province. These included children's books, and books on parts of vocational subjects, such as medicine and engineering.

I was employed by the London office, but my activities became more and more concerned with the kind of books published directly under the authority of the Delegates at Oxford. When the London office agreed that I should go to the Toronto Meeting, the Oxford office arranged for me to have support from Oxford men who were going to it. N. V. Sidgwick, the noted chemist, and an active Delegate of the Press, invited me to join him at a table with two other Oxford men, H. T. Tizard and A. C. G. Egerton, on the liner *Caronia*, which was taking the main party of 500 British members to Canada.

They were three very different people, but there seemed to be a distinct similarity in their outlook. As Oxford scientists they seemed to be on the defensive, but as Oxford men, they were assured of their positions. Sidgwick and Tizard were rather sharp in their division of people into those of whom they approved, and those they did not. They both had a profound and proper regard for Rutherford, but they seemed to me not always to be unerring in their judgment of other remarkable men.

At the Toronto Meeting, D'Arcy Thompson gave one of the evening discourses. Sidgwick and Tizard suggested that we should go in the spirit of barrackers; we should sit in the front row, and make ironical applause. We found the place, a large hall, which was packed. Presently the tall bearded figure of D'Arcy Thompson walked on to the platform with the majesty of a Chaliapine. I could feel my companions bristle. D'Arcy Thompson began to deliver his lecture on the *Nautilus*, the marine organism that drifts across the Atlantic. With complete command of every art and gesture of the theatre he unfolded the enthralling story. After about a quarter of an hour, Sidgwick and Tizard, like the rest of the large audience, were fascinated. D'Arcy Thompson finished his discourse in a tense silence, followed by great applause.

At Toronto I met Rutherford for the first time. A cricket match was organized between a British Association and a local eleven. As one of the youngest members of the Association, I was sent in to bat first; I was out first ball in both innings. As I walked off the field at the end of the match, I became conscious of a tall figure bending over me from behind, and a heavy breathing down my neck. I heard a deep voice say: 'Hugh, Crowther, see you bagged a brace!'

Two of the most interesting figures at the Toronto Meeting were Julian Huxley and F. A. E. Crew. They were the spokesmen of the younger biologists, and protagonists of the then comparatively new subjects of experimental biology and genetics. They travelled steerage in the *Caronia*, and their doings sounded to me, in the cabin class, very interesting. I had heard that they had given some kind of theatrical performance, in which Crew had particularly distinguished himself.

At about the same time as the Toronto Meeting, the International Mathematical Congress also met in Canada. Through this, I happened to see G. Peano, the great Italian mathematical logician; he was a small, dark figure, dressed in frock coat with stiff collar and shirt, which appeared to my Anglo-Saxon eyes most unsuitable in a Canadian summer with a temperature of about 90° F in the shade.

Some members of the Association went on a trip through Western Canada after the Toronto meeting. I judged that it was an extravagance I could not expect my employers to approve, so I did not go. When I got back to England I painstakingly wrote a detailed report of such information as I thought would be of publishing interest. This was sent on to Oxford, and the only comment on it which reached me was: 'Why didn't he go to the West?' It appeared that the Press was engaged in some negotiations there about books, in which my presence might have been of some assistance. I learned from this that it is often better to do things, and leave the explanations until afterwards.

During the next two years I obtained a broad view of the state of technical education and scientific research in Britain. I visited many of the colleges, universities and research laboratories.

It became evident that there was deep ignorance in most sections of the population of what was going on in science and technology. This caused me to think that a new weekly or monthly journal was

Technical and Scientific Publishing 35

required which would present science in all its aspects in intelligible terms to intelligent non-scientists. *Nature* was doing an excellent job for scientists in keeping them informed about general scientific developments, but it was not quite what was required for non-scientists. I wrote to the editors of the *Spectator, Nation, New Statesman*, and the London Editor of the *Manchester Guardian*, for interviews to gather their opinion on the desirability and prospects of such a journal. The last three, who were respectively Leonard Woolf, Clifford Sharp and James Bone, asked me to come and see them. When I made my calls, I was quite taken aback, for all three took it for granted that the real object of my visit was to become a contributor to their journals. All three said they would be glad to read anything I sent them.

I had never imagined myself as a writer, and a lack of facility in expressing myself in an acceptable form on paper had been one of my severest handicaps in my youth.

During my new job of visiting colleges and laboratories I had come across a man who was writing a scientific article for a well-known weekly journal. The thought flashed into my mind, that if that man could write such an article, so could I, but it evaporated as swiftly as it had arisen.

The distinguished editors made only perfunctory remarks about my idea for a new journal, but launched into discourses on their difficulties in getting suitable articles on science, their unfortunate experiences when they had tried, and the kind of article they would like to have. There had recently been some spectacular thunderstorms, and Leonard Woolf said he had looked for years for a good article on thunderstorms, but had never got it. This put the idea into my head of trying to write one. I remember tackling the job in a theatrical boarding house in Birmingham. In the room above were two members of the Lupino family, who had noisy differences of opinion. I found the task of expounding the mechanism of thunderstorms in simple terms desperately difficult, and after spending many of my non-working hours on it, gave it up as a bad job.

Clifford Sharp had told me with a kind of deprecating modesty that he had in fact been trained as an electrical engineer. This disposed me more to the *New Statesman*. For me, however, James Bone was the easiest to work with. He was keen on paragraphs for his *London Correspondence*. The first paragraph of mine to be published in the press was for his correspondence, on 8 February 1926. It was on W. N. Haworth's discovery that the accepted chemical structure of glucose was incorrect. This was part of the work that subsequently led to Haworth becoming the first English organic chemist to receive a Nobel Prize; this field had hitherto been monopolized by the Germans.

I first met Haworth when he was professor of organic chemistry at Newcastle. He had been in Sir James Irvine's laboratory at St Andrews, where he had participated in the line of research on the chemistry of cellulose and the sugars, to which St Andrews chemists had made notable contributions. Irvine had started as a laboratory assistant, and had gradually worked his way up from the lowest to the highest position in the university, ultimately becoming Principal. He had other qualities besides being a good chemist, which had enabled him to accomplish so much. He was a determined and tortuous fighter for his aims. Haworth succeeded in carrying the investigation of cellulose and sugars beyond Irvine. It was not in Irvine's nature to allow himself to be overtaken in anything. He had difficulty in conceding the superiority of Haworth's chemistry, and he tended to use qualities other than chemical to retain the precedence of his own ideas. Haworth was an extremely different man, exceptionally neat, precise and straightforward in his thinking and behaviour. His eyes were very blue and his hair prematurely grey, which gave him an extra impressive look. Considerable tension arose between them. Haworth was proved right, and Irvine concerned himself more and more with academic administration, and less and less with chemistry.

Haworth presently became professor at Birmingham, where he had a larger staff and resources. He planned and organized in a meticulous manner, more like a German than an English organic chemist. In the end, he succeeded in just beating the Continental chemists in the race to the chemical synthesis of vitamin C. He and his team won by three months, but the tension severely taxed his nerves.

After I wrote my first short paragraph on Haworth's work, I still had storms in my mind, and later in 1926 a cyclone killed 650 Cubans and sank a British man-o'-war off Bermuda. I wrote an article on *Hurricanes and Tropical Cyclones* for the *New Statesman*, which was illustrated with a diagram of the parabolic path of such storms, which enables their tracks to be forecast. Clifford Sharp told me that the *New Statesman* had never printed an illustration before; he would have to find out how it was done, but he did not suppose it would be difficult. The article, with the illustration, appeared on 4 December 1926.

The first full-length article which I wrote for the *Manchester*

Technical and Scientific Publishing

Guardian was on the *Coolidge Tube*. This had recently been developed by W. D. Coolidge of the General Electric Company of America. The tube produced very strong beams of electrons, which had direct applications in research and industry, and could also be used for producing intense X-rays for deep treatment in medicine and in the testing of materials. This article appeared on 28 October 1926.

None of my first efforts was rejected, and these beginnings opened the prospect of devoting my main effort to writing. I was not interested in business as such, and without such an interest a primarily publishing career could scarcely be attractive or possible. The atmosphere of British technology and technical education in the second half of the 1920s was repellent and depressing, and I was heartened by the possibility that I might be able to escape from being dependent on it.

Meanwhile, I made every effort I could in the publishing field of pure science. After my attendance at the British Association meeting at Toronto, I went to the meeting at Southampton in 1925, and have attended nearly every annual meeting of the Association since.

Two events at the Oxford meeting in 1926 were particularly important for me. Eddington gave his famous evening discourse on *Stars and Atoms*, to a packed audience at the Oxford Union. It was a splendid piece of exposition for a general audience. The summit of the discourse came when Eddington threw on the screen a fine cloud-chamber picture of the track of an α-particle. A magnificent thumb-mark had been left on the plate by a technician. It was almost as striking as the α-ray track. Eddington turned from his notes, looked at the picture on the screen, and paused as if suddenly arrested by what he saw. It was impossible to say whether he did this by accident or design, but from whatever motive, it was a perfect theatrical gesture. He said that the photograph was of an alpha particle, no, not actually of the particle, but of its track; the track had the same relation to the alpha particle as the thumb-mark had to the technician's thumb. I have never seen a piece of illustration go home with greater effect.

Though one could not tell whether he had carefully thought out the whole situation beforehand, he probably had, for while he was one of the most brilliant of expositors when he had time for preparation, he was one of the worst extempore speakers. He sometimes completely spoiled the effect of a magnificent prepared lecture by consenting to answer questions afterwards. He would then search with long pauses for thoughts and words, and through intellectual caution express himself inconclusively, disappointing most of his listeners. When he had time for preparation, he thought everything out to perfection, and illustrated his argument with original similes of a poetic quality.

After this lecture I went to see Eddington, and asked him whether he would expand it and publish it as a book. He said he would not expand it, but he had some other lectures which he would combine with it and form a book. This was his *Stars and Atoms*, the first of the Eddington and Jeans best-sellers, and in my opinion also the best.

Years later, when I called in to see Rutherford in the Cavendish Laboratory one day, he was examining the apparatus with which he and Oliphant were extending atomic disintegration experiments. When he saw me, he called out: 'Sit down.' He and I then sat on laboratory stools, and he asked: 'How's the writing business?' He continued without pausing: 'Jeans said to me "that fellow Eddington's written a book which has sold 50,000 copies. I will write one that will sell 100,000," and by God he did.' I was told that after the appearance of *The Mysterious Universe*, Rutherford called his 'boys' together and said that if any one of them was thinking of writing a book on the Atom, and calling it *The Mysterious Atom*, he had better come to his study for a talk before going to a publisher.

The Cambridge University Press did not like Eddington's publishing at Oxford, and in spite of further attempts, I did not succeed in persuading him to give another book to Oxford. I also tried to persuade Jeans to write for Oxford, but did not succeed. Jeans could be brusque with people he did not want to see; I was therefore much surprised when he most politely asked me to see him, and even more politely discussed publication with me. I learned later that the real object of his conversation was to see what terms he might get from Oxford to bargain with other publishers.

When the Cavendish Laboratory was extended in the 1930s Eddington wrote a beautiful pamphlet about it, as part of the publicity material in the appeal for funds. During the Second World War, I asked him to expand it, especially for distribution abroad, as a simple exposition of British scientific activity. In his letter of refusal he asked how could one expand a finished work of art? It was impossible to add a piece to an artistic whole.

The other important event for me at the British Association meeting in 1926 was a conversation with Sidgwick. I asked him about scientists who might be valuable authors, and he recommended me to note a Russian named Kapitza, who was working on magnetism in the Cavendish Laboratory; a book from him on magnetism would be very worthwhile.

I accordingly went to Cambridge to see Kapitza. He was then living in Whewell's Court, and when I entered his room, my attention was seized at once by a huge Russian bear-skin rug, beyond which a dark-haired, smooth-shaven Russian with a fine complexion was lounging. He greeted me in a friendly voice, which broke occasionally into falsetto notes, and he spoke an English which contained a number of unintelligible noises. This did not in the least impair his exceptional intelligibility, which was a product of his whole personality. It was said of him that 'he spoke all languages equally well'.

I asked Kapitza if he would write a book on magnetism. He said he might, sometime, but he characteristically gave his own turn to the conversation. I imagine he did not want to write a book on magnetism just then, because he was in the midst of his exciting researches on the production of very strong magnetic fields.

He said the whole question of the publishing of advanced books on physics was very interesting. He would like to think about it, and invited me to call to see him again in about a fortnight's time. When I called on him again after that interval, he told me that he had given further thought to our conversation, and had come to the conclusion that if he were to be associated with any publishing project, it must have two qualities; firstly, it must be international, and secondly, it must be big. He said that R. H. Fowler ought to be brought into it, and he would discuss the idea with him.

The Press's *International Series of Monographs on Physics*, with Kapitza and Fowler as editors, was the outcome. Fowler approached his former pupil P. A. M. Dirac for a book on quantum mechanics. Dirac wrote his famous *Quantum Mechanics* forthwith. Consequently, the series began with a classic, and for a number of years was the leading series of its kind in the world.

When I first called on Dirac he was living in a simply furnished attic in St John's College. He had a wooden desk of the kind which is used in schools. He was seated at this, apparently writing the great work straight off.

The success of Eddington's *Stars and Atoms*, and of the *International Series of Monographs on Physics* made my proposals more welcome to the Oxford office, and I devoted more and more of my efforts to the kind of book in which they were concerned.

In 1939, after the outbreak of the Second World War, I transferred to the Oxford office.

During 1927–28 I was very busy with the publishing of scientific and technical books, but at the same time, the scope for scientific articles for the *Manchester Guardian* grew rapidly.

CHAPTER FOUR
1927-28
INVENTING SCIENTIFIC JOURNALISM

I FIRST BEGAN to write articles on science for the press in the spare time left to me from publishing, as a means of expression that I could not find in business. I did not immediately have the idea of making a living by writing on science.

During 1927-28, I published about sixty articles in the *Manchester Guardian*. One of the earlier ones was on Thunderstorms, for Leonard Woolf's suggestion was evidently still with me. I had by that time learned something of the theories of G. C. Simpson and C. T. R. Wilson.

The researches under Sir William Hardy on the preservation of meat directed my attention to experiments on the freezing and thawing of living organisms, which led me to speculate on the prospects of frozen man in an article on *The Man in Cold Storage*, which was published on 26 November 1928.

'A kind of refrigeration mausoleum might be conceived, in which persons could be preserved for any stipulated period.' An inmate might be thawed out for a time and then frozen again. By reducing the intervals of thaw to one month every 300 years, he could extend his survival for 180,000 years, assuming that he could have a life of seventy years altogether when he was in thaw. What would historians not give to be able to take Rameses the Great, Archimedes, 'Junius', Shakespeare, Newton, Marx, even the meanest peasant of any century from cold storage, thaw him, and question him? Bertrand Russell had quoted the lament of the Chinese poet:

> My eyes saw not the men of old;
> And now their age away has rolled.
> I weep—to think I shall not see
> The heroes of posterity.

But the progress of science had suggested that perhaps men of the future may not always have to weep with the poet. Advances in biochemistry might show how people could be preserved over vast periods, and be recalled at intervals to ordinary life.

My early articles dealt in the main with experimental biology, cosmology, and physics. I went to Giggleswick in North Yorkshire to observe the total eclipse of the sun on 30 June 1927. The sky had been almost continually overcast for a fortnight, to the despair of the astronomers. Then to their joy, the sky cleared five minutes before totality and excellent photographs of the corona were obtained. They were particularly important in their bearing on E. A. Milne's theory of the calcium lines in the spectrum of the corona, which he explained as due to a peculiar feature of the structure of the calcium atom, which enabled it, as it were, to dance upon the sunbeams. The eclipse was watched by many thousands of people spread on the surrounding hills, and the spectacle was greeted by triumphant applause, as if the Almighty was being congratulated on having delivered the goods.

On 28 March I published an article on *Ball Lightning*. I found that natural phenomena of this kind often drew correspondence, whereas articles on the science of the laboratory and the study rarely did.

I published my first article on *Cosmic Rays* on 25 October 1927, and on 4 April 1928 my first on *Smashing the Atom*. The papers of Eddington and Jeans provided material for several articles on the evolution of the stars and planets.

In March 1928 I published a paragraph on Kapitza's method of producing intense magnetic fields, and later in the year an article on C. V. Raman's discovery of his famous effect, under the title: *A Really New Ray*. Among other articles of this period were several on *Colloids*, which I felt were among my best as pieces of exposition. I had met N. K. Adam, which led to his publishing his classic book on the *Physics and Chemistry of Surfaces* with the Clarendon Press. Adam had exceptional powers of exposition, and could cause his hearers to see in their imaginations the positions and movements of molecules on surfaces, which provided the explanation of many fundamental properties of matter, both non-living and living.

I also met Pavlov for the first time. This was on the occasion of the delivery of his Croonian Lecture for 1928 to the Royal Society. Pavlov was then seventy-eight. He was a small, very alert man, walking with a limp caused from a fall years before on the Leningrad ice. His son acted as interpreter, and read his lecture in English. A. V. Hill had originally suggested to me that I should see G. V. Anrep, who was translating Pavlov's book on *Conditioned Reflexes* into English. I brought this to the attention of the Oxford Press, who published it. I had only a few words with Pavlov on this occasion, but six years later had a conversation with him in Leningrad, to which I shall refer later.

I published an article on *Biophysics* in February 1929, in which I suggested that it was one of the sciences of the future. 'One of the next words to fly from the Ark of Science to the Mount Ararat of Fleet Street may confidently be prophesied to be biophysics.' It drew an angry criticism from a well-known anatomist at Liverpool University, who asserted that biochemistry and biophysics 'are wholly foreign to biology, and I cannot imagine their ever being brought into any relation with biologists'.

Quite a number of people expressed interest in my *Manchester Guardian* articles. Among these was C. K. Ogden, the linguist and philosopher, and inventor of Basic English. He published a collection of my articles under the title *Science for You*, and he used this title for a series of books of collected articles by various writers. I contributed three collections to the series, the other two being called *Short Stories in Science* and *Osiris and the Atom*.

I concluded *Science for You*, which was published in March 1928, with an essay on *Science and Journalism*. I said that the communication of the results, and something of the temper of scientific research, to the general public was of immense importance, because to a large extent modern industrial society was the outcome of the application of science to traditional husbandry. As a consequence, in Britain alone, the population had been more than quadrupled. Most of the people living in Britain in fact owed their existence to science. From this it might fairly be deduced that science would have an equally extraordinary effect on social life in the future. The effects of the application of science to industrial and social problems were often not immediately obvious. Consequently, the majority of people who owed their existence to science, were not aware of it. They were nearly all oblivious of the fact, especially if they happened 'to utilize the privilege of their science-begotten existence for the study of the ancient classics'.

The public should be made to realize that social organization is in the long run very sensitive to the application of science. 'Such convictions, if they came to be at all widely held, would exercise an astonishing influence on the future of human society. The public interest in science would beget a public scientific understanding and conscience, and just a little of these, by themselves, would have remarkable effects.' The public would then, for instance, perceive that the scientific experts on coal-mining were on the whole second-rate. The mental atmosphere of the industry was 'repellent to first-class scientific brains', which preferred 'to be happy and poor in the Cavendish Laboratory rather than miserable and rich in the coal industry'.

Assuming that a public scientific conscience was very desirable, how could it be created? I said that it was a mistake to depend for this on school education, and pointed out that 'science is the art of doing precise thinking with approximate facts', and that a great deal of precise thinking could be done with very rough facts indeed. Here was 'the key to the problem of making the public a lot more scientific than it is'.

I remember my sensation when I received the first review of my first book. I was staying in a small boarding house in an industrial city, listening to the provincial chatter at breakfast. When I read the review cutting from one of the leading national newspapers, a window seemed to open for me on an extended world. This first review brought an inquiry for articles from a leading South African paper, which wanted to reprint my *Manchester Guardian* articles.

One of the first scientists to take my scientific journalism seriously was Kapitza, who had an exceptionally wide range of interests and imaginative understanding. These Russian qualities contrasted particularly with most of his British colleagues, who were extremely able men, but in comparison with him, conventionally British. In fact, it seemed to me that Kapitza's contribution to Cambridge of a different intellectual outlook and tradition was even greater than his scientific discoveries. His widening of the Cambridge outlook perhaps contributed more than his actual researches.

One day he said to me that my work was important, and that something should be done about it. Rutherford had vaguely associated me with J. A. Crowther, who, unfortunately in my opinion, had become Rutherford's example of a bad physicist. J. A. Crowther had produced experimental evidence in favour of J. J. Thomson's conception of the atom as a sphere of positive electrification in which electrons were embedded. Rutherford had had to demolish this conception in his creation of the nuclear theory of the atom, so he did not appreciate those who had supported it. J. A. Crowther was never elected a fellow of the Royal Society, in spite of doing very good work in other aspects of science. The eminent French medical radiologist A. Lacassagne told me that he regarded J. A. Crowther as the founder of scientific radiology when he demonstrated that the destruction of bacteria by radiation was due to its physical effects, and obeyed the laws of statistics. There was no specific kind of radiation disease, as medical radiologists had hitherto been inclined to believe. This cleared the way for an objective explanation of the interaction between radiation and living cells, in terms of known principles.

While the similarity of my name to that of J. A. Crowther helped to secure immediate notice in some quarters for my first books, it also had its drawbacks. Kapitza said he must explain to Rutherford that I was in no way connected with J. A. Crowther. Whether he did so or not, Rutherford expressed most encouraging appreciation of my work, and ever afterwards treated me with kind consideration. J. A. Crowther reported to me from time to time that he had had letters from people about my books.

In the last chapter attached to my second collection of articles, *Short Stories in Science*, published in 1929, I dealt with the relation between science and literature. I argued that both depended on the exercise of the imagination, the scientist imagining how things work, and the writer how people behave; both were rooted in the same kind of imaginative faculty. Shakespeare and Bohr, for example, were of comparable imaginative power, one applying this power to the evocation of people, and the other to that of the quantum operations of the atom. It followed, of course, as had been said long before, that science and literature are different aspects of one more fundamental thing, and at a sufficiently deep level there is no fundamental difference between them.

C. K. Ogden was one of the most original, interesting and entertaining men of his period. Besides his own work on linguistics and the philosophy of language, he persuaded Wittgenstein to publish his *Tractatus Logico Philosophicus*, and J. B. S. Haldane to write *Daedalus*, the first of his pieces of general exposition for the public, and encouraged and published the works of many other authors. As a contribution to the propagation of Basic English, I wrote an article on it for the *Manchester Guardian*, which was transcribed into Basic. The experience was instructive, as only a few words of the original draft in ordinary English had to be changed. It appeared that if one was engaged in expounding science to the public, one automatically invented something like a Basic English of one's own.

Ogden had several houses in London, Cambridge, and Brighton. Each had its distinctive and original character. He had a wonderful collection of ancient musical boxes and toys, which he delighted to play, and sometimes to play all together. He collected 'ullages', that is, the remnants of unfinished bottles of wine; he delighted to try with his friends the wide range of well- and little-known drinks he had acquired. He decorated his walls with communications from the great, from Franklin Delano Roosevelt and Winston Churchill downwards.

He was always happy to entertain one's friends from abroad, who were astonished and often captivated by his unique habitations. He produced the reverberating effects of a cathedral by declaiming like a priest on a deep staircase in his house in Montague Street. He wore masks as he gave imitations of famous characters, his favourite being an imaginary speech by George Lansbury.

As the *Manchester Guardian* took more and more of my articles, I conceived that I might make writing on science my main occupation. I began to see that the informing of the public on the progress of science was acquiring new aspects. The amount of science, and its effect on daily life, had greatly increased. In the past, much of the exposition of science for general audiences and readers had been conceived as a contribution to education and to intellectual entertainment. In the latter part of the eighteenth century, and the first half of the nineteenth, there were able men, such as Waltire and Spurgeon, who made their living by lecturing and writing for the public. In the second half of the nineteenth century there were scientists like T. H. Huxley and John Tyndall, who devoted regular effort and attention to writing for the public on science. They regarded it as a social as well as an educational necessity that the public should have a more correct understanding of such fundamental matters as the evolution and origin of living things. They were followed by such men as Sir Richard Gregory, an astronomer who became the editor of *Nature*, which had been founded in 1869 precisely to widen the understanding of science, and H. G. Wells, who dealt with scientific themes. *Nature* acquired a unique position, but it became more a journal for the scientific profession than for the public. Gregory was a statesman of science rather than a writer. In the second quarter of the twentieth century the books of Eddington and Jeans were in the forefront of the works for general readers.

Huxley and Tyndall, Gregory and Wells, and Eddington and Jeans were not, however, primarily writers on science for the public.

Some of the leading newspapers commissioned regular articles from scientists, who were described as their Scientific Correspondents. One of the most noted of these was Sir Ray Lankester. He republished many of his articles under the title *Science from an Easy Chair*. This perfectly described his approach, primarily one of information and intellectual entertainment.

In Britain, the question of the exposition of science was changed fundamentally by the First World War. It had then become evident to a much larger number of the population that the safety and survival of the country depended on the proper utilization of science and technology. A new motive now appeared for the exposition of science.

Profound comments on the social significance of science had been made for centuries before 1914–18, from Bacon to T. H. Huxley, but they had mainly affected the educated classes, and those in search of education and philosophic enlightenment. The new concern with science was inspired by political rather than educational motives. It seemed that there was a need for a new kind of writing: regular exposition of science and its developments, with systematic emphasis on their social significance. Such writing could not be undertaken by scientists in their spare time, however brilliant their occasional contributions in this direction might be.

In the 1920s the prejudice in scientific circles against scientists who wrote in the press was severe. I remember going to interview the eminent physiological chemist Sir William Hardy, who became such a close competitor of Sir F. Gowland Hopkins for the presidency of the Royal Society. He told me that J. B. S. Haldane, and others of the then younger biologists, had done their scientific careers great harm through writing scientific articles for the daily press.

Some years later I saw Hardy again. He had in the meantime become a staunch supporter of the exposition of science in the press. He was now Director of the Department of Scientific and Industrial Research's Low Temperature Research Station at Cambridge, concerned with the preservation of food. The station carried out first-class research of great practical value on the preservation of apples, and other problems. Hardy had become aware of the necessity for securing public interest in laboratories supported by public money.

Hardy's earlier attitude was a striking illustration of a characteristic weakness of specialism and professionalism, the tendency to esteem only discoveries of abstract knowledge. Scientists concentrating their efforts in a particular field must necessarily be prone to it, and can guard against it only by taking thought, as it is inherent in their activity.

It gradually became clear to me that the regular exposition of science and its social significance to the public could not be satisfactorily carried out by scientists in their spare time. It became evident that it would have to be the main activity of the writer, if he was to carry it out adequately. As the scientist was unable to do this job adequately in his spare time, so also was the working journalist, who had to cover the whole of the news.

The parallel development in the United States had rather different features. The opening-up of a rich continent, in which there was an extreme shortage of workers, gave a big stimulus to invention. A strong tradition grew up, of which Edison was the outstanding exponent and symbol, for mechanical and scientific inventiveness. The *Scientific American* was founded in the middle of the nineteenth century to provide a medium for the expression of this interest. The American daily press took note of inventions, to meet the wide public interest in their practical and commercial possibilities.

In the 1920s the American press also began to feel the growing need for systematically covering scientific news. On one side, scientific organizations created a *Science News Service*, and on another, the American tradition of interest in inventions and marvels was applied to daily reporting of science news by working journalists. Able reporters, who might previously have been covering politics, crime or sport, began to concentrate on science. These men created the typically American invention of Science Writers, primarily non-scientists expounding and commenting on science. Fine work has been done by a number of these men; at press conferences a scientist might be very effectively interviewed by one of these able American non-scientific journalists, who had had a wide experience of men and things. Such a journalist may not be aware of the finer prejudices and inhibitions of the scientists, and quite unconsciously his shrewd questions may bring these out in an illuminating way. Latterly, with the extension of science coverage, American science writers have been recruited more from scientists, and their approach and that of the European scientific journalist have become closer.

Science writing has its place, and it may be very good. But it is not quite what I had in mind as scientific journalism. I wanted to see the exposition of science carried out by writers with considerable scientific knowledge, and not depending only on natural shrewdness. Putting the problem another way, it might be approached from the side of the public, to which the science writers belonged, or from the side of science, to which the scientific journalists belonged.

In the course of forty years, these approaches have tended to merge. But new differentiations and limitations have arisen, which will be commented on later. In the late 1920s, when I first began to think and act about the problem, the two approaches were fairly distinct.

By 1928 my conception of scientific journalism had become

Inventing Scientific Journalism

sufficiently definite. I felt that I could now attempt to establish it on a more professional basis. I sought an interview with C. P. Scott. The News Editor of the *Manchester Guardian* informed me that if I would call in Manchester on a certain day and time, Mr Scott might be able to spare me 'two minutes'. I arrived at the *Manchester Guardian* office in Manchester shortly before the appointed time, and asked the porter for Mr Scott. The porter was a big, forbidding man, who looked at me with a withering suspicion. I was ushered through passages into a library furnished in dark wood, and asked to sit and wait. Opposite me was a bust of Scott by Epstein; it looked like the bust of an Old Testament prophet.

I was kept waiting a long time, and then someone came and told me Mr Scott was now ready to see me and I was taken along a corridor to his room. The door was open, and I walked in. The room was rather small and dark. I was arrested by an extraordinary face, on which nearly all the light was focused. It shone out against the dark background like the face of Jehovah.

Scott said, without asking me to sit down, 'Well, Mr Crowther, what can I do for you?' 'I wish to become a scientific journalist,' I said. Scott replied: 'The trouble is, Mr Crowther, there isn't such a profession.' To this I answered: 'I propose to invent it.' Scott then asked me to sit down. I elaborated the views on science and journalism that I have already mentioned, and Scott listened, interposing comments from time to time. I remember his saying that he used to meet Rutherford on the Senate of Manchester University, and never could regard him as other than a great lively schoolboy.

The two minutes lengthened into more than an hour. Every few minutes, men came in with copy for his inspection. He ignored all these interruptions, and after this had happened several times, I noticed that some of the men stared at me hard.

I said to Scott that I wished to be *Scientific Correspondent* of the *Manchester Guardian*. He said that they had never had one, that he would have to think about it, and that he must discuss it with his colleagues. He got up from his chair and said he would have a word with them. Meanwhile, he handed me a copy of *The Times*. The fact that he did not give me his own paper to read impressed me. After ten minutes or so, Scott returned, and informed me that my being appointed *Scientific Correspondent* involved a question of policy, as they had not had one before. He would write to me in a fortnight's time, giving his decision.

I returned to London, and about a fortnight later, on 7 December 1928, Scott wrote to me that 'We are quite willing to appoint you as our scientific correspondent, if that will assist you in your work for us.' So far as I know, this was Scott's last innovation in the *Manchester Guardian*, for he retired soon afterwards.

During these negotiations I was asked to put my ideas on paper, and in these I said that: 'I should hope in time to make the *Manchester Guardian* easily pre-eminent among daily papers for its information and attitude to science. I think the attitude to science is more important than mere scientific information. The public should be helped to realize the greatness of science, and its significance for society and the mind.'

Scott appeared to me to have a higher order of greatness than any other man whom I have met in journalism. He was more perceptive and courageous about progressive causes than many who prided themselves on their advanced views. He was a survivor from the powerful constructive social forces of nineteenth-century Manchester, which had fostered Joule in science, and Engels in sociology. I found him more understanding and helpful than twentieth-century men who considered that they understood twentieth-century society better than he did. Scott's deeper penetration and strength were an inheritance from mid-nineteenth-century Britain which was still in an ascending phase, and contrasted with the superficiality of a country now in a decline.

One of the fundamental strengths of the *Manchester Guardian* at this period was that it was consistently run at a loss; commercial considerations scarcely entered into any decisions affecting moral and intellectual attitudes. This was of enormous help in writing on the significance of the news, and it applied just as much with regard to science as to other subjects. One was not highly paid, but was received with openness and confidence in scientific institutions in countries all over the world. One was favoured with much good information, and able to expound it.

The encouraging reception of my two collections of scientific articles, *Science for You* and *Short Stories in Science*, attracted the attention of W. S. Stallybrass, the manager of Routledge and Kegan Paul, who published the books on C. K. Ogden's backing. He invited me to call. When I arrived, he had the books on his table. He picked one of them up and said: 'You have put a good deal of work into these; why not write a book which could have a wider sale and bring in some money?' He told me that he had approached Jeans for his *Universe Around Us*, and had offered him £750 in advance royalties for it. He had been prepared to go to £1,000, but Jeans had only used this offer to improve the offer from the Cambridge University Press.

Inventing Scientific Journalism

I had thought of writing *An Outline of the Universe*; my idea was to write a brief outline of the physical universe for the general reader. I accordingly drew up a suitable synopsis, which Stallybrass accepted. I thought that the agreement should be drawn up on a business basis, so I went to the agents, Curtis Brown, and asked them to arrange it. The member of their staff who dealt with it was David Higham, who has handled all my books since then, both while he was with Curtis Brown and afterwards, when he founded his own firm.

With my *Manchester Guardian* connection, and my contract for *An Outline of the Universe*, I felt I could now risk devoting myself mainly to writing, so I proposed in 1930 that my engagement by the London office of the Oxford University Press should be converted into a half-time appointment, to which Sir Humphrey Milford agreed.

From this time, I had half of the year entirely at my own disposal. This enabled me to undertake sustained scientific writing, and it also enabled me to visit scientific institutions abroad systematically. In April 1930 I went to live in Berlin for a time, to visit scientific institutions and to begin writing *An Outline of the Universe*.

CHAPTER FIVE
1929
SCIENCE IN SOVIET RUSSIA

BEFORE MY PUBLISHING engagement was made half-time in 1930, I travelled abroad very little. In the summer of 1929, however, I spent my vacation in Leningrad and Moscow. I went to visit my old school-friend Ralph Fox, and to write articles on Soviet scientific institutions. Ralph and I had met when we were pupils at Bradford Grammar School, at the time of the First World War. He was then about sixteen years old and had already shown literary ability. He later became a revolutionary writer, and ultimately died fighting for the Republicans in the Spanish Civil War.

Ralph had gone to the Marx-Engels Institute in Moscow to edit the collection of articles which Marx and Engels contributed to the New York *Tribune*. His presence in Moscow decided me to go there; I had meals with him in his apartment, and he gave me advice in finding my way about.

When Ralph and I first met as schoolboys we were both opposed to the existing social order, but he was far better intellectually informed. He was already an admirer of William Morris, whose verse he advised me to read. I tried, but could not make much of it. He read history, and was awarded a demyship at Magdalen College, Oxford, where he gained a first-class degree in modern languages. Before he entered Oxford he was called up for the Army, and started cadet training for an officer's commission. He met some Australians on the same course, and was influenced by their independent attitude, especially to authority. When the war ended, he had still not been engaged in fighting. He secured an early release, and went up to Magdalen. There was an able and energetic group of students deeply influenced by the Bolshevik Revolution and the establishment of the U.S.S.R. Ralph became a Marxist and Communist. I had no clearly formulated ideas about the social order, and through his influence I began to perceive the depth and power of the Marxist analysis of society.

Ralph was early in touch with Soviet developments. He told me of the founding of the Society for Cultural Relations with the U.S.S.R., and at his suggestion I went to one of its earliest meet-

Science in Soviet Russia

ings. I became a member of the Society, and later of its executive committee, and was active in its work for scientific exchanges. Consequently, when I went to the U.S.S.R., the officials of the Society in Leningrad and Moscow were particularly helpful in arranging visits to scientific institutions.

Besides these connections, Kapitza gave me a personal introduction to his father-in-law, Academician A. N. Kriloff, the eminent applied mathematician, naval architect, and translator of Newton's *Principia* into Russian. He was director of the mathematical physics section of the U.S.S.R. Academy of Sciences. Further, as Scientific Correspondent of the *Manchester Guardian*, to which the Soviet authorities were well disposed, visits to institutions and laboratories were welcome, and facilitated.

I travelled from London to Leningrad in a Soviet ship. Andrew Rothstein had asked me to take a parcel of books to his father, the noted journalist and historian of the Chartists, who had been appointed a member of the U.S.S.R. Academy of Sciences. When I got to the cabin on the ship in the London Docks, I found not a parcel but a large trunk of books; its transport was quite a considerable task. When I arrived at Moscow I was met by Ralph. We secured a droshky and took the trunk to the apartment building where Rothstein lived, and found we had to roll it up several flights of stairs. We did this, placed it in front of the door, pressed the bell, and retreated a few yards, to watch Rothstein's expression when he saw it. However, he viewed it and us without batting an eyelid.

In Leningrad I stayed in the hotel generally used by foreigners. I called on Kriloff, who received me very helpfully, and enabled me to visit several of the Academy's laboratories. In the following year, he arranged for me to meet Kulik, who had recently returned from an expedition to Siberia, to investigate the site of the great meteorite which fell there in 1908 and caused extensive and long-lasting destruction. He had remarkable photographs of the region, prints of which were given to me. The articles that I wrote on this meteorite, illustrated with Kulik's photographs, attracted considerable attention in England and the United States.

I wrote a series of four articles on *Science in Soviet Russia*, which appeared in the *Manchester Guardian* in November 1929. They were the first review by an English scientific journalist of the development of science in the U.S.S.R., and of the projected development of science under the First Five Year Plan. I had not previously grasped the concrete reality of the Plan with regard to science, and even after my visit I did not fully realize many of the implications of it. However, I reported what I saw, and accepted at its face value everything that was said to me by responsible scientists and officials. The result was that some of the things I wrote were wiser and more important than I myself realized at the time. I accepted the reality of the planned development of science under the First Five Year Plan, which had begun only in 1928, and was just getting under way. It was evident that some of the older scientists who had survived from pre-revolutionary times, and held responsible positions in the new Soviet administration, did not believe in the Plan, though they repeated the official views.

My first impression was that scientists and authorities were only too glad to assist me to see everything I wanted to see; this was before the rise of Hitler. The scientists had met few Englishmen since the Revolution. The director of one of the chief laboratories told me I was only the third Englishman with scientific interests to visit his institution since 1917. The degree of isolation of Russians was such that I met educated people who immediately assumed I was an American because I spoke English, and that the *Manchester Guardian* was an American newspaper.

The isolation at that time was not due to military security precautions. It was easy to obtain a visa to visit the U.S.S.R., and travelling was cheap. No scientists went because they assumed that there was nothing to see. After I had been in Leningrad a few days I found I could ring up the directors of laboratories out of the blue. As soon as they understood there was an Englishman on the phone, they found a member of staff who spoke fluent English, and often they came as quickly as possible to bring me to see their institutions. In this way, I met Professor Egiazaroff, the hydroelectric engineer. He had studied at the Metropolitan–Vickers works at Old Trafford in Manchester. He told me about his researches, and that he was consultant for what was believed to be the largest projected hydroelectric plant in Europe, being erected on the River Dnieper. In Moscow I saw the hydrodynamical model of the Dnieper Dam and power station, and I reproduced a photograph of it. Nevertheless, I did not fully grasp the implications of this information until I saw, and walked across the Dnieper Dam, several years later.

On that occasion I saw, as I reached the middle of the Dam, a small party approaching from the other end. There was an important-looking figure in front, followed by several colleagues a yard or two behind. I had heard that A. P. M. Fleming, the director of research and education in Metropolitan–Vickers, where Egiazaroff was trained, was visiting the U.S.S.R. I suspected it might be him, and indeed it was. I had not met him before, and so we first became acquainted in the middle of the roadway across the Dnieper Dam.

On my visit to the U.S.S.R. in 1929, I was impressed by the State Optical Institute where I first met S. I. Vavilov; the Department of Applied Botany directed by his brother, N. I. Vavilov; the Physico-Technical Institute under A. F. Joffe; the Electrotechnical Institute; and the Hydrodynamical Laboratory in Moscow.

The State Optical Institute had been one of the first and most important elements in the planned development of science. Its aim was to make the U.S.S.R. independent in optical instruments, both for civilian and military purposes. The early foundation of a first-class optical research institute later became very important in preparing the way for development in astronomy and space research, and it also stimulated the kind of research that ultimately led to the invention of lasers.

At the Physico-Technical Institute, Joffe explained to me how Lenin had called upon him to produce the physicists needed to carry out the projected development of Soviet physics. In 1918 there were only about forty effective physicists left in the whole Soviet Union, covering one-sixth of the globe. The only way of obtaining the new physicists was to concentrate the forty in one place, and organize them to train promising young men. The Physico-Technical Institute was accordingly founded to conduct research and train professional physicists. Joffe told me that, by 1929, they had already trained two hundred good new physicists. As capable men emerged, institutes in various places in the U.S.S.R. were founded for them. Joffe impressed on me that, according to the plan of scientific development, the men were produced first, and then the institutes were built for them; not the other way round. By creating new institutes in various regions after the talent had been discovered and trained, they were able to contribute to the extremely important object of creating new centres of culture in the main provinces of Russia.

Joffe had with much foresight concentrated attention on solid state physics. He had promoted research on the strength of solids, thin films, and electron spraying processes. Consequently, thirty years later, the Soviet Union was particularly strong in this branch of physics with its fundamental industrial as well as scientific applications.

I visited N. I. Vavilov's headquarters, which were housed in the former Italian Embassy and an old church. The scale and vision of his researches in the origin of domesticated plants and their improvement for agricultural purposes was already evident. I was taken to see two of the department's sub-stations in Detskoye Selo. One was the genetics section, in an old palace or hall, which Queen Victoria had presented to the Grand Duke Boris; one of the palaces belonging to Prince Yussupov, the assassin of Rasputin, had been turned into a flour-testing station.

At the Pulkovo Observatory, research on seismology was being vigorously developed. Prince Galitzin had been one of the founders of the science, and the development of the country's enormous natural resources required its application in mineral and oil prospecting. This energetic development of seismology was also in the future to prove of vast military importance in such matters as the detection of ordinary and nuclear explosions.

In 1929 Moscow presented a much less advanced appearance than Leningrad. The headquarters of the U.S.S.R. Academy of Sciences and other major institutions had not yet been transferred there. The most interesting development was the half-built new Experimental Electrotechnical Institute. A laboratory for high-voltage testing, larger and more advanced than anything of the kind in England at that time, was being built. Another large laboratory was to be devoted to research on physico-electrical work, vacuum tubes, radio valves, X-ray apparatus, and what is now called electronics. Two large blocks of flats for staff were being erected. Accommodation was provided for foreign as well as Soviet scientists and engineers, who were generously treated. I heard that Lange, the German expert invited to advise on high-tension electrical apparatus, found it necessary to buy a suitcase to hold the roubles given to him for incidental expenses.

Another significant institution was the Aero- and Hydro-dynamical Laboratory. This was being energetically developed, but was still accommodated in makeshift buildings. It was providing the foundation for the development of Soviet aviation and hydroelectric engineering; it was under military direction, but I was shown over it in an open manner.

Among the Soviet scientists with whom I became acquainted in 1929 were Engelhardt and Frumkin. To have met these and other scientists early, and then have seen their rising achievement, has been instructive.

I was greatly impressed by the amount and variety of new scientific institutions being built. It seemed that the scientific staffing problem was the most difficult part of it. I wrote in the *Manchester Guardian* that 'It is quite possible the problem of discovering and manufacturing specialized ability is the greatest confronting the Soviet, greater than political problems, which cause much more noise in the world.'

In one of my 1929 articles I asked the question: 'What will come out of the Soviet's integration of science, the State, and industry? In twenty years' time quite possibly the most powerful country in the world. In many parts, at present, only the scaffolding of the integration exists, but the idea seemed to me to be there.' However, the outcome did not appear to me to be certain. In Moscow especially, the old Russia was still heavily present. Moscow was still like a vast village of large country houses in gardens, together with peasant slums. The new clearings and construction were only scratches on the surface. There were many private shops, and a good deal of leisure. There was time for long café discussions, and, as I have already mentioned, it was evident that a number of older experts in responsible positions were at least sceptical of the reality of the Five Year Plan, and were merely conforming with instructions in which they had not much belief. It appeared to me in the summer of 1929 that the U.S.S.R. might relapse into social democracy and a mixed economy; the issue had not yet been decided.

I had seen the scientific situation for myself and had described it in the *Manchester Guardian*. I expanded my articles into a book, which was published in 1930 under the title *Science in Soviet Russia*. It was translated into Spanish by F. Giral, the son of the last Prime Minister of Republican Spain.

Ralph Fox took me to have tea with Rose Cohen, who had been secretary of the remarkable summer schools run by the Labour Research Department at Cloughton near Scarborough in the early 1920s. She had married a Russian official, and was now settled in Moscow.

Quite a time after we had finished tea, her husband, D. A. Petrovsky, came in from his office. He was tired, and took little notice of our conversation, until technical education was casually mentioned. Petrovsky's attention was instantly aroused, and Ralph told him I was acquainted with technical education in England. Petrovsky had recently been appointed Chief of Higher Technical Education under the Supreme Economic Council, with the task of producing the engineers required to carry out the First Five Year Plan. He asked me many questions about British technical education, so I suggested that, if it would be of assistance, I would form an advisory committee of British authorities on technical education, to furnish advice and information. He warmly welcomed this idea, and said he would send one of his assistants to Britain to confer with them.

Petrovsky subsequently told me that G. M. Krzhizhanovsky, the chief founder of the First Five Year Plan, had referred with approval in the Soviet press to my *Manchester Guardian* articles on Soviet scientific and technological development. Petrovsky had outstanding ability. He came from a Ukrainian village, studied in Paris, edited a German socialist journal in New York, and returned to Russia after the outbreak of the Revolution. He soon became a member of the military council under Trotsky, and I was told that he was at one time Trotsky's chief of staff. He became director of education of officers in the Red Army. In memory of his military days, he used to keep his enormous ceremonial sabre propped up in a corner of his sitting-room. He spoke English, French and German excellently, and had been described as the best chairman in Russia.

The launching of the First Five Year Plan in 1928 created a crisis in the supply of engineers. Petrovsky told me that Stalin sent for him and said: 'You trained the officers for the Red Army, go and train the engineers for the Five Year Plan.' Petrovsky had immense administrative capacity, and was free from the common Russian habit of wasting time. I once asked him what made Stalin such a great man, and he answered with one word: 'Brains.'

After I returned to England I formed a committee consisting of B. Mouat Jones, then Principal of the Manchester College of Technology, and later Vice-Chancellor of Leeds University; Dr Herbert Schofield of Loughborough College, now Loughborough University; Professor William Kerr of the Royal Technical College, Glasgow, now the University of Strathclyde; and Dr S. J. Davies of King's College, London, later Dean of the Royal Military College at Shrivenham. This was during the winter of 1929-30, in the period of the second Labour Government. Some of the members felt that official approval of their participation was desirable, and made inquiries in Whitehall. Sir Charles Trevelyan was President of the Board of Education, and they were informed that their participation would be regarded as a useful contribution to the improvement of Anglo-Soviet relations in the educational field.

Petrovsky sent S. P. Tambovtzev to confer with us. He was an able mechanical engineer in his early thirties, with a solid and attractive personality, who came from a village in South Russia. Petrovsky had planned to send him to the United States, Germany and France, in that order, to study technical education, and bring back such information as might be useful in the Soviet development. As a consequence of our intervention, Tambovtzev came to England first.

Our committee, with the goodwill of the Board of Education, made an excellent programme of visits for Tambovtzev, designed to show him examples of different aspects of the British system of technical education. Tambovtzev was shrewd and industrious, and returned with much information. He afterwards went to the United States, where he stayed longer; the rise of Hitler prevented him from going to Germany. Petrovsky did not conceal in his conversation with me that he considered American and German technology and technical education far in advance of the British.

When we had given Tambovtzev all the information which we thought might be useful, I informed Petrovsky that we could not do more until we had visited Soviet technical colleges and schools, and seen on what points we might be able to make further useful suggestions. Petrovsky accordingly invited us to visit the Soviet Union.

This was in the autumn of 1930. In the meantime, in the spring of 1930, I took advantage of my new half-time working arrangements to live for about three months in Berlin.

CHAPTER SIX

1930
BERLIN

KAPITZA'S WORK ON intense magnetic fields had attracted international attention. Editors wanted to have information about it, and Kapitza himself discussed his researches with leading colleagues in other countries. With his energetic and imaginative personality, he made many friends in the centres where magnetism was studied. Among these were Langevin and his pupil Pierre Biquard in Paris, and Dr Paul Rosbaud in Berlin. It was through Kapitza that I met both Biquard and Rosbaud.

When I arrived in Berlin in the spring of 1930, I took a room in a small hotel near the Friedrichstrasse Station. I got in touch with Rosbaud, who was then editor of *Metallwirtschaft*. He had an exceptional interest in, and insight into, scientists as people as well as discoverers. He greatly admired Kapitza. He told me much about science and scientists in Berlin.

As Scientific Correspondent of the *Manchester Guardian* I went to see industrial and academic laboratories that he and others suggested, and wrote articles about them. The Siemens and Halske research organization was particularly helpful. The head of their public relations department called for me with one of the firm's fleet of Mercedes cars. He was a clean-shaven, middle-aged man, who spoke English fluently, and had an authoritarian manner. I was taken round the laboratories at Siemensstadt and entertained there. The most interesting work I saw was on the metallurgy of beryllium, which had converted this very light, brittle, and difficult metal from a laboratory into an industrial product. Among other things, it was being used in an alloy for making springs which were extremely resistant to fatigue.

The public relations man insisted that I should see a film in which the firm was interested, as a partner in the film company that had produced it. I was not in the habit of going to cinemas, and demurred at first, but he insisted. It turned out to be the first run of the *Blue Angel*, through which Marlene Dietrich became world famous; I was unable to deny that the star was unparalleled, and the picture fascinating.

My Siemens and Halske host also took me out to lunch outside the firm on one of our trips. After we had sat down at our table, I noticed that a badge had appeared in his buttonhole, which had not been there before. I asked what it was, and he explained that it was the badge of the *Stahlhelm*. I encouraged him to talk about this organization, and it was soon clear that he was an enthusiastic member of it. He said that he did not wear the badge during the firm's time, but only in his own; I supposed that the workers of Siemensstadt would not have taken kindly to it.

I consulted Rosbaud about getting lodgings, and he suggested putting an advertisement in the paper, and helped me in assessing replies. There was one from the wife of a German doctor in the Motzstrasse. He knew of them, and that they were worthy people. I lived comfortably with this family until the end of my stay. Some of my friends were amused at my staying in the Motzstrasse, which contained the notorious homosexual dance hall, Eldorado, but this was lost on me.

Rosbaud still put in occasional hours on research on the crystallography of metals in the Bodenstein Institute of physical chemistry. He had a bench in F. Simon's laboratory, who was then extraordinary professor. He was conducting his notable researches on low-temperature physics, and had started a new tradition of small-scale experiments. Low-temperature physics had become dominated by Kamerlingh Onnes and the Dutch school, with their large-scale engineering technique. Simon succeeded by great ingenuity in obtaining similar results by small-scale and much less expensive experiments. This made the subject practicable for laboratories with small resources, and enabled research in it to be greatly extended.

At this time, Simon was much concerned with the bearing of the properties of substances at very low temperatures on the structure of cosmic bodies. He had shown that helium could be kept solid at temperatures 40°C above its freezing point, if subjected to a pressure of several tons to the square inch; this led him to believe that all matter is solid under sufficient pressure, and that the centre of the earth is solid, in spite of the high temperature.

I saw Simon from time to time after he came to England, where he participated in the development of low-temperature physics at Oxford. I retained the impression that he was at his very best in the Bodenstein Institute. He then had no large administrative burdens, and was at home in the scientific environment. In England he was always striving to adjust himself, and also to rectify some of the notorious British failings in fuel economy. His thermodynamic soul was scandalized by British waste. In the

Bodenstein Institute there were no serious distractions to the play of his brilliant creative imagination and experimental ingenuity.

One day, after a walk in the Tiergarten, I happened to meet P. M. S. Blackett. He and his family were then staying in Berlin. At one of the parties in his house I met Houtermans, and Martin Ruhemann who was one of Simon's most talented colleagues. He had invented an ingenious small liquefier, and was forging ahead in research at very low temperatures.

At about this time I was much struck by Blackett's use of terms from ball games in his interpretation of his photographs of the tracks of atoms. This made me wonder whether a tradition of ball games had assisted the British in their conspicuous success in particle physics. They have generally been apt in the interpretation of the particulate aspects of phenomena, whereas the Germans, who have a strong musical tradition, were apt to interpret phenomena in terms of waves. I later noticed that Rutherford was fond of analogies from ball games, and that Isaac Newton had suggested that his light-particles swerved in the manner of a spinning tennis ball. I also noticed that F. Joliot-Curie the outstanding French experimental atomic physicist, was a footballer of international standard.

Rosbaud was acquainted through *Metallwirtschaft* with the Bauhaus architects and artists, who were keenly interested in the non-ferrous alloys which they were utilizing for furniture and works of art. Through him I became acquainted with Gropius, Moholy-Nagy and Kepes. After talks with Gropius I wrote an article on his ideas and work for the *Manchester Guardian*, which was illustrated with a photograph of his own modern house at Dessau.

I commissioned Moholy-Nagy to design the dust-jacket for my forthcoming *Outline of the Universe*. I went to a small flat in Siemensstadt, which was being used as a studio. Kepes was then his assistant (he is now professor at the Massachusetts Institute of Technology), and they concocted the design. My name and the title of the book were painted on a glass sphere; this was placed on a large sheet of paper, on which a large number of concentric and intersecting circles were drawn, and then photographed from an appropriate angle. I thought it the best dust-cover design I had ever seen.

Also through Rosbaud, I met his teacher and friend, Hermann Mark, who was professor in Vienna. I asked Mark what had prompted him to study the chemistry and physics of fibres. He pointed to the influence of shortage of raw materials on invention in the textile industry; for example, the British dominated the wool market and monopolized the best grades of raw wool. The textile industries of other countries had to try to make good products from the inferior grades; the industrialists had therefore looked to the scientists to help them, and this had led to general fundamental research on fibres.

I met Mark from time to time over the years, but the most memorable occasion was shortly after the Nazi occupation of Vienna. Rosbaud was visiting England and called to see me at my flat in Russell Square. He was very much depressed about the prospects of his scientific friends in Europe, and especially about his revered master and friend, Mark. While he was saying that he would never see Mark again, for he would never be able to escape from Austria, my telephone bell rang. I picked up the receiver, and the voice said: 'This is Mark speaking, I am staying at the Regent Palace Hotel.' Within half an hour he was in my flat, and he and Rosbaud embraced. Mark, though of Jewish descent, had very fair hair and complexion, and an athletic figure; he had packed his Alpine climbing equipment into his car, and with his family driven to the Swiss frontier. He succeeded in getting through in the guise of a most experienced and obvious Alpinist. Mark became a professor in the Brooklyn Institute in New York, and contributed greatly to the development of polymer chemistry in the United States.

Kapitza had told me to make a point of trying to see Fritz Haber. I was put in touch with M. Polanyi, the deputy-director of Haber's laboratory for Physical Chemistry, one of the institutes of the Kaiser Wilhelm Gesellschaft. He arranged that I should go to tea with Haber in his house at Dahlem. It was not very large, but furnished with Oriental works of art, exquisite lace, and other choice things. He had the thick neck and shaven head of the Briton's idea of a typical Prussian. He was, however, almost orientally polite. He gave me China tea, but excused himself elaborately for drinking coffee.

We exchanged civilities for some time, and then I put to him my primary question: 'How do you explain the difference in the histories of the chemical industries in Germany and in England?' His deportment changed instantly, and he began a passionate explanation, clenching his fists and waving them above his head, forgetting all about Oriental manners. In the middle of the harangue, one of his senior assistants came in, and stood behind his chair for several minutes before Haber took any notice of him.

The gist of what Haber said was that the British had created the crude chemical industry, and made so much money out of it that they lacked the incentive to proceed to the creation of the fine chemical industry. The German firms could not compete successfully with the established and prosperous British crude-chemical firms, so they sent academic chemists to work in the British factories, learn their processes, and then return to Germany, where they established new firms to manufacture finer chemicals, which the British industrialists did not find it sufficiently profitable to make. Thus a feature of the German chemical industry was that it was founded by chemists, not by industrial and commercial men, as in England. The German chemical industry consequently developed under the ownership and control of men with a chemical training; in it there was no split between the scientific and the commercial sides.

The chemical industry provided the pattern for the development of other German industries. For example, the German metallurgical industry acquired a similar scientific orientation, and became more scientific than the older original British metallurgical industry.

Haber also attributed part of the difference to the British social attitude towards learning and knowledge. He said that when Rutherford went into the Athenaeum Club, the last thing he would think of talking about would be nuclear physics. It was not done to talk 'shop'. In Germany, on the other hand, it was considered polite to talk on subjects on which one was specially informed; for example, it would be polite and proper for himself to speak on chemistry. If he visited the Athenaeum, he would expect non-scientific members to invite conversation on chemical industry. He thought that this social attitude had hindered people in positions of influence in England from acquiring a proper general grasp of, and respect for, science.

I published an account of this interview in the *Manchester Guardian*. The *New York Times* representative in Europe, Harold Callendar, saw it, and asked me to write a similar article for his journal. He said that they had often tried to interview Haber, but had never succeeded in arranging it.

The next time I interviewed Haber was in 1934, after he had been forced by the Nazis to leave Germany. This was particularly poignant, for Haber was a very patriotic German. He had invented the synthetic ammonia process, which made Germany independent of outside supplies of nitrates for fertilizers and explosives, and he endured great obloquy for having led the German introduction of gas warfare in the First World War. When Haber left Germany, he went to Cambridge, which he regarded as the centre of the

British scientific world. He was politely though not warmly received there, and when I saw him, he plainly felt this, though he quite understood the British attitude to himself. The vigorous, almost violent, fist-waving enthusiast I had seen at Dahlem was now transformed into a quiet, very distinguished foreign gentleman.

Not long afterwards, Haber had a heart attack, followed by a second, of which he died. The late Professor F. G. Donnan told me that he believed that Haber's heart attacks were brought on by shock, after receiving a communication from Max Planck. According to what I was told, Planck, who had succeeded von Harnack as President of the Kaiser Wilhelm Gesellschaft, sent Haber a communication in which he was accused of having made unpatriotic remarks in England, and asking for his explanation. No doubt Planck was ordered to do this; nevertheless, the shock to Haber of receiving such a communication from Planck was very great. Donnan had entertained Haber and some friends in a Soho restaurant, and he thought that, possibly, a waiter had reported some of Haber's outspoken comments to the Nazi authorities, who then acted on it through Planck.

The Kaiser Wilhelm Gesellschaft biological laboratories at Dahlem were as distinguished as those in chemistry. Goldschmidt and Stern talked to me on genetics, and Mangold, with whom C. H. Waddington worked, on embryology. My visits to the K.W.G. gave me an insight into the way Germany had planned research. In my article on the K.W.G. organization, published in the *Manchester Guardian* in September 1930, I remarked that it was 'a little frightening'. The first principle of the K.W.G. was 'to search for the newest developments in science and encourage them', and to employ the best managerial ability for the administrative side of scientific organization, and relieve geniuses of all possible distractions. I noted that 'For the first time in modern history Germany's export trade is stated to have exceeded England's. In the last analysis, is this due at least in part to the existence of the K.W.G.?' I thought the suggestion worth pondering.

Besides visiting laboratories in Berlin in 1930, I was also invited to a meeting in the Harnack House of the Kaiser Wilhelm Gesellschaft. This was a club and cultural institution, where the scientists in the various K.W.G. laboratories and other institutions could meet for social and intellectual exercise and recreation. I believe it was here that I first met Szilard, who appeared to me at that time as more interested in general ideas than in physics. I noted that, though he was a young Hungarian in Berlin, he carried a copy of the *New Statesman* around with him, which he read in tramcars with absorbed attention.

Polanyi invited me to a meeting of an informal society for discussion of intellectual and philosophical matters of general interest. Polanyi sat at the centre of a circle in a comfortably furnished room, with twenty or thirty colleagues ranged round the circumference, and led a kind of Socratic discussion. Many of the participants were scientists from other countries. It was an informally managed intellectual discussion of a sort that I had not seen in England. Though the topics discussed were contemporary, I felt that they were viewed from outside; this division of the high intellectual life from the brutal rumblings underneath was one of the most striking features of the Weimar Republic.

I was left with the impression that the brilliant scientific efflorescence in the period of the Weimar Republic had an intellectual life of its own, above that of industry and the people, in spite of the integration of much of the scientific research with industry. The cosmopolitan character of Berlin science emphasized this division.

I heard something of what was going on outside science from journalist colleagues. F. A. Voigt was then at the height of his career, and when he was away, Mrs Wagner, the *Observer* correspondent, sometimes stood in for him. Voigt was of British birth and German descent. He was an Englishman with a thoroughly Teutonic cast of mind, but just when one assumed that he was indelibly German in thought, he could be disconcertingly British. As Berlin Correspondent of the *Manchester Guardian*, he performed magnificently.

The Times correspondent, Norman Ebbutt, who was treated badly by his editor, Geoffrey Dawson, was superb. He was an amiable, unpretentious, competent Yorkshireman, fond of his beer. He would normally have led a quiet, efficient, inconspicuous and comfortable life. But the outrageousness of the Nazis set fire to his soul. His sound knowledge of ordinary men and things made him a formidable reporter, and his profound contempt of the Nazis, and all who fawned on them, German or British, gave his dispatches a bite that raised them to the heights of fine journalism. It was also striking to see Darsie Gillie of the *Morning Post*, regarded in England then as one of the most reactionary of newspapers, as an icy opponent of Nazism through his dispatches.

Mrs Wagner was an Englishwoman who had married a German, and had settled in Germany; her perspectives were different from those of visitors on a mission to Berlin. Most of the visiting journalists and writers were glad to know her. She had been helped journalistically by Lord D'Abernon, the British Ambassador, and this had been one of the foundations of her success. She explained features of the German scene to me, and how various well-known literary people had reacted to it. She also gave me some of the press tickets she received for the theatre. In this way, I came to see the performances of Blandine Ebinger and Gründgens in the political revues, which were such a striking feature of the Berlin artistic life at the time.

Voigt told me that he once received instructions from C. P. Scott to see Einstein, and commission him to write an article on the theory of relativity. He was to offer him a fee of, I think, five guineas. He called on Einstein, and had some difficulty in persuading him to write an article for the press, because he had not done so before. He made no comment on the fee, and was apparently not interested in it. After a week or two, the article was produced, but it was three columns long. Scott was pleased with it, and sent a cheque of double the agreed amount to Einstein. Voigt heard afterwards that Einstein carried the cheque around with him for some time, showing it to his friends and saying that it was the first time in his life that he had ever been paid more than had been agreed. Many journals would have been prepared to pay Einstein hundreds of pounds for an article.

I began to write *An Outline of the Universe*, and had made progress with it before I returned to England. I conceived it as an exercise in my conception of scientific journalism, which I expounded in a prefatory note. I described the book as 'an essay in a craft still sufficiently new to be ill-defined, the craft of scientific journalism'.

I gave a short account of Dirac's recently published theory of the proton as a 'hole' in a universe otherwise constituted of electrons. Dirac had succeeded in imagining a physical meaning for the negative roots in the equation describing the electron. The positive roots applied to the electron; he supposed at first that the negative roots applied to the proton. When the positron was discovered shortly afterwards, it became clear that this was Dirac's 'hole'. He had conceived a mathematical theory of 'anti-matter', one of the greatest intellectual feats of the century. In his Nobel Prize address in 1933, he extended this notion to the conception of the 'antiuniverse'.

Dirac's conception of 'holes' as an aid to the physical imagination was applied later in the quite different field of solid state physics, and led to the invention of the laser.

As an investigator in magnetism, Kapitza was deeply interested in another of Dirac's profoundly original ideas: the theory of the unit magnetic pole. Dirac suspected that magnetic poles, which hitherto had always been found in nature in pairs, might be able to exist by themselves. Today, in 1968, experimenters are still trying to demonstrate the existence of unit magnetic poles. One day they may well be found, and provide confirmation of yet another of Dirac's extraordinary intuitions.

Dirac became very friendly with Kapitza, and a frequent visitor to the U.S.S.R. The Russian translation of his classic work on *Quantum Mechanics* had a sale several times as large as the world-sale of the English edition.

My *Outline* was published in 1931, in the midst of the economic crisis of that year. In Newcastle I saw a display of it occupying a whole shop window. But the Newcastle streets were empty, and the population struck almost destitute by unemployment; there was not much sale for my book. I sent copies of it to various authorities and friends, in particular, Rutherford and F. G. Hopkins. Rutherford, as was his habit, was friendly and encouraging; he said I must have put some work in it. Five years later, Hopkins in his oration at Birkbeck College made some remarks on scientific journalism, which appeared to have been influenced by my *Note*.

The *Outline* was chosen for reissue among the first *Pelicans*, and it was translated into Chinese. The Pelican issue was arranged through Krishna Menon, who was an editor of the original *Pelican* series. He called on me to invite me to contribute to the new series. The next time I met him was about fifteen years later. Frederic and Irène Joliot-Curie were staying in London, and Krishna Menon, now High Commissioner for India, called on them while I was there. There were only two chairs in their room, so Menon was put on one, and I on the other, while Irène Curie sat on one side of the double bed, and Frederic Joliot on the other, back to back. Krishna Menon had a very elegant attire, staff and gloves, and the conversation was rather disjointed.

When the *Outline* and other of my books were later reissued in the *Pelican* series, I found that one of the happy results was that all kinds of people, including some of my friends, began to read them. I had already observed philosophically, while staying in expensive hotels during my travelling days, that men who were spending pounds on a meal would sit with a sixpenny *Penguin* or *Pelican* propped up on the table in front of them.

At the University reception for the British Association at Cambridge in 1938, Seward was receiving guests on behalf of the University. I took my place in the queue of members, all with white waistcoats or dinner jackets. When my name was announced and I stepped forward, Seward seized my hand warmly, and began to tell me he had been reading a book of mine with interest. The queue began to pile up behind me, and I detached myself as quickly as I could. After taking a few steps, I found myself talking to a friend, who had with him A. B. Cook, the eminent historian of religions. He introduced me to him, and at once Cook said he had recently been reading a book of mine, which he had found very interesting. I was surprised, and very much pleased, for Cook was distinctly outside the usual circle of readers to which I could aspire. After a short and lively conversation I moved on, and ran into the physicist A. F. Ferguson, a very big, tall, and amiable man. 'Oh, Crowther,' he began, 'I have recently been reading a book of yours . . .' Before he could finish I said: 'What is all this? It is the third time that this has been said to me in ten minutes.' I then learned from Ferguson that Seward, Cook and himself had been on a holiday voyage to the West Indies. After a while they had been driven to seek entertainment in the ship's library, and the only book that they could find which appeared possibly to be worthwhile, was the Pelican edition of my *Outline*. So the book went round, and I had three readers who might not otherwise have ever looked at any of my books.

CHAPTER SEVEN

1930
PRODUCING THE ENGINEERS FOR THE FIRST FIVE YEAR PLAN

MOUAT JONES AND I were the only members of our committee of advisers on technical education who were able to accept Petrovsky's invitation to visit the U.S.S.R., because the others felt unable to ask their respective institutions to grant them the necessary leave.

The chairman of the Manchester Education Committee, under which the College of Technology then came, was Councillor Wright Robinson, an enthusiastic socialist and educationist. He was very keen that Mouat Jones should accept the invitation, and when it came up for consideration on his Committee's agenda, he said he took it that their Principal accept this highly important invitation. It was immediately agreed, and the Committee passed on to the next business, before some members had fully grasped what it was.

Mouat Jones was a Conservative in politics, though liberal in education. He told me that after he accepted the invitation, one of his oldest Manchester friends, an ardent Conservative, never spoke to him again. As I have experienced on several occasions, men like Mouat Jones proved in practice bolder and more determined on matters of principle than some who made a parade of principle.

Mouat Jones was an Oxford chemist by training. As a chemist, he revised Roscoe and Schorlemmer's celebrated treatise on chemistry, and in the First World War was involved in the chemical side of gas warfare. He was a rather retiring bachelor, whose quiet manner made his brilliant and entertaining wit as an after-dinner speaker all the more effective. He was one of the most charming of friends and travelling companions.

Mouat Jones and I set off across Europe to Moscow in the autumn of 1930. I had with me a copy of Jeans's *Mysterious Universe*, which had just been published, for review, and spent half an hour or so in the train between The Hague and Berlin looking through it, and then wrote the review. Mouat Jones was scandalized that I had not read it from cover to cover, but I had read *The Universe Around Us* and other works, and assured myself that Jeans had not said anything very different from what he had

Producing the Engineers for the First Five Year Plan 71

said before. The Cambridge University Press seemed pleased with my review, for they quoted from it constantly in their advertisements for about twenty years.

When we arrived at Moscow, there was a small party of senior officials to welcome us at the station. They were much older than I was, and the interpreter told me afterwards that they were surprised, for they had assumed that we would be greybeards. Petrovsky himself commented on my age, and I think one of the reasons why Mouat Jones joined my committee was because it appealed to his sense of humour. He once said to me: 'Fancy a young journalist turning up from Chile or Peru, and telling the U.S.S.R. how to run its system of higher technical education!'

The aspect of Moscow in November 1930 was profoundly different from that of 1929. In the meantime, Stalin had launched the collectivization of agriculture and the industrialization of the country. The leisurely, social democratic, suggestibly unstable atmosphere of 1929 had been replaced by a purposeful revolutionary drive. Everyone seemed to be preoccupied with an important task or journey. In the Department of Heavy Industry, which was looking after us, hundreds of men and women in uniformly dark winter clothing and long felt boots, were streaming through the corridors. A man stepped out of one of the streams, and began talking to me; at first sight, he looked like any other, but after a moment I recognized him as Academician Lazareff, whom I had met before. Food was very short, and there were queues outside the shops. Even as guests of the Department of Heavy Industry, there was difficulty in feeding us; we were given dishes which were often difficult to identify.

A packed programme of conferences with officials and visits to technical schools of many kinds had been arranged for us. We were asked to give lectures on technical education in England. Mouat Jones and I turned up punctually for the first of these, but when we arrived not one member of the audience had yet come. We waited for a while, and then went for a walk without saying where we were going. We heard later that a considerable audience had assembled at the usual time of one hour after the announced time. After this, strict orders were given that audiences had to arrive punctually, and they did so.

Petrovsky told us that in developing technical education he had acted on the principle enunciated by Marx, that the factory system 'is the embryonic form of the educational system of days to come, when, for all children above a certain age, productive labour will be not only a means of increasing social production, but also the only method of bringing about a many-sided development'. (*Capital*, Vol. I, p. 489, English translation.) Technical colleges had accordingly been organized in every case in connection with a factory. We were, of course, familiar with Dr Herbert Schofield's work at Loughborough, where he had developed the system of 'training on production'.

In order to produce specialist engineers, capable of doing useful work, in the shortest possible time, higher technical education had been reorganized on the monotechnic principle. There were consequently an extraordinary number of colleges, with virtually a single course, of university standard. There were monotechnics for turbine engineering, for building, for margarine manufacture, etc., each giving an advanced course in its speciality.

Mouat Jones saw the point of the system in the existing situation, though he believed strongly that technical education should be on a liberal basis. He thought that the Soviet Union would have to modify their system later; in fact, this is what the Soviet authorities did. Mouat Jones had had a distinguished military career in the First World War. He fully grasped the military importance of the prodigious effort of industrialization that we were witnessing. He said the Soviet leaders were putting first things first, that is, they were creating an industry which could support a major armaments industry. Unlike nearly everyone at that time with his political opinions, he perceived at once that the Soviet industrialization was a major phenomenon which would almost certainly succeed.

The atmosphere in November 1930 had the purpose and discipline of military Communism. The only other occasion in my experience when I saw something of the same spirit, though on a much less intense scale, was in London when it was bombed during the Second World War.

In the drive for technical education every qualified engineer in the U.S.S.R. was ordered to devote a part of his time to teaching. In our visits to the various technical colleges and schools we met high civil servants with engineering or science degrees, whom we had previously met in conferences in government offices, doing their stint of teaching. We visited schools where raw young peasants were being taught to use simple machine tools by methods like those of a gymnasium instructor teaching a class to swing clubs in unison. We saw schools and colleges of every grade and variety, from simple craft training to the most advanced abstract and applied research. All were working on a multiple-shift system, in order to make the fullest use of the equipment available.

I saw what could be done to develop higher technical education when there was the will and the power. Scientific and technological development need not be slow: whether it is or not, is a matter of social decision. The change in the Soviet Union between 1929 and 1930 impressed me more deeply than any other event in my experience. I saw how much could be done, and how quickly, to create a socialist state. It seemed very probable that if the industrialization had not been carried through, the Soviet Union would have disintegrated; those who had carried it through had saved the first socialist state. It appeared to me that Stalin, as the leader of the industrialization and collectivization policy, was the saviour of the Soviet State. I thought it a pity that he used such sanguinary methods, attended with so much suffering and hardship, and I indicated in my writings that this was my opinion. I thought it a mistake not to be more open about the social costs of the policy, for in my opinion its positive altogether outweighed its negative aspects. Comparing the U.S.S.R. in 1929 and in 1930, it seemed evident to me that the administrative contribution of Stalin, in laying the foundations of a socialist industrial state was an absolutely major social contribution. I subsequently heard the question posed, whether the U.S.S.R. owed more to Lenin or to Stalin. I think it is a good question. Lenin died early, so we do not know whether he could have actually carried out the industrialization of the U.S.S.R. No doubt he could have done, but Stalin did it. Starting a revolution is not the same thing as consolidating it, and it is conceivable that the consolidation was in fact the harder task.

The price of Stalin's achievement was, however, high. In the social tension, which was vastly increased by the rise of Nazism in Germany, suspicions were multiplied and deepened. Petrovsky himself, and Rose Cohen, were among the many who later fell under suspicion, and disappeared.

On returning to England in 1930, I wrote two articles on *Russia's Technicians*, which were published in the *Manchester Guardian* in December of that year. These gave a concentrated account of what Mouat Jones and I had seen. Mouat Jones told me that a friend of his had said to him that he had read his articles in the *Guardian*, and had been surprised at his having such views. Mouat Jones replied that he had not written them, but his friend said: 'Nonsense, my dear fellow, I recognized your style at once.'

I expanded my articles and notes into a book, *Industry and Education in Soviet Russia*, which was published in 1932. Petrovsky advised me to put 'industry' into the title, as that was more likely to attract interest than 'technical education' by itself.

Twenty years later I asked Tambovtzev of what benefit our

activities in Britain had been to the development of technical education in the U.S.S.R. He said that he found my father's views on evening class and further education the most useful in the Soviet situation of 1930. They gave help on the most pressing problem of improving people's technical knowledge while they were also engaged on a full day's productive work.

Russian higher technical education before the Revolution had in fact been of a high standard; higher than in England, especially in its training in mathematics and fundamental science. The defect was that there was so little of it, and it was associated with the ruling classes. Many of the most highly qualified engineers in Russia had fled during the Revolution. The Soviet Union was not short of ideas for higher education, but of qualified people to carry it out.

After his early visits to England and America, Tambovtzev became a professor in an institute of research and training on machine tools, a kind of work which now, thirty years later, is being more energetically developed in England. He was in charge of the manufacture and erection of the famous metal statue on the top of the Soviet Pavilion at the World Exhibition of 1937 in Paris. In the Second World War he became an important officer in connection with tanks, and was in charge of the disposal of the German tanks captured in the Soviet zone of Occupied Germany. For these services he received the appropriate decorations.

After my return in 1930 I saw British technical education in a new perspective. It appeared extraordinarily lacking in size, plan, and drive. The scores of institutions, founded at different times during the previous hundred years in response to differing regional and social conditions, had grown up like trees in an unplanned and unpruned forest. There was a lack of vision and material resources. The British scene in technical education caused me to suggest that all existing British technical colleges should be marked down as institutes of the second grade. The technical faculties of the universities should be closed, for they were under-equipped, and working in an inappropriate atmosphere. I proposed the design of 'seven resident higher technical colleges, or technical universities; one for London, one for the Midlands, one for Lancashire, one for Yorkshire, one for Wales, one for the North-East Coast, and one for Scotland. The professorial chairs should bear prestige and emoluments to attract the best technical brains in the country. These men should be given freedom to search for new processes which would relieve us of our increasing dependence on foreign patents. This is not an impracticable dream. In Soviet Russia today,

Producing the Engineers for the First Five Year Plan

schemes which make this look negligible are devised almost daily, and occasionally even launched. There is no shortage of brains or skill in Britain, but our ability is jammed by an antiquated constitution. If we had a Five Year Plan we could probably execute it very much more effectively than the Russians are executing theirs, but, we have no Five Year Plan!'

Thirty-eight years later, there was still no thoroughgoing creative plan for science and technology which arrived at fundamental construction and reconstruction rather than extension, multiplication, and modification. By the 1960s the Imperial College of Science and Technology, and the former technical colleges such as those at Glasgow, Manchester, and Bradford, had been expanded into technical universities. These were not, however, entirely new foundations thought out from the beginning. The major opportunity to found one was lost when it was decided to make the Churchill memorial just one more Cambridge College, instead of a great and original new technological university.

It appears that it is not possible to have such a plan without prior fundamental social changes.

CHAPTER EIGHT

1931

SOCIAL CRISIS AND SCIENTIFIC INSPIRATION

News of the development of the U.S.S.R., and in particular of planned science under the First Five Year Plan, stimulated interest in England in Soviet science. The growth of unemployment and financial crisis caused an increasing number of people to look to the Soviet Union to see whether new ideas could be learned for dealing with the menacing problems of an industrial and technological society that was becoming out of joint.

The leaders of Soviet Science were made aware that there was a growing desire to know and understand what they were doing, and they themselves began to feel the need for being better and more widely understood. They saw that grotesque misconceptions of their philosophy of science, and their planning of scientific research, were being expressed abroad. Many persons in foreign countries believed that the Soviet development of science was a fantasy; they thought that the Soviet Union had neither the scientists nor the resources to carry out scientific work which would contain anything of significance to scientists in Western European countries. Many people believed that the Russians were congenitally lacking in understanding for machines; their highest, and unattainable, aspiration must be to catch up the Western Europeans and Americans.

In this atmosphere, the Soviet scientists decided to send a strong delegation to the International Congress on the History of Science, starting in London in June 1931, with the aim of explaining their philosophy, and describing how they were carrying it out. They wished to show how a socialist state conceived and developed science, in order to persuade scientists to see that, as scientists, their best hope for the future lay in socialism.

The decision to send the delegation was made at short notice. Eight scientists were appointed to deal with different aspects of the subject, under the leadership of the eminent Communist N. I. Bukharin, who was a member of the Supreme Economic Council. He had recently been appointed head of scientific research in industry, and was chairman of the U.S.S.R. Academy of Sciences' section dealing with the history of science. The other members of the delegation were the applied botanist and agriculturalist N. I. Vavilov; the zoologist Zavadovsky; the physicist A. F. Joffe; the mathematical physicist B. Hessen; the mathematician Kolmann; the technological economist and planner M. Rubinstein; and the electrical engineer W. H. Mitkevich.

They brought substantial papers on a well-balanced and unified range of topics. Owing to the short time for preparation and organization, some were still finishing their papers up to the last moment before leaving Moscow. Bukharin left in such haste, that he was in the air for some time before he discovered that he had left his manuscript behind, and the aircraft had to return to Moscow so that he could retrieve it.

The impact of the Soviet delegation was remarkable. The Congress was virtually the first to be held on the history of science. The subject had hitherto been pursued mainly in an antiquarian spirit. A number of the participants were elderly scientists, who had taken up the history of their subject as a pleasant diversion during their retirement, while others had worked at some historical aspect of science in their spare time; some were wealthy amateurs amusing themselves with studies in the history of science. The President of the Congress, Dr Charles Singer, who was the most eminent British historian of science, was one of the few devoting the whole of his efforts to the subject.

The speakers and papers presented an extraordinary medley of topics and approach. The organizers had asked for preliminary notice of papers, to assign time for their delivery during the sessions. The Soviet delegates had not sent any, and when they arrived with their lengthy and carefully organized papers, they learned that no time on the programme had been assigned for them. When it was explained to them that, accordingly, they would unfortunately not be able to be heard, they were astonished. They did not expect that any considerable notice would be necessary for an official exposition of the Soviet views on science and technology to be heard.

Their conception of values became all the more strained as they listened to reminiscences from the elderly, and trivia from obscure amateurs. They naturally felt that their lack of opportunity to speak was due to reactionary suppression of Soviet views. Ultimately, the Congress was extended by half a day, the whole of which was to be devoted to the Soviet communications. As even this would not provide time for more than summaries of their papers, the delegation decided to have all their papers translated into English, printed and published, in five days, ready to be handed out at their special session at the end of the extended Congress. James Bone headed the paragraph which I sent to him on this: *A Five Days' Plan*.

The Soviet Embassy became the headquarters of this unique publishing venture. Several expert translators worked almost continuously, turning the Russian texts into English, while printers' boys stood by, to rush with copy and proofs between the Embassy and the printing works. Authors, translators and readers worked through the night, and compositors came into the printing works at six in the morning to proceed with composition.

I reported that 'Doubtless this production of a book on the history and philosophy of biology, physics and economics in five days will be one of the most surprising feats of this Congress and its successors.' Actually, it was not found possible to present bound copies of the collected papers in book-form at the special Soviet session, but prints of all the papers were available and distributed to members of the audience. The bound volume, *Science at the Cross Roads*, running to about 200 pages, appeared a week or two later.

The extraordinary contrast of the Soviet approach to the Congress with that of the other participants caused many members to become more conscious of their own attitude to the history of science, and consider whether it was satisfactory or could be improved; was it to be taken for granted that science and its history had nothing to do with society and politics? The ferment stimulated by the Soviet participation made many members study speakers' attitudes to the history of science, as well as the technical contents of their papers. Looking at the speakers in this way, one noticed that several of the best were working scientists, who seemed to have paused for an hour or two to consider the history of their subjects. Though pregnant with ideas, some of these speakers did not appear to have acquired much technique in historical studies.

The Soviet scientists' uniform philosophical approach brought out the extremely individualistic, not to say anarchic, approach of most of the other contributors.

In the stimulated atmosphere, some participants with the older views became bewildered and angry at what they regarded as illegitimate intrusions into their subject. When the discussion on the Soviet contributions was opened, there was a long disapproving silence. This was broken by a young student named David Guest, who was then twenty years old. He suggested that, in keeping with the Soviet view of the relation between science and society, it was possible that the source of some of the profound contradictions in the logical basis of mathematics was not to be found in terms of mathematics alone, abstracted from society, but was to be sought in prior contradictions in the society from which mathematics had arisen. Thus, argued Guest, the separation of the history of mathematics from the history of society and politics might be the reason why the fundamental contradictions in the logical basis of mathematics appeared to be insoluble.

David Guest did not live long enough to make the great intellectual contribution of which he was capable; he was almost at once involved in the political struggles of the rapidly growing social crisis. He went to fight on the side of the Republic in the Spanish Civil War, and was killed in 1938; one of the most gifted men who died in that struggle, a loss to British and world intellectual life.

The most remarkable paper delivered at this congress was that of B. Hessen, on *The Social and Economic Roots of Newton's 'Principia'*. It had never occurred to me, or to most other people, that Newton's *Principia* had had any social and economic roots. I knew from Marxist thought the principle that all science had its roots in society, but I had not perceived or looked for this in any concrete and particular case, especially not in the greatest case of all: Newton's *Principia*. I had assumed in the conventional manner that it was a purely intellectual creation.

Hessen sketched an outline of the social origins of this greatest of scientific works. The argument of his paper was quite simple, and it was obvious to British historians that there was much relevant information which he did not quote and, indeed, of which he appeared to be ignorant. But the limitation of the scope of Hessen's knowledge was irrelevant; he was a professor of physics in Moscow University, not a British historian. It was the penetration of his thought, arising from his command of both mathematical physics and Marxism, that enabled him to reach new depths in the understanding of Newtonian science which intellectually superseded other historical analyses, however much more learned.

Hessen's paper revealed to me a method of prosecuting the history of science which was more profound than the conventional one. I thought at once of applying it to other periods in British science. After reflection, it seemed to me that the British science and scientists of the nineteenth century would be the best to start on, because the subjects were sufficiently recent for the technical content to be familiar. It seemed such an obvious thing to do that I started on it precipitately, for fear of being forestalled. In fact, it took four years to complete and publish, appearing in 1935 under the title *British Scientists of the Nineteenth Century*.

When it was in page proof, I lent it one afternoon to B. Finkelstein, a Soviet theoretical physicist who was at that time in London. About midnight on the same day, I was knocked up, and when I went to open the door, he was standing before me with the proofs in his hand. I asked him in, and before I could say that he certainly need not have bothered to bring the proofs back so quickly, he told me that he considered that my book was better than Hessen. Of course, I knew exactly how much I owed to Hessen: both the idea and the method. Hessen could not be expected to know the details of English history as an English writer might. Nevertheless, after all allowances, I was gratified by Finkelstein's remarks.

During the Congress of the History of Science, I had many conversations with Hessen and Bukharin. Hessen was most anxious to visit the Patents Office Library, to see historical material bearing on the influence on British science of patents of the Newtonian period and earlier, and I took him to the Librarian there.

On another day, I arranged for Bukharin and Hessen to visit the National Physical Laboratory, which had to be done through official circles, by application from the Soviet Embassy. The Embassy provided one of its cars to take us from London to the Laboratory at Kingston-upon-Thames. When we were about a quarter of a mile from the Laboratory, the car, a large Daimler, broke down, so we had to proceed on foot. We passed a number of shops on the way; one of these was a hardware store, containing tools, taps, utensils and a variety of nuts, bolts and oddments of various kinds. Bukharin became unduly fascinated with the contents of this shop-window, pondering on them with deepening absorption. Hessen began to be uncomfortable, evidently feeling that this was not a suitable occupation for a possible head of state on the way to an official engagement. The breakdown of the car had already detracted from the dignity of our progress. Bukharin was slightly nettled by Hessen's reproof, and smiled rather sharply; then he took hold of Hessen's nose, and pulled it firmly, making some jocular observation in Russian.

This, and other incidents with Bukharin made me like him as a person, but prepared my mind to see that he could never be a suitable head of state. He was too much of the playful intellectual to become the kind of realistic administrator required to consolidate the Soviet State.

Within a few weeks of the visit of the Soviet delegation to the Congress on the History of Science, three very notable centenaries were celebrated: of the birth of James Clerk Maxwell; Faraday's discovery of electromagnetic induction; and the foundation of the

British Association for the Advancement of Science. They coincided with the national economic crisis, the collapse of the Second Labour Government, and the devaluation of the pound.

The Maxwell centenary inspired several profound essays on physics and the history of physics, especially those by Einstein, Planck and J. J. Thomson, which gave me most valuable insight into the importance and intellectual characteristics of Maxwell, Faraday, Davy and Galileo, and were of great help to me in writing *British Scientists of the Nineteenth Century*. Einstein gave me a graphic idea of the relative intellectual weights of Galileo, Newton, Maxwell, and Faraday; Planck gave me an impression of the stature of Maxwell, in comparison with other physicists of the nineteenth century, and J. J. Thomson made me grasp that Davy and Faraday knew no higher mathematics. Reflecting on this, I perceived that it was because they came from the lower social classes, who in their time were very rarely able to enter the English universities.

The emphasis in the Faraday celebrations was more on the significance of his discoveries for industry: electromagnetic induction for electrical engineering, and benzene for the chemical industry.

Among the foreign delegates who attended the Faraday celebrations were the Japanese physicist Nagaoka, the German physical chemist Bodenstein, Niels Bohr, and W. Heisenberg. Bodenstein made a quiet, academic impression. I was glad to have seen him, as he was the head of the laboratory in Berlin where F. Simon worked. Bodenstein's researches were the starting point for the modern study of chemical chain reactions. Nagaoka was taller and tougher in personality than I had expected. He made a particularly interesting remark on Faraday, speaking with feeling about what he regarded as the British common sense, which had enabled Faraday to be given opportunity and encouragement. He spoke as if he were acutely aware of the absence of this kind of common sense in other quarters.

The centenary meeting of the British Association began late in September. I had commented in an article a week before it started, that the objects and history of the Association were peculiarly relevant to the present circumstances. It had been founded in 1831, after the Napoleonic wars, when debt and strain had caused the British to neglect science. The period 1918–31 had features in common with 1815–31. In the latter period, though Britain was benefiting from the application to industry of inventions made by ingenious workmen, her leaders lacked a conception of the possibilities of science applied to the development of civilization. The improvements in manufacturing processes made the social chaos worse by increasing unemployment. In an atmosphere of Luddite disorders, post-war finance, and neglect of the scientific attitude, David Brewster, Charles Babbage and others campaigned for a better utilization of science. Brewster vigorously expounded this need in the *Edinburgh Review*. He complained of the lack of plan in the co-ordination of British scientific institutions, and compared their casual organization with the comprehensive schemes of the French. He proposed a British Association for the Cultivation of Science which was to stimulate the British to understand the importance of science. The suggestion that it should meet in places outside London was inspired by an idea in Bacon's *New Atlantis*.

Brewster and his colleagues succeeded in forming and launching the British Association, but it soon lost the features of an instrument for organizing an appropriate system of science, and became more of a propaganda and holiday institution; it was assimilated to the Victorian outlook. At the higher end it formed an agreeable meeting place for leading scientists, where they engaged in yearly semi-private conversations, and exchanges of information on their latest researches; at the lower end, a scientific circus for non-scientists. From the latter aspect, it acquired the nickname of the 'British Ass'.

The centenary meeting was held in London, though this was contrary to the original custom, which had specifically laid down that it should meet in the provinces, in order to stimulate scientific activity outside London. The reason for holding it in the capital was that no other place had the facilities for the large attendance expected, or was sufficiently convenient to facilitate the attendance of the exceptionally large number of distinguished persons invited.

The speakers from abroad included Köhler the psychologist; the chemists von Euler, Karrer, R. Kuhn, Wieland and Windaus; the physicist R. A. Millikan; and the cosmologist W. de Sitter. I witnessed an amusing incident between Millikan and the fourth Lord Rayleigh, who was presiding at a lecture given by Millikan in the Royal Institution. Millikan broke the Institution's strict rule of not exceeding the appointed hour, and Rayleigh looked tauter and tauter. Presently, Millikan sensed the situation, and turned to Rayleigh with an excuse. Rayleigh said: 'You have. already had minus fifteen minutes.' Millikan then stopped abruptly.

The President of the British Association in 1931 was General Smuts, who was interested in the philosophy of science, and a leading statesman of the British Empire. His presence manifested the Association's concern with science in the British dominions.

The holding of the meeting in London did not have the intended effect. Though it was large and distinguished, it was not physically noticeable in the vastness of the capital city. Intellectually, it was almost swept out of the public consciousness by the national economic crisis, when Ramsay MacDonald betrayed the Labour movement, and formed the National Government.

Against this background, the Victorian form of the British Association seemed rather irrelevant. In the situation, the institution had little to offer. In fact, Brewster's British Association of 1831 appeared more modern. The government of the Association, and many individual scientists, felt uncomfortable, and began to discuss what science could do to help the nation in the crisis, and what action the Association and scientists should take to make themselves more effective in this direction.

The crisis unfortunately deflected attention from the large number of excellent expository papers, both non-technical and authoritative, which would ordinarily have been widely noticed. One was by A. V. Hill, who explained that the famous Hill–Meyerhof theory of muscular contraction had been shown by Lundsgaard to be incomplete. Great self-confidence had always been a characteristic of Hill, but on this occasion he indicated that experience had once more demonstrated the need for humility in undertaking to explain nature.

Another important biological paper was given by J. B. S. Haldane, on the mathematical analysis of the distribution of genetical characters in a population. It was pioneer work of fundamental importance. Haldane, a strong personality, with a big figure, a bold and almost exasperated expression, worked out his results by primitive and long-winded mathematics. J. B. S. Haldane was a Dr Johnson of science. He dominated parties and small circles of conversation. He once asked me to be his guest at the University College of London Fellows' dining club. His conversation consisted of intellectual fireworks in various languages, and was a kind of performance. When nuts and wine were served, he placed a row of walnuts on the table, and then cracked one by banging it with his forehead. 'You see,' he said, 'my skull is very thick.'

I did not really enjoy this, nor did I like very much his highly esteemed expository writings; they were full of sparkling ideas, but without adequate logical development.

Ramsay MacDonald appeared on the platform at one of the meetings, with Sir William Bragg at one side of him, and Rutherford on the other. I believe it was about his first public appearance after the formation of the National Government. There was a general air of bewilderment, and nobody quite knew what to do. It was evident that everybody expected MacDonald to speak. Bragg and Rutherford looked distinguished, grave and detached, as if they were waiting for important news of something about which they were not informed.

MacDonald rose and made some sententious remarks about science. It was not his speech, but his deportment, which was remarkable. He was red-faced and stooping, the acme of a hypocrite. The contrast with Bragg and Rutherford beside him made it all the more conspicuous; never were the straightforward characters of Bragg and Rutherford more impressive. There was a mixture of transcendent ability and innocence in their expression.

While I was deeply struck by this demonstration of the difference of these scientists from MacDonald as men, I was also concerned by their tranquillity, so attractive in itself, but, I thought, not what was required for dealing with the MacDonalds and the political and social problems of the time. It was a demonstration of the limitation, as well as the virtue, of the natural scientist as then conceived. I felt that a new type of scientist would be required to cope with a society in which science and technology were becoming politically important, and in which scientists would have to perform their part as scientists as well as citizens in social and political affairs.

The economic, financial and political crisis of 1931 placed British science and scientists in a new perspective. News of the First Five Year Plan, and the planning of science and technology incorporated in it, aroused increasing interest in Britain. It was an example of socialism in action with regard to science, and it became a source of ideas for dealing with British social and scientific problems.

In this general situation the scientific and medical panel of the Society for Cultural Relations with the U.S.S.R., in which I was active, organized visits for two large parties of British scientists and doctors to the U.S.S.R. in the summer of 1931. The two parties, both of which I accompanied, had altogether more than sixty members, about twenty scientists and forty doctors. They included J. D. Bernal, W. Le Gros Clark, H. D. Dickinson, J. D. Cockcroft, J. B. S. Haldane, Julian Huxley, P. G. 'Espinasse, Glen A. Millikan, J. Pilley, N. W. Pirie among the scientists; and E. Mapother, A. Salter, M.P., and Somerville Hastings, M.P., among the medical doctors. For most of them, it was their first visit to the U.S.S.R.

The parties travelled 'hard' at a low price, and we sailed to Leningrad by Soviet ship. In my four-berth cabin were two scientists and a medical doctor from a mental hospital, besides myself. My scientific companions, with whom I became acquainted for the first time, were P. G. 'Espinasse and W. Le Gros Clark; we have ever since remembered this journey with pleasure.

The ship called at Hamburg, where it took on German machinery for Siberia, a sign of the Soviet industrialization, the Soviet market for engineering products, and the economic relations between Germany and the U.S.S.R.

When we arrived in the Soviet Union, I observed that a good deal of progress had been made since I had been there in the previous year. The Soviet scientific and medical authorities painstakingly showed us an illustrative selection of their institutions, and expounded their policy and organization in explanatory discourses.

N. I. Vavilov, who had recently been in London at the International Congress on the History of Science, received some of us in his office in the Department of Plant Industry, formerly the Stroganoff Palace. He had a huge long room, with tables covered with papers, and the floor with large maps of various countries, which he was using in connection with his researches on the world origins and distribution of domestic plants.

N. I. Vavilov was a man of middle height, dark hair and moustache, with a handsome figure and charming voice. His manner was enthusiastic and imaginative and yet practical. He was one of the best scientific men whom I have met. Culturally, he was one of the finest persons produced by pre-revolutionary Russian society. He had worked at the John Innes Horticultural Institution under Bateson before the First World War, and had a splendid command of English. Vavilov did not, however, belong to the revolutionary society, even though he enthusiastically agreed with its aims. In a conversation with me on what kind of society is best for the scientist, he paused and pondered, and then lent over to me and said in a confidential way: 'Of course, all scientists are anarchists at heart.' In my opinion this view was one of the causes of his subsequent difficulties with the Soviet authorities.

As the majority of our parties were medical doctors, there were many medical visits. These included a large and properly equipped clinic for keeping the health of the population in a large section of the city under continuous observation and record. It was a development in preventive medicine which indicated the pattern of the future, the provision of centralized and adequate facilities for the general practitioner; it gave the doctors on the staff more opportunity for specialization according to their talent, and more efficient treatment for patients.

We also saw an abortion clinic, and maternity, cancer and mental hospitals. One of the women doctors in the abortion clinic gave us an exposition of their views and policy on it. The British visitors were surprised at the frankness of the exposition. They were also surprised that they should be admitted to see women in labour, and a very bloody cancer operation. One or two members of our party fainted. The Soviet authorities were open in showing the good and not so good in their medical organization.

When the parties arrived in Moscow, the Society for Cultural Relations became embarrassed when it realized we were travelling 'hard'. Some of our members were already well known; the members of the parties were themselves not at all embarrassed, because the conditions of travel had been fully explained and agreed in London.

However, as N. I. Bukharin and his colleagues had only recently been welcomed and entertained in London, more lavish attention was now given to us, though we had not expected it. Bukharin entertained us on behalf of the Soviet Government. We were invited to a party, which was held in the Dynamo Stadium, the Wembley of Moscow. When we arrived, the huge arena was empty, and we wondered exactly how we were to be entertained. We were conducted to the Government box in the stadium seats, and invited to sit down. The summer evening was warm, and we surveyed the seventy thousand empty seats. We heard vaguely that there was to be a banquet, and we wondered where. Presently we saw some scores of soldiers assembling in immaculate order on the grass arena, in the part nearest to us; they turned out to be the band of the G.P.U., the military security force. They played soft music, which floated to us on the warm evening air. For a long time we saw no sign of Bukharin, or any other authoritative person who might be our host. But at length he appeared, together with Karl Radek, whose pockets were stuffed with newspapers half hanging out. Hessen, the author of *The Social and Economic Roots of Newton's 'Principia'*, appeared, and became involved in a lively discussion with a scientific colleague. They were arguing whether a differential equation could be a reflection of social and economic conditions; Hessen for, and his colleague against. The music murmured on, and the argument became sharper.

There was plenty of opposition to Hessen's views. Sometimes

they were in the form of ridicule. When L. D. Landau was still a very young man, the service for telegraphing pictures between Leningrad and Moscow was installed. Landau and some of his friends in Leningrad succeeded in booking the first telegraphing of a picture by the new public service. It consisted of a mathematical diagram caricaturing Hessen's ideas.

While Hessen was in the midst of his reply to his critic, the banquet was announced. We were led through the well-appointed dressing-rooms, and into a large hall, where the most lavish variety and quantity of food any of us had ever seen was piled up. Many in our party looked at it in silence, thinking of the tins of biscuits and packets of chocolates they had brought with them from England to face the straitened conditions they had expected.

Besides this Bukharin gave a large reception in one of the city buildings which was notable for its size and informality. Many of the people there were not particularly associated with science. One was Olga Knipper, the widow of Chekhov, whom Bukharin introduced to me. She was a dark-haired, well-built matronly lady, not at all a Chekhov character.

During this visit I witnessed an example of the immense international respect for the Huxley family. Professor and Mrs Kohts, the noted biological evolutionists and investigators of the behaviour of anthropoid apes, had created a fine Darwin Museum, illustrating every aspect of Darwin's work and ideas. They had had a large picture painted from the famous photograph of Thomas Henry Huxley with his grandson Julian, then aged seven, sitting on his knee. Julian and Juliette Huxley and I visited the Museum together. Professor Kohts was delighted at being able to welcome the Huxleys, and insisted on the three of us being photographed sitting underneath the picture.

One of the most striking impressions that I brought back from this visit was the uniformity of outlook of Cambridge scientists, in spite of wide variations in personality and political beliefs. John Pilley, who was an Oxford man, also commented on it.

Several members of these parties became famous; Cockcroft received a Nobel Prize. In 1931, Cockcroft was a quiet, capable Manchester electrical engineer who had taken a first-class degree in mathematics at Cambridge and then started on research in physics. I had first seen him three years before in Kapitza's laboratory. This was in a shed which had formerly been part of the department of physical chemistry. Kapitza's specially designed dynamo for producing intense magnetic fields was being erected. I noticed a young man with a measured manner at the farther end of the laboratory, looking at working drawings on a table. Kapitza did not introduce him to me, and I took him to be a laboratory assistant. Cockcroft was helping with the design and operation of the dynamo which had been made at Metropolitan–Vickers in Manchester where he had served his engineering apprenticeship.

I did not learn until two years later that Cockcroft had begun interesting experiments of his own. Blackett advised me to seek him out. I found him in yet another corner of the department of physical chemistry, with two large glass tubes from a petrol pump and some other equipment. He aimed at accelerating particles to very high energies, for bombarding atoms. Most of the room was filled with physical chemistry material, and the set-up looked very preliminary indeed.

When the party of scientists arrived in Leningrad and was taken to the Europa Hotel, there was a long wait in the lobby, while rooms were allotted. Cockcroft sat down on a large chair, and pulled out a sheaf of papers. On one was a graph of a curve; he pondered this, and then began to write out underneath it calculations in very neat numbers and words. Presently he was told the number of his room, and no time had been wasted.

A year later, after he had made his apparatus work and had succeeded in achieving the first disintegration of atoms by machinery, he was world-famous, and naturally attracted close attention.

In June 1931 I published what I consider the worst article I ever wrote. It was entitled *Exploring Space: Dreams of Visiting the Moon*. I coldly criticized this as a fantasy, without any reasonable basis. The article was a reflection of the attitude of conventional science to the idea of space exploration, which it held in contempt.

In 1957 I went to a meeting called by the Royal Society to discuss space research. It was evident that the British were taken by surprise by the first Sputnik. They had not believed that the construction of a sufficiently powerful and reliable launching rocket was practicable at that time. It seemed to me a reflection of the same scientific conservatism which had distorted my judgment in 1931. Such experiences teach how dangerous it is to say 'no' to anything in science.

CHAPTER NINE

1932
THE NEUTRON

When Rutherford was elected Cavendish professor of experimental physics at Cambridge he had already made his crucial experiments, at Manchester, proving that it was possible to disintegrate the atom by utilizing the naturally provided rays from radium. After he had settled in Cambridge, he completed and published an account of the experiments in 1919. He was forty-eight years old, an age after which great achievement in physics is rare. But he was in splendid psychological and physical health. He organized the exceptionally large number of gifted young men around him, rather in the spirit of a hero prepared to rush forward at the head of his troop, brandishing his sword against the unknown. His grandfather came from Scotland, and with his big figure, flashing eye and drooping moustache, he was evidently not far removed from Viking ancestors. He always preferred to take the lead with his own hand, and was not very comfortable when he had to depend on others for special skills or knowledge which he did not possess. He was the opposite of his great predecessor, J. J. Thomson, who was a general of the mind and worked largely through other people, to whom he gave intellectual directions. Rutherford was an imaginative Achilles, and personally dealt with difficulties on the spot; J. J. Thomson was an Agamemnon, who commanded the armies, and achieved great victories by strategy.

As a hand-to-hand fighter Rutherford had intense interest in people. He had a prodigious memory for personal details of anyone whom he regarded as in his area of concern.

As a scientist he had a very strong physical imagination; he could see how nature worked better than he could express it in words. When his four papers on the demonstration of the disintegration of the atom are coolly looked at, the reader is apt to marvel at his confidence in his results.

I asked Joliot-Curie, one of the greatest experimental physicists, what he thought of this aspect of Rutherford's published papers. He said that one had to remember that an experimenter probably had a hundred times as much supporting material as he actually published. A great deal of this might be imperfect and incomplete, and unsuitable for publication, but its total effect on the judgment might be decisive. The experimenter who has worked for years in the laboratory on a problem acquires many impressions of its nature which are not sufficiently definite to be put into words, but create an intuition on what kind of conception or explanation is probable or improbable.

The great mathematician G. H. Hardy once remarked to me that it took about twenty years to develop an intuition as to whether a problem is worth tackling or not.

In 1920 Rutherford was invited to give the Bakerian Lecture to the Royal Society. This is often devoted to a summary of some important series of published investigations and Rutherford chose to summarize his researches on the disintegration of the atom. In his lecture, however, he went far beyond this, and this departure from tradition was to have a profound and entirely unforeseen effect on the pattern of the international development of nuclear physics.

He applied his great imagination to the implications of his disintegration of the atom, showing that its nucleus had a structure, and was built of particles to be identified, which were held together in ways that had to be found out. As his experiments had shown that protons could be knocked out of the nucleus, it was evident that protons were a constituent. One would expect that electrons would also be a constituent. Now the ordinary hydrogen atom consisted of a proton with an electron circulating round it at a comparatively great distance. Was it possible to get them closer together, or even united? If so, then the positive charge on the proton and the negative charge on the electron would cancel each other. One would have a neutral particle of about the same mass as a proton; in fact, there might be a whole series of neutral particles of mass 1, 2, 3 and 4. Rutherford pointed out that such a neutral particle would have remarkable properties. It should be able to enter the nuclei of atoms easily, because, having no electric charge, it would not be repelled by the nucleus upon which it impinged. It should produce nuclear transformations, and its existence would offer an explanation of how the heavy elements had been built up from the elementary protons and electrons.

Thus, in 1920, Rutherford had in his mind, and had published, a conceptual blue-print of the future of experimental atomic physics. It is one of the most remarkable imaginative efforts in physical science.

With these ideas in the minds of himself and his collaborators, he desired to pursue them with all speed and power. Rutherford sought for some intense form of energy which might compel protons and electrons to combine to form neutral particles. Experiments were tried on the discharge of intense electric sparks in hydrogen gas. He thought that perhaps a proton and electron in some hydrogen atoms might be made to coalesce, but the results were negative. They were an attempt at what is now called fusion research, the combination of nuclei and particles to form new kinds of atomic nuclei.

Rutherford became interested in obtaining equipment which would produce intense electrical effects for accelerating particles, in order to imitate and, indeed, surpass nature. He had used the radium atom, which spontaneously ejects very energetic particles; but these are only of certain specific energies, which limited the range of experimentation with them. He wanted very energetic particles, but of a flexible range of energies, so that more versatile experiments could be made with them.

I think that one reason why Rutherford gave such thoroughgoing support to Kapitza's experiments on intense magnetic fields was because he thought that they might provide a technique he could use in his experiments on the atomic nucleus. As Rutherford was not an engineer, he was dependent on others for the invention and construction of physical equipment on an engineering scale.

Two of the main types of experimentation in Rutherford's laboratory were the classical small-scale experiment that could be done in a room no larger than a domestic sitting-room, with apparatus about the size of cooking utensils; and the new engineering-scale experiment using machinery of factory size, which Rutherford saw was necessary, but with which he was not personally at home.

His chief collaborator in his own classical type of experimentation was J. Chadwick, who had come with him from Manchester. Chadwick had been educated in one of the municipal secondary schools in Manchester, and had gone to Manchester University, where he became one of Rutherford's pupils. In 1912, at the age of twenty-one, he made the important discovery that atoms struck by heavy particles ejected by radium might emit a wave-radiation.

He went to Berlin to continue his studies, and was interned in the prison camp at Ruhleben near Berlin. There he and other internees discussed science together, and organized a little laboratory. Under the less uncivilized conditions of the First World War, they were befriended by some German scientists who helped them with equipment. Also interned in the camp was a British regular army officer, who had been sent before the war to pursue military studies in Germany. This was C. D. Ellis, who as a boy at Harrow School had shown scientific interest and ability, but had chosen to go into the Army.

Chadwick, Ellis and some others, including the electrical engineer H. S. Hatfield, kept their interest in science fresh, and Ellis presently decided to leave the Army and devote himself to science. After the war he entered Rutherford's laboratory, where Chadwick was now deputy director of research to Rutherford.

The two friends were of very different types. Chadwick was dark, thin and pale, with a pessimistic temperament. He was often anxious, in pain, and exasperated. These characteristics, which were so prominent, were in fact superficial; beneath his apparently delicate and strained exterior there was a wiry tenacious physical constitution, which enabled him to pursue researches with the utmost thoroughness and ruthless criticism. Nothing came from Chadwick that was not doubly or trebly proved. He was Rutherford's right-hand man, who kept up the standards and saw that none of the younger men were allowed to fall below it. I remember an occasion when, while Chadwick was talking to me, a young experimental physicist came to show him some photographs of α-particle tracks in photographic emulsion. The young experimenter spoke of the possibilities of the technique with great enthusiasm, but Chadwick criticized the accuracy of the method severely. The young man went away with a sombre expression, and Chadwick remarked that it did not do for young men to be too optimistic. I met the experimenter nearly thirty years later, and asked him whether he remembered the incident. He laughed, and said it had left no hard feelings.

Ellis, who had the deportment of a regular officer in civilian clothes, wore a military moustache and had the easy manner, within a certain formal social framework, of an army man. He became Scientific Adviser at the War Office in the Second World War. Ellis had a quick and lively scientific mind and liked to speculate about the possibilities of a scientific idea.

From 1920 at least, Rutherford and Chadwick had been looking out for a neutral elementary particle in nature, and had made experiments that had yielded no results. In fact, after 1919, the researches of Rutherford and his colleagues did not make such swift progress as he desired. The concentration of genius and ability in the Cavendish Laboratory was, as it were, dammed up. After a few years Rutherford's impatience for swifter progress became noticeable.

The Neutron

One of the few considerable advances in the subject was made in Paris by S. Rosenblum, who used the big electromagnet constructed by Cotton to analyse the stream of α-particles from radium. By submitting it to a strong and also very constant magnetic field, the stream could be split into several rays, each of which consisted of α-particles of a particular energy. This gave some accurate detailed information bearing on the structure of the nucleus which had ejected the particles.

Rutherford welcomed and esteemed this work, which he felt was a considerable advance at a difficult stage in nuclear research. It has been said that he favoured the consideration of Rosenblum for a Nobel Prize, but this was strenuously opposed by Madame Curie. Rosenblum, rightly or wrongly, suspected that the opposition was due to her prejudice in favour of her own family.

Rosenblum was a scientist of great ability who did not have the good fortune to achieve the complete expression of his talents. He was a small, mobile, gentle man, not fitted to impose himself in any situation. When France was overrun by the Nazis in the Second World War he was in imminent danger as of Jewish descent. He was one of those smuggled out of France by the rescue system organized by Louis Rapkine; he was packed into a trunk at the southern frontier of France, for transport into Spain. The trunk was successfully carried through, but on the journey it was turned up and down, and when it was opened in Spain Rosenblum was found to be almost dead. Fortunately, he revived.

While Rutherford and Chadwick continued to keep neutral particles in their minds, research with fast particles, which had been Rutherford's classical instrument, continued. Meanwhile, the search for artificial or engineering methods of accelerating particles was promoted. The physicist E. T. S. Walton, who came to Cambridge from Trinity College, Dublin, and Belfast, made an important theoretical analysis of possible methods. Among others, he discussed the cyclotron method, worked out by E. O. Lawrence in California.

It was clear that the cyclotron method of whirling particles around, like a stone at the end of a string, presented technical difficulties that could be evaded by accelerating particles along a straight line. T. E. Allibone and J. D. Cockcroft, both Manchester-trained electrical engineers, attacked this problem. They were at that time junior and obscure, compared with Rutherford and Chadwick; they were being backed on the outside chance that they might produce something which would work in the vague future.

The main line of nuclear physics in the Cavendish Laboratory was still dominated by the classical style of experiments with small sources of natural radioactive material. And it was from this line of research that the next major advance in nuclear physics was to come, not from large-scale engineering nuclear physics which was still a minor sideline that had not yet been shown to work.

I first heard the rumour that the neutron had been discovered in experiments at the Cavendish Laboratory at a meeting of the Tots and Quots in London, when J. D. Bernal mentioned it in casual conversation. This society, founded by Solly Zuckerman, consisted of a group of scientists and others interested in science, which flourished in the 1930s and during the Second World War. It was created and held together by Zuckerman's unique combination of gifts. No other scientist had the same range of sympathies and understanding, and was appreciated by so wide a circle of persons of differing opinions. The motto of the society was an abbreviation of *Quot homines, tot sententiae*, which Solly translated as 'Many men, many opinions'. The society usually entertained a chief guest, who was invited to open a discussion after dinner, on any subject he chose. The society met in Soho restaurants, one of which included the house in Dean Street in which Karl Marx had lived. His ghost may have had the curious pleasure of listening to the wisdom of Lord Cherwell and other eminent guests, besides the views of numerous young men, several of whom acquired varying degrees of fame. Among those who came in the earlier years was Hugh Gaitskell; he appeared not to be very much at home with the atmosphere and views of most of us. I will say more about the Tots and Quots, especially during the war years, later.

Having heard the rumours of the discovery of the neutron I went to Cambridge to investigate. I learned that Chadwick had indeed discovered the neutron and was to give his first account of it at a meeting of the Kapitza Club, to which I was invited. This club, founded by Kapitza, was another unique social invention that flourished around 1930. With his other gifts Kapitza combined sociability, a wide range of interests, and a love of discussion. He found the English and Cambridge social and intellectual habits very formalized. He held them in high esteem, but felt their limitations. He was one of the few foreigners who considered the English system of higher education superior in some respects to the continental European. He thought its severe social and intellectual discipline advantageous, though it gave the young student less freedom; as an undergraduate he was treated more like a schoolboy and less like a mature man.

With personal knowledge and experience of the indiscipline of the old Russian higher education, he valued the English discipline all the more. But he felt that in some ways it went too far. He noticed that young English research students, who were more uniformly talented than the corresponding groups in other European universities, were excessively afraid of expressing their opinions for fear that they might appear ignorant or stupid. One of Kapitza's motives in founding his club was to reduce this inhibition by encouraging the young men to express their thoughts, which are sometimes highly suggestive; they reveal a perspective that had not occurred to those who had long been working in the subject. At the first meeting of his club, the young members behaved in the conventional English way, saying nothing until they were asked, and then being very non-committal. Kapitza broke this down by opening the discussions with more or less fallacious assertions about the subject. He made conscious howlers that were so obvious that the shyest young research worker had no hesitation in correcting them. In this way he created in Cambridge a new standard of general discussion of physics, and engendered a freer exercise of the scientific intellect.

It was to this club that Chadwick gave the first account of his discovery. He described anomalous observations by scientists in several countries, which could not be explained by current conceptions. He then showed how, with the addition of further rigorous experiments, all the observations could be explained as due to neutrons. With his unrivalled technique in this kind of experimentation, his proof that the neutron existed was conclusive. He answered completely all the objections raised by his own ruthlessly critical judgment, and every hearer was convinced at once that a new fundamental particle of nature had been discovered. A flood of new ideas and experiments immediately arose in the minds of his audience, and in the hour or so of inspired discussion which followed, several important new lines of research were conceived.

Blackett, who at that time had a house in Bateman street, had given me a shake-down on his sofa for the night. In the morning I woke up at about eight o'clock, and in forty minutes wrote an article on what I had heard at the Kapitza Club. I showed it to Blackett, who confirmed that it contained no howlers. It was published by the *Manchester Guardian* on 27 February 1932, under the title *The Origin of Matter*. This became the best-known of my articles. Rutherford, and Chadwick especially, were at that time particularly chary of the press. Consequently, the newspapers found it difficult to get information on the great new discovery. After a day or two, James Bone rang me up, and said that there was a story in Fleet Street that the discovery of the neutron was a hoax, as it was impossible to get adequate information of it. I told him that I had heard the story from Chadwick himself. Bone said: 'All right. If you are sure of it, send me something about the neutron every two days for the next fortnight.' Actually, the *Guardian* published ten pieces of mine, on various aspects of the neutron, during the next four weeks.

As Rutherford and Chadwick were not very forthcoming, the American press especially made use of my article on *The Origin of Matter*, which was widely reprinted and quoted in the United States. When I first visited the United States five years later, American science writers spoke to me of it as if it had been published yesterday.

It had even been cut out of a copy of the *Guardian* and pinned up on the notice board in the Cavendish Laboratory. When I went to the next meeting of the Kapitza Club I happened to sit in a corner of the settee, waiting for the proceedings to begin. McLennan, the senior Canadian physicist and close friend of Rutherford, of whom he had been a colleague in Canada, came in and sat in the other corner of the settee. He did not know me, and began to discourse on the merits of the article on the Origin of Matter in the *Guardian*, and said: 'The man who wrote that knew something.' These words provoked polite laughter among other members of the company, and McLennan looked disconcerted, wondering what was amiss. R. H. Fowler then said to him: 'The author of the article is sitting beside you.' McLennan turned, and addressed a number of appreciative remarks to me. They were all the more interesting because of McLennan's well-known bluntness as the dictator of Canadian physics.

After the discovery of the neutron, James Bone never queried anything I suggested about science. If I rang up the London Office, I was immediately switched over to a stenographer who took down what I dictated. Bone said to me: 'You have a hunch for science.'

Bone's request for continuous news of the neutron prompted me to look up the main papers of the chief contributors to the discovery. Like all major scientific discoveries it was the culmination of advances in many laboratories in different countries. The origin of any discovery is an arbitrary point in the history of science, at which something new comes in that makes the ultimate discovery probable. In the case of the neutron, one might say that

The Neutron

the critical innovation was the use of polonium as a very strong source of α-particles. This radioactive element, discovered by Marie Curie and named after her native country, can be used to produce strong beams of energetic particles. The German physicist Bothe, a small, mild, unassertive man, discovered that if certain substances were bombarded with these beams of particles from polonium, a somewhat anomalous radiation was produced, which seemed to have an abnormal penetrating power.

R. A. Millikan, the American physicist then interested in cosmic rays, while giving a lecture in Paris, compared the effects of Bothe's rays with those of cosmic rays, and suggested that this should be explored experimentally, to see whether they would help to explain each other's nature. F. Joliot and Irène Curie started experiments on these lines. They bombarded the light metal beryllium with polonium rays, and observed the disintegrative effects of the very penetrating radiations so produced. They then placed a barrier of paraffin wax between the beryllium and the apparatus for observing the penetrating radiation. They discovered that the barrier, instead of decreasing, greatly increased the number of observable disruptions.

This extraordinary result transformed the psychological situation, presenting a startling phenomenon which could not be overlooked or evaded. They showed that the increase in disruptions was due to protons knocked out of the paraffin wax by the very penetrating radiation from the light metal beryllium. The Joliot–Curies suggested that the very penetrating radiation was of a wave nature, and the interaction between this wave-radiation and matter was governed by 'a new type of interaction between waves and matter'.

Chadwick, following Rutherford, had sought for the neutron since 1920, and was prepared in thought to recognize it. He belonged to the British and Rutherfordian tendency to interpret phenomena in terms of particles rather than waves. He proved that the new radiation did not consist of waves, electrons or protons. On the other hand, all observations were consistent with their being streams of neutral particles of unit mass. In that case, all the phenomena could be explained in terms of known physical laws, and the invocation of a new law was unnecessary. Such particles should be expected to produce collision effects in Wilson cloud chambers, which could be photographed. Chadwick's colleague N. Feather made the experiment, and got the effect.

About a fortnight later, Rutherford lectured at the Royal Institution on the discovery of the neutron, and referred to the possibility of the building-up of new atoms by the entry of a neutron, effecting the transformation of one kind of element into another.

Bone asked me for some personal notes on Chadwick for his *London Correspondence*, which were published in the same issue as my main article. He wanted to know whether the discovery would have any practical application, and I notice that I said that 'their practical application will doubtless be discovered before long'. Within eight years the release of atomic energy had been accomplished.

In 1945, on her first visit to England after the Second World War, Irène Curie told me that the reason why she and her husband, Frederic Joliot, had not interpreted their famous experiment with the paraffin wax correctly and failed to discover the neutron, was because they had never read Rutherford's Bakerian Lecture of 1920; they had not done so because they had assumed that, like most of its predecessors, it would merely be a summary of earlier published work.

In Cambridge the discovery of the neutron had an effect which was of a higher order than just one more discovery. It relieved a period of thirteen years' frustration. The discovery of the neutron rejuvenated Rutherford's spirit. He was sixty-one, but he regained the force and certainty of touch of his earlier years; he spoke as a man inspired. The Cavendish Laboratory was raised, as it were, to an enhanced or excited state. Discoveries became a daily expectation, and in the heightened atmosphere investigators had increased confidence and insight.

Shortly after Chadwick's discovery I went to Bohr's Easter conference on theoretical physics. After arriving in Copenhagen I called on him to pay my respects. We immediately fell into a conversation about Rutherford and the great event in the Cavendish Laboratory. In the course of the conversation Bohr remarked: 'Rutherford is not a clever man, he is a great man.' Towards the end of our conversation he said that he believed he had discovered one of the reasons why the neutron had remained undiscovered for so long: the extreme rarity of its interaction with the electron; it hit the proton comparatively easily, but the electron very rarely. This arose from the wave-structure of matter. Bohr explained that the mathematical analysis of the effect was analogous to that worked out by Rayleigh, in his explanation of the blueness of the sky as due to the scattering of the waves of the sun's light rays by molecules in the atmosphere.

I asked Bohr where he was going to publish this. He said he had

The Neutron

made no arrangement, and thought he might some day write a short paper on it for the Royal Danish Academy of Sciences. I asked him whether I might publish an article on it in the *Manchester Guardian*. He said he would like to see the article first. I wrote it on the next evening and handed it to him the following day. After keeping it for several days he handed it back to me unaltered, saying that, in consideration of his former days at Manchester, he agreed to its publication. The article appeared on 19 March 1932. Shortly afterwards, I received a note from W. P. Crozier, the news editor of the *Guardian*, saying that they were always glad to publish my articles, but would I please bear in mind that sometimes there were very few of their readers who could understand them.

Thirty years later, on the publication of Bohr's Rutherford Memorial Lecture, I learned that Rutherford had first heard of this work of Bohr's from my article, and had written to him about it. So there had been at least one reader who had understood it. On 21 April 1932 Rutherford wrote to Bohr:

> My dear Bohr,
> I was very glad to hear about you all from Fowler when he returned to Cambridge and to know what an excellent meeting of old friends you had. I was interested to hear about your theory of the Neutron. I saw it described very nicely by the scientific correspondent of the Manchester Guardian, Crowther, who is quite intelligent in these matters. I am very pleased to hear that you regard the neutron with favour. I think the evidence in its support, obtained by Chadwick and others, is now complete in the main essentials ...

Bohr's 1932 conference was attended by Heisenberg, Dirac, Kramers, Kronig, Bloch, R. H. Fowler, C. G. Darwin, and about a score of other leaders of theoretical physics.

The fame of Heisenberg and Dirac was already taken for granted. The personalities who were new to me, and made a particular impression, were Kramers and Bloch. Kramers seemed to have a great resource of mental power on which, for some reason, he never drew to the full; he always seemed to have something in hand. I believe that if he had concentrated his powers more, his contribution would have been even greater than it was. Bloch seemed to me to show wonderful brilliance and enthusiasm. I thought he would emerge quite in the front rank very soon. In fact, he did not receive a Nobel Prize until more than twenty years later.

A fascinating event at this conference was a musical parody of Goethe's *Faust*, in which H. Delbrück was the most active writer and producer. Kramers played Faust, and Mme Kronig, a very good singer, sang Gretchen's songs, accompanied by Heisenberg, an excellent pianist. Her voice broke achingly each time she came to a fundamental contradiction in the quantum theory. Kramers lay on a couch with his trousers in disarray, and meditated on the vanity of science and life; his performance was more than an amateur amusement.

A very popular turn in this entertainment was a parody of Bohr lecturing, with Bohr himself sitting in the middle of the front row, convulsed with inarticulate laughter at the reproduction of his mannerisms of thought and expression.

While I was still in Copenhagen I heard that there was to be a special meeting at the Royal Society under Rutherford, and I suspected that this portended another major discovery.

CHAPTER TEN

1932
THE TOP OF THE WAVE

WHILE I WAS still in Denmark, the news came through from the Cavendish Laboratory that Cockcroft and Walton had succeeded in disintegrating the atom by machinery, the experiment that may be regarded as the beginning of modern large-scale science.

The discovery of the neutron had inspired the Cavendish scientists, who swept from one triumph to another. Observing the progress with wonder, someone asked Rutherford: 'How do you manage always to stay on top of the wave?' 'I made the wave, didn't I?' he replied.

On arriving back in England, I immediately went to Cambridge to see Cockcroft and Walton, and their apparatus. They were still very much the first-rate young Cavendish research men who had done a good piece of work, and it seemed to me that Rutherford still saw them in that light. They had not yet reached the scientific seniority accorded to Chadwick. I was shown P. I. Dee's wonderful cloud-chamber photographs of the disintegration of lithium atoms by protons accelerated in Cockcroft and Walton's apparatus. I had the impression that these were esteemed at least as much as the Cockcroft and Walton experiment itself.

Cockcroft saw and felt that his experiments portended the utilization of nuclear energy in the future, and showed signs of wanting to say so. Rutherford was fiercely opposed to this; he said that the experiment had no useful purpose whatever, and had only to do with the revealing of the innermost secrets of nature. In 1933 he made a much-quoted remark that 'anyone who expects a source of power from the transformation of these atoms is talking moonshine'. I believe that this and similar remarks have been supposed to indicate that Rutherford thought that nuclear energy would never become available. My impression was that he was primarily concerned to stop premature speculation and hopes. The press was pining for him to make some forecast on the imminent release of atomic energy, and he was determined that no premature claims should come from his laboratory. He had a fierce contempt for scientists who claimed more than they could perform. The clue to his behaviour is to be found, I think, in his attitude to the quantum theory before, and after, Niels Bohr had successfully applied it to his nuclear model of the atom. Before this, he was apt to regard departures from classical mechanics as 'moonshine', but afterwards he took the quantum mechanics into consideration, though he continued to view it warily.

The first practical release of nuclear energy did not come from the new large-scale engineering physics, but from the neutron, a product of the small-scale physics. In 1920 Rutherford had first envisaged the transformation of atoms by neutrons that penetrated their nuclei. It was a particular transformation of this kind, discovered in 1938 by his old pupil and colleague of his Canadian days, Otto Hahn, which led to the first practical release of atomic energy.

Rutherford had looked to his engineer-physicist pupils to devise a machine which could be substituted for natural radium as a producer of streams of fast particles. If such a machine could be devised, it would be under control and be much more flexible than those sources, such as radium, which happened to be found in nature. Cockcroft applied an electrical engineering method of multiplying voltages, by which electric pressure of 600,000 volts and more could be given to atomic particles. He utilized transformers of the kind used for raising the voltage of current from power stations to that suitable for the transmission lines of the electricity grid.

He did not proceed very quickly with attempts to disintegrate atoms with his machinery. There was a deep belief, inherited from the principles of the old physics, that it must be extremely difficult for a charged particle to enter and disintegrate an atom. However, in 1928 Gurney and Condon showed that according to quantum mechanics, the chance of penetration was much greater than had been imagined. The Russian physicist G. Gamow independently made the same discovery. He came to Cambridge, and the new theory increased Rutherford's optimism that bombardment with accelerated particles might be successful; he became impatient for results. He encouraged Cockcroft and Walton to try more experiments. They did so, and disintegrations were observed at much lower voltages than had been expected.

Cockcroft had studied electrical engineering under Professor Miles Walker of the then Manchester College of Technology. Miles Walker delighted to give special instruction to young engineers of talent. Cockcroft was one of two who voluntarily came to him for this extra teaching. After the First World War the Institution of Electrical Engineers founded a fellowship to celebrate the victory. It was to assist a young electrical engineer of talent to secure further training or prosecute research. Miles Walker was invited to recommend its first recipient. He told me that he asked the other of his two special students whether he would be interested, but he had been offered a good post which he did not want to refuse. He then sounded Cockcroft, who leapt at the suggestion and used the fellowship to go to Cambridge to take a degree in applied mathematics. He secured a first-class degree, thus adding command of mathematics to that of electrical engineering.

Miles Walker was a pale, clean-shaven, rather bowed figure. He studied law and mathematics before he became an engineer. He was very professorial in his logical attention to detail, and in this aspect seemed the reverse of the practical engineer. Nevertheless, he wrote two of the treatises most esteemed by practical engineers; his specification and design of heavy electrical machinery, and his unique book on the diagnosis of troubles in such machinery. He was elected a Fellow of the Royal Society in 1931, one of the very few engineers to be recognized in this way by the scientists in the early 1930s.

In the economic and social crisis of 1931 Miles Walker was one of the foremost in recommending attention to the social relations of science and engineering. He advocated the application of engineering methods to the solution of social problems.

Before Cockcroft carried out his own famous experiment, into which he was almost pushed by Rutherford, he had helped with the design and construction of Kapitza's dynamo, and he had designed a special electromagnet for C. D. Ellis to analyse the α-ray spectrum of radioactive substances. A very remarkable example of his aid to others occurred in 1935, when Kapitza did not return to Cambridge from the U.S.S.R. The new Mond Laboratory, which had recently been built for Kapitza, and equipped for research in low-temperature physics, was left without a director. Rutherford requested Cockcroft to take over its direction. Cockcroft was now internationally known for his crucial atomic experiment, yet he virtually dropped atomic physics and undertook the direction of research in a very different field. Within two years first-rate new work was coming from his laboratory of low-temperature physics.

Cockcroft's capacity for helping other people was so great that he became a sort of scientific dogsbody of genius. If anyone wanted anything done they tended to go to Cockcroft for help. Shortly after he had taken over the Mond Laboratory, I called in Cambridge one day to see Rutherford. Conversation veered round to the Mond Laboratory, and Cockcroft happening to come in, Rutherford ordered him on the spot to show me round. Cockcroft bridled slightly, said nothing, but quietly and thoroughly showed me round.

Shortly before the Second World War Cockcroft became deeply involved in the mobilization of physicists, in preparation for the impending conflict. In the earlier part of the war he was director of radar research for the Army, again quite a different field from atomic physics. With the development of work on the atomic bomb, he was drawn back to it. He was a member of Tizard's mission which had a crucial role during the war in establishing co-operation between British and American military science. When the comparatively small British effort on the atomic bomb was transferred to Canada, Cockcroft was placed in charge.

The atomic energy establishment in Canada became the source of experience for the construction of the Atomic Energy Establishment at Harwell, which he was appointed in 1946 to create. Shortly after he had begun on this task I asked him what progress had been made; he said that so far, a large and very muddy hole had been excavated on the former airfield where the vast new institution was to be built. He started from practically nothing in the shape of buildings and equipment.

Cockcroft's construction and direction of Harwell was a magnificent feat in the organization and administration of scientific work. The establishment provided the scientific information for the early work on the British nuclear weapons, and for the first large-scale production for civilian as well as military purposes, at Calder Hall.

Another remarkable aspect of Cockcroft's work in the early post-war years was his diplomatic skill in dealing with an exceedingly difficult security relation with the United States and other allies. He kept his temper under great strain.

Before the war Cockcroft had done a great deal of administration for St John's College at Cambridge. When the Mastership fell vacant, it seemed obvious that he would be elected, but he was not. After his strenuous years at Harwell he accepted the Mastership of the new college to be built at Cambridge as a memorial to Churchill. This also he carried through. On the last occasion that I saw him, not long before his death, I asked what he most liked doing, and he replied: 'I like making things.'

He built and established Churchill College, but it is not yet clear where it will rank among his various achievements. Many people think that it should not have been conceived as a Cam-

The Top of the Wave

bridge college, but as an entirely new independent British technical university. Cockcroft was not invited to create such an institution; nevertheless, with his characteristic competence he produced a solid new college. One of its features is the large number of sons or grandsons of famous men among its fellows. As Cockcroft expressed it to me: 'We are very dynastic here.'

Cockcroft's highest genius was for organizing science, a quality of supreme importance in the era of large-scale science which he did so much to found. As such he became a model for his epoch, and his original scientific work continued to preserve its utility. Thirty years after his construction of the first successful artificial atomic disintegrator his type of accelerator was still the most widely used in the laboratories of the world.

It seemed almost impossible that a third major piece of work should emerge from the Cavendish Laboratory in 1932, but this happened. The succession of achievements inspired Eddington to refer to 1932 as the *Annus Mirabilis*, the marvellous year.

In September of that year P. M. S. Blackett and G. P. S. Occhialini published their invention of an automatic cloud-expansion chamber, with which swift particles, such as those from cosmic rays, could be made to take photographs of their own tracks. This greatly increased the efficiency of the investigation of cosmic rays, which Blackett had taken up after he had made the first photograph, in 1925, of the disintegration of an atom, and had completed a long research on the detailed study of the tracks formed by particles in cloud chambers.

The latter unspectacular study had greatly increased the precision and power of the cloud chamber as an instrument of atomic research. In 1928 Skobeltzyn had obtained cloud-chamber photographs of very swift particles connected with cosmic rays. Blackett transferred his attention more to cosmic-ray research, which Rutherford did not altogether appreciate; it was not easy to pursue a line in the Cavendish Laboratory on which he was not enthusiastic. Rutherford's dominance, and the tensions which arose between the members of his large group of extraordinarily able pupils, made the choice and development of any independent line very testing. Blackett chose cosmic rays, and stuck to them.

His automatic cloud chamber soon produced very important results. C. D. Anderson in Millikan's laboratory in California had continuously operated a cloud chamber, and in the vast number of photographs he had obtained he detected several tracks which appeared to be due to cosmic-ray particles of about the same mass as that of an electron, but with a positive instead of a negative charge. Anderson announced this in a note in September 1932, but did not publish his photographs at the time, and did not go beyond suggesting the existence of a particle of positive charge and mass less than that of the proton, as a counterpart to the long-known negative electron. Blackett and Occhialini made it absolutely clear that the positive electron had indeed been discovered.

Shortly afterwards, Blackett became professor of physics at Birkbeck College in London. He was the first of Rutherford's leading Cavendish collaborators to leave the, by then, perhaps excessively intense Cavendish atmosphere. By the end of 1932 the Cavendish Laboratory itself was rather like an atom about to explode. There was so much genius in so small a space, and so much excitement.

The experimental proof of the existence of the positron added a further dimension to Dirac's greatness. He was the only British theoretical physicist of the same order as Planck, Einstein, Bohr or Heisenberg and, it seemed to me, with a genius in his own field comparable with Rutherford's in his. Dirac pointed out that physical meaning could be given to the negative as well as the positive roots of the wave-equation of the electron. The positive applied to the electron, which has a negative charge, so the negative root might apply to a positive particle. He tried to fit the proton to the negative root, but this positive particle is nearly two thousand times heavier than an electron, so he was in a difficulty. The experimental discovery of the positron showed that it was really this particle whose existence he had deduced theoretically.

His general idea was therefore correct. He extended it into a theory of anti-matter, and suggested the possible existence of an anti-universe. It followed, for example, that there should be a negative proton as the counterpart of the well-known proton. He was deeply delighted when, some thirty years later, the negative proton was shown experimentally to exist.

Dirac was elected to the Lucasian professorship, Isaac Newton's chair, in 1932, when he was just under thirty years of age. His father was a teacher of languages in Bristol, of Swiss descent; he was ill-paid, and Dirac was educated in an elementary school and at the Merchant Venturers' College where he graduated with a first-class degree in electrical engineering. He was very quiet, and a good student, but it was not suspected that he was a genius. He became an apprentice at the British Thomson-Houston works at Rugby. After a time it was thought that he had better go back to academic work and qualify as a teacher. He returned to Bristol and entered the University. His exceptional ability was noticed when he was found to be reading original papers on his own initiative. He was entered for a scholarship at Cambridge and was awarded an exhibition at St John's College, but was not at first able to take it up, owing to lack of additional means. However, when his talent became clearer it was arranged that he should go to St John's after all. Dirac's quite exceptional gifts were first fully recognized by R. H. Fowler at Cambridge.

In his earlier days Dirac was extremely reserved and spoke very little. It has been said that when he first went to Copenhagen, someone remarked that his vocabulary seemed to be restricted to five words, 'Yes', 'No', and 'I don't know'. This, and a host of other stories have arisen about Dirac, with varying degrees of authenticity, like the stories of other very remarkable men such as Winston Churchill. They may not be entirely true in fact, but they are in keeping with the personality of the subject. I will therefore tell some of the many Dirac stories I have heard.

When he first visited Copenhagen, his hosts thought he was rather lonely and that he needed some looking after. Mrs Bohr, the very charming wife of Niels Bohr, took him round the fine collection of modern pictures in Copenhagen. They looked at a series of masterpieces, but no comment could be obtained from him. Presently they paused before a picture of sunlight on a boat on the sea, the whole created by a few strokes of the brush. The picture caught Dirac's attention and he commented: 'I like this picture better; it has the same degree of unfinishedness all over.'

Heisenberg, who was an accomplished pianist, once played several pieces, and then asked Dirac which he liked best. Dirac reflected: 'The one in which you crossed your hands.'

Dirac and Blackett were studying at Göttingen at the same time, and met there. One day Blackett, who has a fine head of hair, remarked that he must get his hair thinned. 'One in four, I suppose,' was Dirac's comment.

Another physicist reported that once, when he had a slight cold, he was driving Dirac somewhere. He had taken some aspirins from a bottle, and as they drove along a slight rattle was made by the loose tablets in the bottle. After a while Dirac asked: 'What's that rattle?' His driver explained. After a silence Dirac observed: 'I suppose that the maximum rattle will occur when the bottle is half empty.'

Some years later Dirac read E. M. Forster's novel, *A Passage to India*. His interest in fiction was unexpected, and someone in Cambridge thought that the two great men ought to be brought together. This was arranged, and they observed each other in long silent respect. Presently Dirac asked: 'What happened in the cave?' 'I don't know,' said Forster, which concluded their conversation.

Dirac became very friendly with Kapitza, who perhaps understood his psychology better than most English people. On Kapitza's lawn he began learning to play tennis. Later on he became a keen gardener, cultivating asparagus. Dirac was generally very good at anything which engaged his attention. When occupied with legal business he perplexed the lawyers, and when he took up one of the Chinese board-and-counter games he ruthlessly defeated opponents.

Dirac and Heisenberg were invited to lecture together in Japan. On their journey round the world they visited many places, one of which was the University of Hawaii. It happened that a little while after this there was a meeting of presidents of American universities. The President of the University of Hawaii told his fellow-presidents that it was not difficult to recognize talent. For instance, not long ago two young men had turned up at his university, one of them saying he was Heisenberg and the other Dirac, but he saw through them at once, and sent them packing. 'But,' murmured one of his colleagues, 'they *were* Heisenberg and Dirac.'

I once asked Dirac which he thought was the greatest discovery of recent times. He pondered, and then replied: 'The General Theory of Relativity, because it did not follow from what had gone before. It came out of the blue.' He thought that if Einstein had never lived, fifty more years would have passed before it was discovered.

'Out-of-the-blueness' was, of course, a very characteristic feature of Dirac's own greatest discoveries. Their combination of intellectual boldness and elegance took the breath away.

Dirac told me in 1949 that he considered himself lucky in having been born when he was. He had become a research student looking for something to do just when the new quantum mechanics had been discovered. It was Heisenberg's idea which was the starting point of his own contribution. With his extraordinary intellect he had very human and kind qualities.

In 1932 I reported on other aspects of science besides physics. Jeans, Eddington and Milne were involved in vivid controversies on cosmology. In chemistry Haworth and his colleagues were progressing in the synthesis of vitamin C. Kennaway and Cook announced their discovery of the precise chemical constitution of cancer-producing constituents of coal-tar.

Jeans, Eddington and Milne had strongly contrasting personalities as well as cosmological theories. If one had met Jeans in a club without knowing who he was, one might have assumed him to be a stock-broker with intellectual interests. He was an alert, energetic person of rounded and well-dressed figure, with his mind very much on the matter under discussion. He was able in business matters and left a fortune of £256,000. He reformed the publication of the Proceedings of the Royal Society, bringing them out punctually. He once told me that among his predecessors as editors, Stokes had been dilatory, while Schuster could not make up his mind. Jeans's academic life was comparatively short, as he left university work after his first marriage, to the daughter of Tiffany, the New York jeweller.

Eddington was very much a Quaker intellectual. He was often solitary, and withdrew into long silences. You used to wonder whether he was expecting you to go, and just when you had decided that you had better, he made a remark that showed he had been pondering on what had been said. One might have thought that he would not have paid much attention to finance; however, he left a fortune of more than £40,000.

Milne kept a substratum of Yorkshire accent. He was very quick and active in his movements, and had a concrete kind of mind, even though he was passionately interested in abstract and logical problems. He liked to have a problem stated as quickly as possible, in order to get to work on it; he was always anxious to be getting on. Milne's rapid speech contrasted with Eddington's long silent ponderings. But underneath these differing appearances there was a comparable amount of intellectual activity.

Jeans and Eddington both had strong artistic sensibilities. Jeans was deeply interested in music and played the organ. He loved rising or falling sequences of notes, as in Bach's music, and a similar feature was prominent in his writing. He liked to illustrate by dwelling on the proportions of things, and was essentially a prose writer. Eddington was a poetic writer; he thought out fine similes for natural phenomena.

In 1931 I had become acquainted with the work of W. T. Astbury on the molecular structure of wool and textile fibres. His X-ray studies threw light on the explanation of why rams' horns, which consist of bundles of fibres, are permanently twisted, and why ladies' hair could be given a permanent wave by stretching and heating. Astbury's work was a major step in the elucidation of the structure of animal fibres, and the chemical structure of protein molecules. He helped to prepare the way for the great molecular biological discoveries of the 1950s, though these eluded him.

Astbury was a bluff blond Staffordshire man, the son of working-class parents, and was absolutely direct and plain-spoken, and sometimes this went too far. He was apt to express downright opinions before troubling to check them. A devoted scientist, he insisted on his research being conducted entirely from the scientific point of view, in no way subordinated to the demands of the textile industry. He did first-rate scientific work, but did not find the best working relationship between science and industry.

Astbury used to speak enthusiastically of his youthful memories of the wild annual week's holiday which he and his family used to spend at Blackpool. He told me it gave an intensity of joy that a middle-class person like myself was unable to understand.

Perhaps the most important advance in X-ray analysis since its foundation was, as W. L. Bragg has pointed out, the development of the strategy of X-ray research by J. D. Bernal. It was he who developed the method of obtaining a certain result from a number of individually inconclusive observations, a method essential for analysing very complicated substances such as occur in biological tissues. In 1940 I was entertained by scientists in Paris, just before the fall of France. One of them said that Bernal should have been awarded a Nobel Prize for his demonstration that the accepted chemical structure of the steroids was wrong.

Bernal was an assistant director of research under Rutherford, but his methods of work and thought, which have proved so valuable in the application of X-ray analysis to biological substances, did not appeal to Rutherford. He liked to have everything clear and simple, whereas Bernal was always conscious of the complications. Once, when I was seeing Rutherford, Bernal came in. Rutherford was explosively critical because something was not as clear as he would have liked it. Bernal stood by, hanging his head, but still unable to forget the complications. It was a clash of temperaments and ways of looking at things.

While Astbury was carrying out his X-ray researches on biological fibres, Svedberg at Upsala was developing his ultracentrifuge, by which he was able to separate large biological molecules into simple molecules of one big unit. I reported this in 1931 as a step towards the recognition of a *Unit of Life*. It was another contribution to the founding of molecular biology.

Svedberg was a thin, dark, mobile man of a Mongolian cast of countenance; he may perhaps have been descended from one of the Esquimaux tribes inhabiting the Arctic North of Scandinavia. He had an artistic temperament, and a very beautiful young wife.

The exciting announcements coming from the physicists in

The Top of the Wave

1932 seemed to spur others. In May 1932 Sir Frederick Hopkins, who was then President of the Royal Society, referred to work by the physical chemists, F. P. Bowden and C. P. Snow, in what was later to be called molecular biology. They had applied photochemical and spectroscopic methods of chemical analysis to the study of the structure of molecules of substances of biological importance, such as vitamins. This work was in one of the lines which has since proved to be of major importance. However, not much more was heard of this particular piece of work. Bowden moved into other fields, especially the physics and chemistry of surfaces. Snow gradually withdrew from scientific research and concentrated on administration and literature.

Another of the varied events of 1932 was the International Congress of Prehistoric and Protohistoric Sciences. Leakey spoke on the fossil anthropoid apes he had found near Lake Victoria in East Africa. The Abbé Breuil and Elliot Smith had recently travelled to China to see the Peking human fossils.

Another great Catholic scientist of those days was the Abbé Lemaître, the author of the theory that the universe is the ever-expanding result of a primeval super-atomic explosion. He had been a pupil of Eddington. I joined a party with him at lunch, hoping to hear more about the explosive origin of the universe, but he was far more interested in talking about the secrets of Freemasons. He was a plump, cheerful man, who chuckled with a deep pulmonary reverberation as he exposed one more trick of those naughty Freemasons.

Mr Alexander Keiller gave an account of the excavation of Windmill Hill in Wiltshire to the conference on prehistory; this was on the grand scale. He unearthed many things, including the earliest known skeleton of a domesticated dog. He had a fine private museum, to which he kindly admitted me. I wrote considerable reports of his paper and exhibits, and reproduced a fine photograph of the skeleton of the dog. James Bone pressed me for more personal information about Keiller. I explained that it was his first public speech, as he had not even addressed any of the learned antiquarian societies. He had the slightly hunted look of a Rugger Blue facing his degree examiners and uncertain of the result, but everyone was delighted by his modest demeanour.

Mr Keiller told me he had wished to name his famous skeleton Canis Familiaris Felstedensis in honour of Felstead, the Derby winner. On the day of Felstead's Derby the navvying staff of his excavators at Windmill Hill were whistled together to hear the result. After returning to the digging they discovered the skeleton.

But unfortunately Keiller's learned antiquarian friends would not agree to Felstedensis, so the official name became Canis Familiaris Palustris.

I mentioned that the discovery had been made possible through the consumption of a very well-known domestic product, as Mr Keiller was a member of the family of Dundee preserve manufacturers, and his excavations could not have been made without much wealth. He wrote to me that though my articles were accurate in content, some were execrable in taste.

The centenary of the British Medical Association also occurred in this year. The Association held a huge dinner in the Albert Hall. I found the sight of a crowd of thousands of medical doctors forbidding. Seen in bulk, they looked as tough as an equal number of business men.

I was also engaged in reporting the conference on Biology in Education, organized by the British Social Hygiene Council. Among the speakers were Sir Stephen Tallents, Sir William Hardy, Professor A. V. Hill and Sir Walter Morley Fletcher. Sir Walter said that physical science had greatly changed the world, but biology would change it far more, for it would change man himself, not merely the environment. I was asked to edit the proceedings of the conference, which were published as a book on *Biology in Education*.

One of the most active bodies in this period was the Society for Experimental Biology. James Gray became its leading figure. In 1932 he told me some details about its foundation. The British Journal of Experimental Biology preceded the Society. The leaders in starting the Journal were F. A. E. Crew, L. T. Hogben and Julian Huxley. Later on, Gray was appointed editor.

In September 1932 Zuckerman explained to me how the new approach in biology had penetrated the subject in which he was then engaged: the behaviour of the primates. Biology in the latter half of the nineteenth century had consisted mainly of morphological research directed towards the elaboration of the concept of evolution and to the problems of classification. This century had seen a rapid growth of physiological research, and current views on the theory of evolution were now based very largely on the data provided by genetical experiments. The time had also come when physiology and the critical study of behaviour could be applied in the consideration of classification and phylogeny, particularly in the primates.

Another biological event of 1932 was the publication of Joseph Needham's great treatise on *Chemical Embryology*; it came to me

The Top of the Wave

for review. Needham published this huge work in three volumes, with a bibliography 220 pages long, and a list of more than 4,000 relevant papers, when he was thirty-one. It inspired stories of varying degrees of authenticity. One was that when he appeared at a continental scientific congress, an elder European scientist, who had not met him before but was acquainted with his works, looked at him uncertainly, and then asked: 'And how is your distinguished father?' Another is that when he delivered his huge manuscript, his publisher was intimidated by its size and said that it would have to be subsidized. Needham asked how much, and the publisher mentioned a sum which he thought would certainly put off so young a man. Needham wrote out a cheque for the amount on the spot. So the publisher found he would have to publish the work, and later on was surprised to find that the heavy cost of production was paid off quite quickly.

Reporting the discoveries in the Cavendish Laboratory and other notable scientific advances in 1932 was absorbing, but I was also able to see something of how Soviet science was developing. The Sixth Mendeleev Congress of Chemistry was held at Kharkov in the autumn. I was the only Englishman to accept an invitation, and one of the total of six foreign guests. It was my fourth annual visit to Soviet scientific institutions. The progress had been considerable, and the tension of November 1930 had been reduced. There was some confusion in the Congress itself; the arrangements had been planned for an attendance of 1,500, but 3,200 turned up. With the severe food and housing shortage, their accommodation taxed the organizers. It was difficult for the foreign guests to find their way around or follow the papers; it was possible, however, to talk with participants and others who spoke a common language.

I stayed with Martin Ruhemann, who had recently gone from Berlin to the Physico-Technical Institute at Kharkov. This was the first of the major new regional institutes built under the scheme for the planned development of Soviet physics. The scientific director was Obreimov who had been sent to study abroad. There was a strong low-temperature physics department under Shubnikov, in which Ruhemann and his wife worked. Sinelnikov was building up an atomic physics department and already constructing an accelerator for disintegrating atoms. I learned much more from the scientists in the Physico-Technical Institute than from the Congress, about the progress and problems of Soviet science. The Physico-Technical Institute was new and in process of creation, and was run by men in their thirties. I got a strong impression of the optimism and the difficulties of the new Soviet scientists.

At the Congress itself, I first saw A. I. Oparin. He expounded his now famous views on the chemistry of the origin of life.

I reported in the *Guardian* near the end of December 1932 that all my scientific acquaintances of various nationalities in the U.S.S.R. had told me that they were confident that the Russians would succeed in their construction of a new social order. They 'differed in their opinions of the nature and value of this order, but without exception they believed that the Soviet would have some sort of permanent success'.

CHAPTER ELEVEN

1933
THE WAVE ROLLS ON, ABOVE THE CRISES

THE CAVENDISH STREAM of research continued in spate in 1933. The inspiration and the development went on, apparently detached from the social and economic crises disturbing the country and the world. In February the Royal Society's Mond Laboratory was opened, built at a cost of £15,000, provided from the Ludwig Mond bequest to the Society. It was designed and equipped for research at very low temperatures under the direction of Kapitza.

The new laboratory was the most beautiful in England, and emphasized a change in the Cavendish policy of research. The Cavendish founders had made inexpensive apparatus, often with their own hands, in a tradition deriving from the craftsman. The total annual expenditure on equipment of the Laboratory was measured only in hundreds of pounds. The Mond Laboratory was conceived in a different style. The equipment was more of the engineering than of the craft type, and the general rate of expenditure was on a much higher scale.

The two chief pieces of equipment in the new laboratory were Kapitza's dynamo for producing intense magnetic fields, and new liquefiers for producing liquid hydrogen and helium in quantity cheaply. Another interesting piece was a nitrogen liquefying plant, donated by C. W. P. Heylandt. He was the eminent German engineer who made important improvements in low-temperature engineering. He bought Dewar's patents for vacuum vessels, and other inventions used in low-temperature operations. Dewar's vessels were suitable only for small-scale work, so during the period 1904-13 Heylandt set out to solve the invention of large vessels of up to three thousand gallons, which could be used for handling liquid oxygen in industrial quantities. In a fine piece of industrial scientific research he succeeded, and thus revolutionized the use of liquid gases in industry.

Heylandt visualized insulated ships for the transport of liquid methane, a technique utilized thirty years later in bringing liquid methane from the Sahara to Britain. His system was also used for holding the liquid air used as a fuel in the propulsion of high-altitude rockets. He made many experiments with cars driven along a track by a horizontal rocket, with which very high speeds were attained.

When I met Heylandt through Kapitza he appeared a ruthless, keen, acute German scientific industrialist. He was a very able but coolly forbidding man. He had acquired wealth from his engineering developments, and I heard that he had an extraordinary house, in which internal walls and floors could be moved by pressing buttons. He was the kind of man who was passionately interested in, and proud of, engineering devices for their own sake, besides their exploitability.

Through his liquid gas inventions Heylandt later had an important role in the development of the German rocket missiles, which were the effective beginning of modern space exploration. It was one of Kapitza's strengths that he understood this kind of engineer as well as scientists, and he was able to draw scientific and technical insight from an exceptionally wide range of scientists and engineers.

Kapitza aimed at combining his two techniques of intense magnetic fields and very low temperatures to extend knowledge of the properties of matter. The experiments might also throw further light on the behaviour and structure of atoms, and reveal properties of matter at very low temperatures which might be put to practical use.

The engineering construction and aesthetic design of the laboratory building were advanced and in striking contrast with the rough rooms of the old laboratory, descended from the tradition of the craftsman's workshop. Kapitza had had a plaque of Rutherford placed in the entrance hall, and the figure of a crocodile on the wall outside the entrance. These were both carved by Eric Gill. In Russian 'crocodile' is a nickname for a great man. Besides these imaginative decorations, the director's office had been furnished with modern scarlet-upholstered steel furniture.

Kapitza arranged that I should receive an invitation to the opening ceremony of the laboratory. The Vice-Chancellor entertained one hundred and four guests for lunch at Corpus Christi College, and the Chancellor, Stanley Baldwin, announced the opening at a meeting in the Arts School. Among the guests were most of the leaders of British science. Heylandt was one of the chief guests, and Biquard and Rosbaud were also present.

The most amusing incident occurred during the meeting in the Arts School. Rutherford took the chair, and Kapitza sat on a chair under the platform, facing the audience. Rutherford started the proceedings with a brief explanatory speech. After remarks by the President of the Royal Society and Sir Robert Mond, he called on Baldwin to speak. Baldwin apparently quite unconsciously repeated almost word for word what Rutherford had said. As Baldwin went on, Rutherford, and also Kapitza, looked embarrassed and grew redder in the face; perspiration ran down Kapitza's temples. When Baldwin sat down, there was a distinctly awkward atmosphere. Sir J. J. Thomson, in the body of the audience, was next called on to speak. He soon began to retail interesting reminiscences, among which was the observation that in earlier days, there were two kinds of scientist in the Cavendish, those who made discoveries, and those who got the credit for them. This was an allusion to a foible of his own, for some of J.J.'s distinguished pupils had felt that he had tended to appropriate some of their best ideas.

This, and other stories, soon secured the audience's attention. J.J. then started on a story about a young woman and an egg. He behaved as if he had suddenly realized that it was unsuitable for the occasion and he began to make it more and more complicated and obscure. Everyone became absorbed in trying to detect what he was driving at, but without success. When he sat down, the contretemps at the beginning of the meeting had been quite forgotten, and a happy mood spread to all.

The founding of the Mond Laboratory created some tension, for it had concentrated what were then large resources on one of Rutherford's chief collaborators, of whom there were several of equal merit. Besides this, Kapitza, who was a Soviet citizen, had been elected an ordinary Fellow of the Royal Society. The records of the Society had to go back two centuries for a precedent: scientists who were not British citizens were customarily made Foreign Members. Further, Kapitza was elected one of the first six of the Royal Society's new research professors who had no duties other than to pursue research at a suitable place. Three of them were Nobel Laureates.

Rutherford had been the main influence in securing these resources and appointments for Kapitza, and some felt that he was biased in his favour. Rutherford had a tendency to focus his strong emotional feelings on particular men. He had no son and seemed to need a way of expressing an overflow of paternal feeling. Kapitza was not the only one who appeared to arouse Rutherford's paternalism, but he was the most unusual of these personalities; he knew how to arouse Rutherford's interest.

Kapitza foresaw the growing importance of the explanation of science to the public, and realized that it was not just a matter of intellectual amusement, or of writing down to the uninformed, and he drew Rutherford's attention to this.

I concluded my article on the new laboratory and its prospective programme of research with the observation that for 'the next twenty years Professor Kapitza and his colleagues and followers will have plenty to do'. Things did not quite work out like that. But for a little while longer, the wave of science rolled on, above the crises.

In 1932 W. P. Crozier, the news editor of the *Manchester Guardian*, had asked me to write a regular article on the *Progress of Science*, containing five or six varied items of new scientific information. I did not very much like the idea, as the publishing of scientific information without relating it to its social significance was against my conception of scientific journalism. An article consisting of half a dozen varied items was too mixed, and the items too short for this to be done at all adequately. However, Crozier wanted it, and these *Progress of Science* articles continued until the Second World War.

In 1933 I published as a short item in one of these articles an account of Jansky's discovery of radio waves from outer space, the great discovery that was to become the foundation of the new science of radioastronomy. I recounted how the 'hiss' atmospheric he had detected came from a direction which rotated in space once a day, and was therefore connected with the rotation of the earth. He had also detected a yearly rhythm in the direction, showing that it was fixed in space, and connected with the revolution of the earth round the sun. Thus it might be due to radio-waves coming from some particular star or nebula.

In 1946 an international conference of physicists was held in Cambridge, to restore the international collaboration which is so strong in this field of science. It was attended by Bohr, Pauli, Fermi, and many others, and was almost festive, in the cheerfulness over the end of the war and the prospect of unhindered international collaboration in research into the deepest secrets of nature. The meeting-room was furnished with twin desks; at the desk at which I sat, about three-quarters from the front of the room, there was a pale, neatly dressed youngish man of about forty, who obviously was not accustomed to this kind of conference. I presently asked him where he came from, and he said: 'My name is Jansky; I can't think why they invited me here.' I at once had the impression of a first-class scientist, not in the style of Bohr, Pauli or Fermi, but of the man who makes great discoveries, not by extraordinary brain-power, but by careful and attentive observation, and then the persistent elucidation of the new fact, which turned out to have such enormous consequences.

About ten years later still I wrote a short book on radio-astronomy. I remembered my meeting with Jansky, but had quite forgotten the paragraph I had written in 1933.

I have wondered why I forgot the paragraph I wrote in 1933. I think one reason may be the type of article in which it appeared: mere scientific information does not make such an impression on the memory as that which is related to more general considerations.

Jansky's initiative in founding radioastronomy was not encouraged by the Bell Telephone Company, and his efforts were ultimately diverted to research more directly connected with the company's interests. Jansky most unfortunately died at an early age; he was one of the very best kinds of scientist.

In the course of my work for the Oxford University Press in 1932-33, I met A. W. Nash, the professor of oil engineering and refining at Birmingham University, who told me he wanted to create an encyclopedia of the science of petroleum. The Press took up this suggestion; A. E. Dunstan and H. T. Tizard were persuaded to join in the work, the preparation of which was announced in 1933. As Nash once said to me, he revelled in this kind of work. The *Science of Petroleum* presently appeared in a series of large volumes.

This connection with the science of petroleum caused me to attend the World Petroleum Congress in London in 1933, and subsequently one in Paris, and also to write a small book about petroleum. I learned something of the outlook of the oil scientists and technologists, of the world fuel problem, and of the practical importance of seismology and the earth sciences, and began to understand better the economic stimulus to the earth sciences, and why the Americans and Russians were so prominent in them. I also gained an increasing impression of the remarkably inadequate effort of the British in the scientific development of their Empire.

After I had known Nash for some years he told me that the vice-chancellorship of Birmingham University would be vacant, and he would like to have a name to put forward from the science side; he asked me whether I had any to suggest. I told him that when I was lecturing at Harvard, I had heard that R. E. Priestley had made an outstanding impression there. Nash asked me to make inquiries about Priestley, who was then Vice-Chancellor of the University of Adelaide in Australia. I wrote to Cockcroft and others for an opinion, and they spoke highly of Priestley's academic administrative work at Cambridge. I told Nash, and Priestley was elected Vice-Chancellor of Birmingham University.

Another event of 1933 on which I wrote, and which was significant for the future, was the ascents to the stratosphere by Soviet aviators. Some information from Joffe about their scientific programme came to me, and it was clear that the Russians were already training a considerable team for high-altitude research.

Meanwhile, Rutherford and his colleagues were sweeping forward with majestic confidence and inspiration after their triumphs of 1932. Rutherford had successfully completed the fourth major decade of research in his glorious life. The first closed with his Montreal period, when he and Soddy promulgated the theory of the spontaneous disintegration of radioactive substances and reduced the subject to intelligibility and order. The second was the establishment of the nuclear theory of the atom, after he arrived at Manchester; the third, the artificial disintegration of the atom, at Manchester and Cambridge; and the fourth culminating in the discovery of the neutron and the disintegration of the atom by machinery. In the first three periods he worked mainly by personal effort and direction, in the fourth mainly by direction and personal effort.

At the British Association Meeting at Leicester in 1933 Rutherford made a triumphal appearance. Owing to the proximity of Cambridge, there was 'a descent of the whole group of atom disintegrators headed by Lord Rutherford himself'. He had opened the discussion on atomic structure held at the Meeting in Leicester twenty-seven years before, which had been concluded by Kelvin with the assertion that atoms were eternal and indestructible.

Now he stood there again, with Cockcroft, Walton, Dee, Oliphant and Blackett around him. C. R. Burch and T. E. Allibone were also there, and spoke of the engineering work done in Manchester on the apparatus used in the Cavendish Laboratory. Dee showed his marvellous cloud-chamber photographs of atomic disintegrations, and Blackett skilfully presented the latest information on the positron. 'The morning was thrilling, and the qualities which make Lord Rutherford's genius were plain to the fortunate audience. The multitude of triumphant achievements reinforces belief in the power of humanity to overcome its difficulties.' Never was the positive and optimistic quality of science seen more conspicuously.

As Rutherford warmed to his comments on the recent Cavendish events, he began to speak without reserve. He said something like the following: 'I told my boys I wanted a million volts in a soapbox; none of those great balls of van de Graaff.' Presently he sat down amidst affectionate applause, and then the chairman called on the next speaker: 'Dr van de Graaff'. Rutherford looked as if he had been struck by lightning; he had no idea that van de Graaff, who worked in the M.I.T. at Boston, Massachusetts, was present.

Van de Graaff came forward from an obscure place at the very back of the lecture theatre, a young and very modest American experimental physicist, who had not very long before been a Rhodes Scholar at Oxford. Van de Graaff said he would not be speaking on his accelerator, but on the problem of utilizing very high vacua in electrical machinery, which would enable high voltages to be handled in small spaces. His paper was technically original, well constructed, most modestly delivered, and very interesting.

As soon as van de Graaff had finished, Rutherford stood up swiftly, and congratulated him in most enthusiastic terms, in comic variance with what he had said before.

Among the significant events of the year was Blackett's departure from the Cavendish Laboratory to be professor at Birkbeck College in London. Rutherford took the chair for him at his inaugural lecture, in which Blackett gave a splendid review of the state of research on cosmic rays. He soon created a school of cosmic-ray research at Birkbeck. The laboratory conditions were so bad that it was scarcely possible to walk between the benches, and his study was about the size of a telephone box.

Rutherford once said to a research student who complained of poor research conditions: 'Why, I could do research at the North Pole.' If the Birkbeck Laboratory was not quite the North Pole, it was then about the worst physical laboratory in any English university. But Blackett, like his great teacher, created a new centre of research under difficult conditions. He made use of the then empty university site in Bloomsbury to erect a large cloud chamber surrounded by a powerful electro-magnet, with which important discoveries were soon to be achieved. He quickly attracted an international group of research men, one of whom was O. R. Frisch. The new Birkbeck group produced interesting results, including observations on cosmic rays deep down in the disused British Museum tube station.

Blackett showed the remarkable power of leadership in research and the development of research departments, which he later exercised in such a striking way at Manchester, the Imperial College, and elsewhere.

At Oxford physics had for a long time been in a rather depressed state. The appointment of Lindemann, who had returned to England with a very high reputation, from Berlin where he had been working with Nernst, had not led to an immediate improvement. He became involved in academic and political struggles which had their own significance, but did not immediately benefit Oxford. He decided to develop low-temperature physics and was recommended to invite K. Mendelssohn, a young member of the German school, who was then with F. Simon at Breslau. Mendelssohn came to Oxford in 1933, and early in 1934 he liquefied helium, the first time it had been done in England.

A little later the Nazis made it impossible for Simon, an elder cousin of Mendelssohn, to stay in Germany. Simon had had a distinguished career as an officer in the German Army in the First World War. Such men were not immediately persecuted, but as Nazi power increased even they were forced out. So Simon also came to Oxford. Under their leadership, and that of N. Kurti, Simon's colleague who came with him, Oxford low-temperature physics became very active and the reputation of Oxford physics rose.

While science was advancing in 1933, the social and economic crisis in Britain was still very severe, with millions of unemployed and under-employed suffering severe hardship. One of the signs that the scientific world was beginning to take more notice of these problems was the Royal Society's election of Sir Richard Gregory to a fellowship. Gregory had been made a baronet, on the recommendation of Ramsay MacDonald, for his services in increasing the public understanding of science. Gregory was not a research scientist, and the Royal Society had for a long time been chary of electing scientists who had not an extensive research record. With his contact with politics on one side, and science, as represented by the Royal Society, on the other, Gregory was in a position to make a unique contribution to the relations between science and society. His understanding, common sense and business ability made him an indispensable member of any effective movement in this field.

CHAPTER TWELVE

1933
GERMAN SCIENCE AND CULTURE IN EXILE

THE RISE OF Hitler was followed by an exodus of scientists, scholars and artists from Germany. Many of these came first to Britain, as the nearest convenient place. Their arrival greatly extended the range, and stimulated the interest of British scientific and cultural life. Early in 1933 H. Freundlich the colloid chemist, and M. Polanyi the physical chemist, resigned from their posts in the Kaiser Wilhelm Institute of physical chemistry at Berlin-Dahlem. Polanyi became professor at Manchester. Hevesy, who had developed the use of isotopes in chemical analysis, was expelled from his chair at Freiberg, and went to Scandinavia. I met him in Copenhagen, and we went for a walk along the Lange Linie, a famous walk beside the Copenhagen harbour. Though Hevesy had been so disgracefully treated by the Germans, he had a profound regard for German values. Like many of the scientists who had to leave Germany he had a deeply Teutonic outlook, of which he was probably not conscious.

It happened that in the Copenhagen harbour there were two warships side by side, one a German pocket battleship, the other a British cruiser. They were both about the same size, otherwise they were almost incredibly different. The German ship was extremely clean and smart. The German sailors were always active, cleaning, polishing and drilling, and they all had sharp creases pressed in their trousers. The ship shone like a polished jewel. The British ship was dirty and unkempt. The sailors lounged in deck-chairs, with their jackets unbuttoned, collars unfastened, smoking pipes and listening to the B.B.C. from loudspeakers on the deck.

Hevesy could not conceal his intense admiration for the German efficiency. Noticing what was passing in his mind, I remarked that the German ship seemed somewhat smarter than the British. This released him from the need any longer to conceal his feelings, and praise for the German superiority poured forth. How could the British hope to survive, he said to me with intense earnestness and sympathetic sorrow.

Hevesy's former collaborator F. Paneth, who was dismissed

from his chair at Königsberg, came to the Imperial College in London and presently was appointed professor at Durham. Paneth's scientific and cultural tradition was quite different from the English but he adapted himself most skilfully to his new English environment. Besides being a master of isotope chemistry he was an accomplished geologist and a notable historian of science, being particularly interested in the scientific writings of the medieval schoolman Albertus Magnus. Paneth's range of interests was in marked contrast to that of most British scientists, but he did not, like some of the scientists who had come from central Europe, tell them that they ought to broaden themselves. Instead, he showed them how to do it.

As a former Königsberg professor he was specially acquainted with the tradition of Kant, besides being interested in Kant's evolutionary theory of the cosmos. When he came to Durham he studied the local scientific tradition, and especially the views of the local astronomer Wright, to whom Kant had referred. Paneth demonstrated that Kant was largely indebted to Wright for his theory of cosmology.

The Bauhaus architects and artists whom I had met in Berlin began to arrive in England. Unlike their English colleagues at that time, they were keenly interested in science and technology. Moholy-Nagy, the constructivist painter, film pioneer and producer first visited London at the end of 1933. He had already designed new methods of lighting and scenery for productions of *The Tales of Hoffmann* and *Madame Butterfly* under the direction of Klemperer, and he had made experimental films in which the intrinsic possibilities of artificial lighting were explored, and not used to copy sunlight. He had invented what he called drawn sounds ten years before, a form of synthetic music derived from drawn forms.

Moholy lived in England for several years before he became principal of a college of modern art in Chicago. He worked very hard to adapt his talents and inventions to the very different English artistic and commercial environment and to communicate his ideas. The Bauhaus tradition saw no split between art and science and its creators were as happy to meet British scientists as they were to meet British artists and architects; sometimes, I thought, they were more at ease with the former.

Moholy strove to be polite, and to stand by friends. I noticed this particularly during a preview of an exhibition he had arranged at Fortnum and Mason's. As we went round the exhibits he was making diplomatically kind remarks about everybody. It put me in mind of a story which I immediately told him. Six British naval cadets held a dinner on the evening before passing out of the naval college as officers. At the end of the dinner, they swore to sing each other's praises for the rest of their lives. Today, five were admirals and one was a commodore.

Moholy, whose command of English was imperfect, was puzzled by this. He smiled politely, but clearly had not seen the point. I went on, and shortly afterwards left. After I had got some way towards Piccadilly Circus I suddenly became aware of someone speaking excitedly behind me. I turned round and found Moholy chasing me. He asked breathlessly, with a puzzled expression: 'Why was one only a commodore?' I said I supposed it was a difference between the Continental and the English view of things, the Continental expected a thing to be perfect, while the Englishman assumed it would be more real if imperfect. Then he said: 'I now understand; we shall *all* be admirals.'

Moholy's linguistic difficulties involved him in various little problems. When he was rather superiorly entertained in Oxford, and very much on his best behaviour, he said to his hostess on departing: 'Thank you so much for your hostility.'

Marcel Breuer, the inventor of modern steel furniture, visited England in 1935. He told me that he invented it in 1925, while he was at the Bauhaus in Dessau. He had been devoting much time to the improvement of the design of wood furniture, to make it more elastic, economical, and handsome. One day he bought a bicycle, and, not knowing how to ride, had to learn. The idea of using steel for making furniture came to him while staring at the handle-bars, as he strained and struggled with them.

Breuer went on to the use of aluminium alloys which were light and corrosion-resisting. Some of the first inquiries for his furniture came from the Middle East where wood furniture is attacked by termites. He held that the most important object in a room are the persons who live in it. If the room did not express their personalities, it was badly furnished, whatever the furniture itself looked like. The room and its furniture should be subjugated to the purposes of its occupants.

After Breuer had settled for a time in London he invited me to spend a day with him in the country. The place to which this modernist chose to take me was Stonehenge, which, of course, must have been ultra-modern at the time when it was erected.

Walter Gropius, the founder of the Bauhaus, went into partnership with English colleagues. Unlike several of his chief collaborators in his Bauhaus days, he was completely German in origin and tradition. He carried the traditional German thoroughness and philosophical profundity into his architecture which he described as arising out of the necessity to combine machinery with aesthetics. In the nineteenth century the pursuit of mechanical advance caused construction and art to become separated. The new style aimed at recombining them in a fundamental manner. Consequently, in his view, the relations between art and architecture and science and engineering were complementary. He and his colleagues were necessarily interested in all of them.

Gropius visited me in my flat in Russell Square. We walked past the new monumental building of London University and I noticed that he deeply disapproved of it. He remarked that it had been built to last for five hundred years, and who knew what would be required long before the lapse of that period of time, and how would this building meet it? His comments struck me all the more because the building had been designed by one of the more progressive of the English architects.

Gropius, Moholy-Nagy, Breuer and Kepes all lived in London for some time. Gropius was presently offered a chair at Harvard, which was a more suitable place for him. Moholy-Nagy, Breuer and Kepes also moved on to the United States where the scope for them was wider and easier. I was greatly impressed by the way the scientists, architects and artists from Germany were able to settle for a time in the very different English environment. I often reflected that if the situation had been reversed, and I had had to leave England and seek refuge in Germany, I would have been quite incapable of surviving by my own efforts only.

Among the German scientists who came to Cambridge at an early stage were M. Born, H. Bethe and R. Peierls. They had the thorough, all-round training in theoretical physics characteristic of the German tradition of Planck, Sommerfeld, and Born himself. The English theoretical physicists were inclined to follow their personal line, rather than apply the new theory comprehensively and work over the field systematically. At the stage which atomic physics had then reached, this was not as useful as the German tradition. The vast number of important new experimental facts required to be systematically worked out, rather than attempting to discover still more fundamental information before the new knowledge had been digested and consolidated.

Rutherford was not particularly successful at assimilating the German talent. His own tradition was so strong, and he had an intelligible bias towards other scientists who, like him, came from the Commonwealth. His attitude arose from his conception of science and the way it should be done. He was among the foremost in helping refugees on political grounds, through the Society for the Protection of Science and Learning, of which he became President.

But he did not like German modes of scientific thought. I have been told that Ewald, one of the founders of X-ray analysis, once found himself sitting next to Rutherford at dinner at Trinity College. He had not been introduced to him, and Rutherford presently politely asked him what he did. 'Fundamental physics,' replied Ewald. Rutherford then asked him what branch he worked in, and Ewald replied: 'X-ray crystal analysis.' 'That's not fundamental physics,' said Rutherford in his involuntary style.

Rutherford was also a little conscious of the competition with himself for scientific eminence, which the great work of the Braggs had brought. In a sense, Rutherford was right; the greatest discoveries arising from X-ray analysis were to come in biology rather than physics.

Born was presently elected to the chair of theoretical physics at Edinburgh. I had visited Edinburgh when C. G. Darwin occupied this chair, and Barkla was the head of the department. The change in the department after Born's arrival was startling. It had been difficult to find Darwin in his office, and almost impossible to find Barkla in his. Darwin was always polite, but behaved as if he was under pressure to rush off to do something that was more interesting than university work; he was not really suited for an academic life. When he later became Master of Christ's College at Cambridge, he showed the same amiable impatience with undergraduates. He was not at home with the immature. Barkla used to receive one while he was putting his hat and coat on. He had lost interest in conventional physics, and was much more interested in music and gardening. He had a splendid baritone voice which could have earned him a career as a singer.

His research students had to discover facts about the non-existent J-phenomenon, which he thought he had discovered many years ago. I was fascinated by the spectacle of young Scotsmen confronted by the problem of obtaining their Ph.D., which was necessary for their future, and the necessity of discovering non-existent facts in order to obtain it. I listened to wonderful Scottish metaphysical disquisitions on the possible meaning of obscure observations and measurements. The upshot was that while nothing definite had been found, there was certainly something there. A year or so later I would meet the same men, who in the meantime had obtained their Ph.D.s; their language about the J-phenomenon was now unprintable.

Things were very different after the arrival of Born. It became difficult to find people away from the laboratory. In a large room that had been empty and solitary in Darwin's day, there were now about a score of Ph.D. students sitting at desks arranged in order, like a fifth-form classroom. Born went round, personally helping and encouraging each of them. He performed a great service for Scottish theoretical physics.

At the same time Born most carefully attempted to adjust himself to Scottish life. He made no criticisms of what he found either in the state of science in the country, or of the way of life. He merely did what he thought was the right thing and let it speak for itself. The first time that I called on him in Edinburgh, he invited me to high tea. He and his wife had even adopted this habit, and at about 6 p.m. we stoically consumed cold ham and tea. The sight of these outstanding leaders of German culture going to such lengths touched me deeply. In our talk about science afterwards Born said to me: 'You undoubtedly have a kind of insight into science.'

On another occasion, Born told me about the origin of the famous paper by Heisenberg, Born and Jordan, in which the new matrix quantum mechanics was invented. Heisenberg, who was one of his research pupils at Göttingen, had recently come to him and explained his idea for the new quantum mechanics. The appropriate mathematical apparatus for it had not yet been worked out. Heisenberg had had a classical education, and was not widely read or trained in mathematics and physics. When he had been examined for his Ph.D. he had not been able to explain a simple optical property to his examiners, one of whom was Sommerfeld. Heisenberg was certain he had failed, and did not go to the dinner after the examination, which candidates usually held to celebrate their success. The other examiner concluded that he ought not to be given his Ph.D., but Sommerfeld insisted that he should.

Born and Heisenberg were attending a meeting of a mathematical society at Hanover. In the train between Hanover and Göttingen, they were discussing the mathematics of the new quantum mechanics, and it occurred to Born that the sort of mathematics which was required was of a kind he had learned in lectures in his youth. This was matrix algebra. He had long since forgotten the details, but he suddenly remembered that elsewhere in the train was a mathematician, Jordan, who was expert in matrix algebra. They went down the corridor, and the three of them, Born, Heisenberg and Jordan, collected together in their compartment, and worked out the substance of the great paper before the train reached its destination. It is remarkable that Schrödinger also invented his wave-mechanics, which is mathematically equivalent to the Heisenberg–Born–Jordan mechanics, in a railway train.

Born's contribution and understanding in Britain was most admirable. Nevertheless, he was profoundly teutonic in his intellectual outlook. During the Second World War I discussed with him the idea that was being mooted, for the foundation of an International University. Born said that one of the services of such an institution would be that it 'would tell us what to think'; there would be a supreme authority in international science and learning. There are arguments for and against this idea, but this particular argument of Born's is one that could scarcely occur to anyone brought up in the English tradition.

CHAPTER THIRTEEN

1934
POLITICS CHANGE SCIENTIFIC CIRCUMSTANCES

HITLER'S ACQUISITION OF power changed the atmosphere of Europe, and had far-reaching effects on science, many of which were not immediately evident. At about the same time, also, several of the scientific leaders of the previous period passed away. Haber died in 1934. Later in the year Marie Curie died. Rutherford had written an obituary notice of her twenty years before, which was kept in the *Manchester Guardian*'s file of obituary notices of eminent persons still living. It was sent to me for bringing up to date. I felt rather embarrassed, and decided that the best thing I could do was to leave what Rutherford had written alone, and add an extra paragraph at the end, mentioning important things that had happened since.

I had seen Mme Curie in Brussels during the Solvay Congress of that year, a few months before her death. She was extremely pale and withdrawn, and, I felt, too frail to be approached. I met several of the other participants, in particular Irène Curie and her husband Frederic Joliot. They had just discovered artificial radioactivity, and had emerged as major figures, following the role of Mme Curie, both in French and world atomic physics.

At Brussels they had the fresh and youthful brilliance of highly gifted research scientists, absorbed in the enthusiasm and excitement of discovery, and not yet deeply involved in the wearing problems of policy and administration. They were concerned in getting the importance of their great discovery understood and accepted, which did not happen immediately in all quarters. Joliot was thrilled to receive from Rutherford a letter in which he said that he had been looking for the effect all his life, but it had been left to Joliot to find it.

During this visit to Brussels I saw the fine house for scientists provided by the Belgian National Fund for Scientific Research. After the First World War the Belgian Government had found itself in the possession of a sum of money originally subscribed abroad for the relief of war victims. This became the nucleus of the national research fund. Besides supporting researches such as those of Piccard on high-altitude balloon flights, the fund built this social centre for scientists, which became an excellent focus for international scientific conferences. It contained a library, restaurant and bedrooms, in which scientists could stay under comfortable conditions. I met there E. O. Lawrence and Peter Debye.

Lawrence described his cyclotron and his programme at the Solvay scientific meeting; he manifested the strength of his personality by unconsciously wagging his finger at Rutherford in making his points to him. Lawrence was then a rather thin, tall young man with spectacles, and a naturally confident manner. He was evidently a born leader with something of a Rutherfordian psychology. His intuition was, however, for technique rather than physical ideas. He had a genius for making apparently impossible devices work.

Debye was a squat, diplomatic Dutchman. He had just accepted a very high and well-paid position under the Nazi régime in Germany, and he was very careful what he said. I gained an impression of his enormous competence, and also that he was very self-centred. Before long he found that not even he was clever enough to enjoy life under the Nazis, and he went to the United States.

Joffe also attended this conference. He was my guest at lunch in a small Belgian hotel, where he ate cooked horse with immense gusto. Joffe greatly enjoyed food. I have thought that his temperament, which enabled him to enjoy such things as eating and swimming so much, must have helped him in the very difficult and important work he did in planning the creation of Soviet physics.

In 1934 the Russian crystallographer Fredericks invited me to an evening in his flat in Lesnoe, where Joffe's institute was situated. Joffe was one of the guests. His enjoyment of the splendid supper provided was expansive. The most remarkable dish consisted of about half a dozen different kinds of smoked fish, probably from the Gulf of Finland; they were delicious, and one especially was incomparable. Joffe did more than justice to it.

Another of the scientists who entertained me at Lesnoe was N. N. Semenov. He was one of the most fascinating scientists in the world. Besides his great scientific gifts, he had an artistic quality. His wife was a small, pale, delicate lady, who spoke the kind of good English which is the result of first-rate early instruction. Semenov was thin and tallish, with dark hair, glistening eyes, narrow jaw, and a rather oriental cast of expression; he was fond of wearing beautiful Tartar skull-caps.

Semenov and Hinshelwood shared a Nobel Prize for their

researches on chemical and chain reactions. Hinshelwood was more of a learned scholar and less of an imaginative creative scientist than Semenov.

Hinshelwood was more conservative in outlook. It seemed that Oxford scientists of his and earlier generations tended to be generally conservative and conventional, partly because science there for so long was accorded a much lower place than classics and the arts. It has been said that an Oxford scientist and some arts friends were having tea in a house in Oxford. The scientist looked out of the window, saw a figure going by, and exclaimed: 'There's Sherrington!' His friends inquired: 'Who's that?'

Sherrington was a quiet and retiring man, but he ought to have been known. The first time I met him, he received me politely in his office. I had recently been concerned in the publication of the English translation of Pavlov's book on *Conditioned Reflexes*, and I enthusiastically expounded conditioned reflexes to him. Sherrington listened with keen interest, but after about twenty minutes he began to ask one or two searching questions very quietly, which interrupted my flow of enthusiasm.

I last saw Sherrington when he was living in Caius College, Cambridge. We discussed the remarkable work accomplished by Joseph Needham in China, and in the history of Chinese science. Sherrington remarked that Needham's endeavours had the quality of the romantic style.

Some scientists were not prepared to accommodate themselves to the Oxford scale of values. When Lindemann became professor there, his cultural background was German, and the idea that pupils of Planck, Einstein and Nernst should take a back seat appeared to him ludicrous. The disputes in which he was involved were not due merely to his disputatious nature.

If Lindemann revived physics in Oxford, Joffe revived it in the whole Soviet Union. Two men of more contrasted temperaments would have been difficult to find. Lindemann was cool, and his social habits did not coincide with those of the majority of people. He was a vegetarian, a teetotaller, and virtually a non-smoker. He was a well-built man, pale, dark and saturnine, who had been athletic in his youth. Whatever he did seemed to be involved in difficulties. He always seemed to be spoiling for a fight.

As scientists, Lindemann and Joffe were rather similar. They were very quick and versatile, rather than deeply original. They had qualities necessary in planners of science.

Joffe was also a big man, but what he did seemed not to be involved in difficulties. Of course, he was surrounded by them but he was very diplomatic; when they became very awkward, he was apt to disappear.

Joffe survived difficulties which would have defeated most men by evading, or just not recognizing, their presence, and being able to become entirely absorbed for the moment in the amenities of life. He was always the same, in the early 1930s in Leningrad, in Brussels in 1934, and in Moscow in the tense days before the death of Stalin. The last time I saw him was in Moscow in 1950. He looked extraordinarily bronzed and well, and at ease. He told me he had just returned from a holiday on the Caucasian shore of the Black Sea, where he had spent a lot of time bathing.

Following on their researches which contributed to the discovery of the neutron, Irène and Frederic Joliot-Curie observed that when boron was bombarded with the radiations from polonium, positive electrons appeared to be ejected from it. They also observed that positrons appeared to be ejected from beryllium by intense wave-radiations.

Blackett suggested that the positrons were not ejected from the beryllium, but were formed by a condensation of the impinging waves in the neighbourhood of the nuclei of the beryllium atoms, according to the theory already proposed by Dirac. The Joliot-Curies called this transformation of waves into matter 'materialization'. Pursuing these researches, they discovered that the positrons from the bombarded boron were not produced by the 'materialization' process. They were emitted for a considerable time after the bombardment of the boron had ceased, and their intensity decayed in the same way as that of a disintegrating radioactive substance. The bombardment of the boron had converted it into a new radioactive substance.

This discovery that it was possible to make new radioactive substances in the laboratory was of immense importance, for it brought within the reach of experimental investigators a process which must be occurring in the stars and be playing a primary role in the evolution of matter in the universe. Joliot immediately foresaw wider implications of their discovery. He spoke in his Nobel Prize address in 1935 of the conception of a nuclear chain-reaction which might, he suggested, release nuclear energy on a large scale. At the time this was regarded more as an imaginative exercise than a concrete contribution to physics. The quick succession of important discoveries by the Joliot-Curies gave a splendid stimulus to French science just at the time when it was to lose Mme Curie; they carried on the Curie tradition at its highest intensity.

The two Joliot-Curies were extremely contrasting, but equally charming and equally remarkable. Joliot was of lower middle-class origin. His grandfather had been Louis Napoleon's pastry cook. Joliot was an electrical engineer by training, and had been educated at the Paris municipality's technical school, the École de Physique et de Chimie, whose director was Paul Langevin. The great friend of his boyhood was Pierre Biquard, and after completing their early schooling together, they went to the École, where Joliot graduated as an engineer.

He then got a job in an electrical engineering firm in Alsace-Lorraine but he did not like the industrial life. He wanted to leave it and enter science, but he was unable to pluck up the courage to go to see his old master, Langevin, so Biquard went on his behalf and explained the situation. Langevin said he would look out for something suitable. Presently he informed Joliot that Mme Curie required a junior laboratory assistant. The post really was junior; Joliot started right at the bottom of the staff. His extraordinary experimental gifts soon became evident; he began to participate in research, and then to lead it.

Joliot in daily life was a man of quite outstanding charm. He understood ordinary persons and joined in their interests, and loved explaining scientific things to them. Wherever he was, he attracted people; porters, waiters, fishermen, were equally fascinated by him. This human quality was the basis of the general attraction of his personality. But in order to sense his scientific greatness it was necessary to see him at work in the laboratory. He became completely absorbed in manipulating the apparatus, which appeared to have the effect of operating his scientific thought and intuition. He became grave and weighty, and a profoundly different personality from the one generally known in ordinary life outside.

Irène Curie was four years older than Joliot. She was very different from a daughter of the lower middle-class, a kind of princess of the intellectual world. She had been educated by her eminent parents on original lines devised by them, and she had not been sent to school. Her manners were not those acquired in the usual seminaries. She had been brought up to behave with extraordinary directness; the great eminence of her parents made this possible for her. Her outspokenness could be devastating, and it was combined with a wiry figure, a penetrating voice, and a face that most people would not have found beautiful.

Irène Curie was, however, a most charming person if one could accommodate oneself to her original mode of behaviour. For instance, she would pick up a newspaper and start reading it in the middle of lunch and become oblivious to her guests. This and similar habits greatly disturbed some people. However, if one took no notice of it there was no need to worry. After a time, she would become aware of a remark someone was making and make an interesting comment on it. She liked ski-ing, singing folk songs in a rather cracked voice, and reciting in English verses from Rudyard Kipling. She was completely without fear and behaved particularly well when she found herself surrounded with Nazis after the fall of France.

The Joliot-Curies discovered artificial radioactivity through the bombardment of light elements such as boron and aluminium with charged particles. This method did not promise similar results with heavy elements such as gold or uranium, because the electric charge on their nuclei would be so large that it would repel any charged particle projected at it. Fermi obtained a bewildering number of transformations, which seemed to suggest that uranium and thorium could be transmuted into still heavier elements, not previously known on the earth, or in the cosmos.

As Fermi was a theoretical physicist, these remarkable experimental results excited particularly great interest. They were broadly in line with the sort of changes which Rutherford had foreseen the neutron might cause. Fermi followed this with his concentration of neutrons by surrounding them with water or paraffin wax. In this way he increased their concentration by more than a hundred times. Later on this method of concentrating or bottling neutrons was to be of decisive importance in controlling the release of atomic energy and thus making atomic power possible.

Fermi was a man of very extraordinary ability. He accomplished major works in both theoretical and practical physics, and in their application. When he delivered papers, he would lean against the table, with his legs crossed, and one elbow on the bench, as if he were chatting in a bar. His informal manner made scientific thinking appear effortless and amusing. Nevertheless, he seemed a lesser personality than Joliot-Curie. His scientific genius did not seem to raise his general outlook above a conventional level. Fermi was not, like Rutherford, a dominant personality. He left the impression that so long as he could get on with science he was not very deeply concerned about anything else.

Not the least of the extraordinary events of this year was Rutherford's last major piece of experimental research. With Oliphant and Harteck, he had personally entered the new field of accelerator physics. I had the impression that he particularly enjoyed being able to catch up Cockcroft and the engineering 'boys', in their own field. He and his collaborators produced experimental evidence for the production of hydrogen atoms of mass 3. Rutherford was then sixty-three years old, an age at which important discoveries in physics are rarely made.

He was full of health, enthusiasm and good nature. Once when I called on him at this period he asked me cheerfully what I was doing, and I said I was writing a book. 'When I have to write a book,' he remarked, 'I collect the material. Then, in July, I clear the decks, put everything else on one side, and start on the manuscript. By October it is finished, and I send it to the publisher, *for better or for worse.*' I have never forgotten this excellent procedure and the final comment; it is about the best advice to an author that I ever heard. Rutherford was ruthlessly critical and accurate in his work and was hard on people who he thought did not do as well as they could, but he had no use for perfectionism.

My visit to Kharkov in 1932 had made me acquainted with the Physico-Technical Institute of the Ukraine and its staff. Through this, I was invited to attend the International Conference on Theoretical Physics at Kharkov in May 1934. The guests included Niels Bohr, who was accompanied by his wife, Rosenfeld, Waller, Plesset, E. J. Williams, and J. Solomon, the son-in-law of Langevin, who was later tortured and murdered by the Nazis. Landau took the chair at the meetings, and the Russians included Frenkel, Fock and Tamm.

Rosenfeld, who spoke Russian and was a very close friend of Bohr, acted as his interpreter. Bohr did not take a particularly active part in the scientific discussions. He seemed to be treating his visit as an occasion for learning more about the U.S.S.R., the nature of Soviet life, and the Soviet attitude to science. He also regarded it as, to some extent, a sightseeing holiday. I remember more clearly these sides of his activities than his interventions in technical discussions.

While we were there, the weather became very sunny and hot, and Bohr needed a sun hat. There was no sun hat in the Kharkov shops big enough for Bohr's head, which was narrow and very long. Furnished with the largest obtainable straw hat perched on the top of his head, we went off to visit a collective farm. The car got stuck in a country track, and had to be pushed out; Bohr insisted on pushing it very energetically from behind.

Bohr and his brother, the pure mathematician Harald Bohr, had both been considerable athletes and footballers. Harald played for

Politics Change Scientific Circumstances

Denmark, and it was said that once he got away with the ball, it was very difficult to stop him. His only drawback was that he did not distinguish the sense of the direction in which he was running, for or against his own side.

The most interesting feature of the conference from my point of view was the emergence of E. J. Williams as a physicist of international standing. He dominated the discussions, showing an equal command of experimental and theoretical physics. A few months before he was little known, except in a few places in Britain. I first heard of him years before in Swansea; I had asked the professor of physics there, E. J. Evans, whether he had had any particularly good students. He said that by far the ablest was a young man called E. J. Williams, then at Manchester. The next time I was in Manchester I called in the physics department and asked for E. J. Williams. People seemed a little surprised that I asked for him, and he could not be located at once. I went round the laboratory, and someone went with me to the basement, where it was thought he might be. We poked around in the dark passages, asking for Williams, and presently a very red-faced young man came out of what appeared to be a coke-cellar, with his mouth covered with crumbs.

Williams was the son of a Welsh stone-mason. He published some important work on the theory of the atomic structure of metals while he was at Manchester, but he did not seem to fit in there very well. He went to the Cavendish Laboratory and did good work on the experimental and theoretical investigation of atomic collisions but acquired a reputation of being slightly aggressive. It may have been a reaction to the Cambridge social atmosphere.

Then he went to Copenhagen for a short time. In one of the theoretical discussions in Bohr's institute, progress was held up because a point appeared not to have been investigated. After the discussion had gone on for some time, Williams, who was sitting at the back of the theatre, shyly produced a reprint of one of his Royal Society papers, saying that perhaps this might be of use. It was found that Williams had dealt with this particular point years before.

From that moment, Williams became internationally known, and a transformed person. His lively manner now appeared to be the pointed expression of a powerful intelligence, which suddenly blossomed at the Kharkov conference with fascinating variety and poise. I heard him draw attention with unconsciously masterful certainty to a mistake in Dirac's reasoning, the only time I ever heard that done. Williams had a perfect balance between experimental and theoretical reasoning, like that of Clerk Maxwell.

Williams had not seen much of the wide world before his visits to Copenhagen and Kharkov. He was paralysed with embarrassment at having to share a sleeping compartment in a Russian train with a lady, and when he was moved to another containing men only, was almost equally embarrassed at seeing in one of the berths a very big Red Army officer with a shaved head. He was at first very chary of Russians in the restaurant car, but I presently discovered him sitting between two Russians with his arms round their shoulders, a long row of empty beer bottles before them, and singing lustily.

Williams was elected professor of physics at Aberystwyth, where he started excellent atomic research. I have been told, however, that he broke the College's rule that private telephone calls should be paid for, and not left on the books of the department. After he had a particularly long personal call, he was rung up, and a female voice inquired: 'Was that not a private call, Professor Williams?' He firmly replied: 'No, it was not.' 'And,' Williams added, in his description of the incident to his friends, 'what could she say then?' He heard no more about private calls.

When he set about reorganizing the research laboratory, he found that the chief technician was somewhat argumentative, and he could not understand why. Presently he discovered that he was also Mayor of Aberystwyth, and as such *ex officio* on the governing committee of the College. 'After discovering that,' said Williams, 'I always took my hat off when I met him in the street.'

Williams exercised his versatile ability in the Second World War, when he assisted Blackett, and worked out the mathematics of the operational research into the protection of convoys of ships, which was an important factor in the survival of Britain.

Williams died of cancer at the age of forty-three. His early death was a serious blow to British physics, and the British themselves must also take some blame for not fostering his great talent sooner.

Another remarkable performance at the Kharkov conference was the interpreting of the proceedings by the theoretical physicist Frenkel. He exhibited an extraordinary facility for immediately and completely understanding everything, and expounding it in various languages. He was like the interpreters at the United Nations, except that he was dealing not with ordinary things, but with abstract arguments in mathematical physics.

The Physico-Technical Institute at Kharkov had developed very much in the two years since I had first visited it in 1932. I was struck by the work on a 5-million-volt Van de Graaff accelerator, and by the construction of a large rotating anti-cathode X-ray tube, capable of producing very intense beams of X-rays. These were evidence of the rapid development of engineering skill in technical physics. They were an indication that the widespread belief that the U.S.S.R. would never equal the best science outside it was wrong.

The papers by Landau and Lifshitz on the production of electron and positron pairs by the collision of nuclei, by Tamm on the exchange effect between neutrons and protons, and Frenkel on electro-dynamics were a demonstration of the first-rate competence of the Soviet theorists. The foundation was being laid for the future Soviet achievements in the release of atomic energy.

While in the Soviet Union I visited Joffe's institute again. The administrative director arranged the details of this visit, and I told him I wished to write on the institute and other branches of Soviet science at length. To do this, I would have to spend at least three months in the U.S.S.R. collecting the latest material. He at once gave me an invitation to do this. I returned to England and prepared plans for staying in the U.S.S.R. through the late autumn and winter to carry out this project.

Early in 1934 Kapitza's new helium liquefier in the Mond Laboratory at Cambridge went into service. It was a reciprocating engine which cooled gas by expansion. The lubrication of such machines at temperatures near absolute zero is technically very difficult. Kapitza avoided the difficulty by running a loosely fitting piston in a cylinder without a lubricant, but so fast that the gas had no time to escape before it had been cooled. His invention helped to restore the British to a leading place in research in low-temperature physics.

In the summer of 1934 Kapitza went to the Soviet Union for the annual visit he had made for several years. It was presently heard in England that he was having difficulties in securing permission to return to England. Rutherford, the Royal Society, Cambridge and Trinity College, and the Department of Scientific and Industrial Research, which had given Kapitza substantial financial support, were much perturbed at the news.

When I went back to the U.S.S.R. I took with me a letter from A. V. Hill to hand to Pavlov, explaining the concern felt, and asking that everything possible would be done to facilitate Kapitza's return. When I arrived in Leningrad, Miss Wilm, the representative of VOKS, the Society for Cultural Relations with Foreign Countries, whom I had met several times before, took over the practical arrangements for my visit. Miss Wilm was a most experienced person. She had acted as interpreter for Beaverbrook and Arnold Bennett on their celebrated visit to Leningrad in Beaverbrook's yacht. She had been shocked and mystified by Beaverbrook's behaviour to Bennett. She told me that Beaverbrook made jokes during meals about Bennett's stutter, while Bennett sat silent, and apparently passive. Possibly Beaverbrook and Bennett understood each other, but to the Russian admirer of the great novelist such behaviour appeared absolutely barbaric.

On another occasion she was interpreter for C. V. Raman during a visit to Leningrad. Raman was a brilliant, largely autodidactic genius, who thought very much on his own lines. He had a high opinion of his achievements, in which he was justified. I once asked him to write a book on his speciality, the scattering of light. 'What, one book,' he replied, 'why, *ten* books would be necessary to describe *my* work.' I first met Raman in 1924. It was on a tug in Montreal Harbour. A party from the British Association was being taken round the harbour and generously entertained. The skipper of the tug had prepared a large bowl of very strong punch and was hurt by Raman's refusal to drink any. He was a strict Brahmin and shoved the skipper away with repelling gestures with his hands, saying, 'No, no; soda-water, soda-water.'

As soon as possible after my arrival in Leningrad I called on Weinstein, the chief representative of the Soviet Foreign Office in the Leningrad Province. I told him of the disquiet in England about Kapitza, and showed him A. V. Hill's letter for Pavlov. With him was Oumansky, then a young diplomatic official. Weinstein, who had the short beard and decisive manner of a revolutionary of the Lenin period, had only one eye, which, as is often the case, was exceptionally piercing. He treated me with the utmost politeness and I have no doubt that this early visit to him facilitated the rest of my visit.

I called on Kapitza's mother. This excellent lady was a lecturer on anthropology and education in a teachers' training college. I gathered that her expert tuition had greatly helped in the early education of her gifted son, who had an impediment of speech. Kapitza's father had been an Army general of engineers, who had had an important part in designing and building the fortifications of St Petersburg. Mme Kapitza was also a striking personality, of level temperament. On the wall of her sitting-room she had the magnificent painting of Kapitza and Semenov discussing a scientific problem in their research student days. The artist was one of their friends, a man of comparable talent. He had depicted their

Politics Change Scientific Circumstances 141

personalities, and shining glass apparatus with its reflections, brilliantly.

Kapitza was staying in her flat at the time. There were two bodyguards, who seemed to camp in the staircase, and followed Kapitza wherever he went. So far as I could see, they did not interfere with him in any way. At this time, when I first saw Kapitza in Leningrad, he still believed that if he defied the Soviet authorities, they would be forced to allow him to return to England, because of the strength of his English and international connections.

In the intervals of my visits to scientific institutes to collect notes for my book I was taken to various entertainments. One of these was a concert. In the midst of the performance, the manager came on to the platform and announced that it must cease as Kirov, the governor of the Leningrad Region and one of Stalin's chief colleagues, had just been assassinated.

My Soviet hosts were dumbfounded by the news, and so was Leningrad as a whole. On the next day columns of silent people walked through the streets, at a loss to understand what had happened. The people who had been looking after me became deeply preoccupied, and more or less forgot my presence.

I was much struck at the time, and even more so since, that though I was a foreigner in Leningrad, carrying a letter of protest against a Soviet Government action, no one appeared to be the least concerned about me; one might have thought that when one of the heads of state had just been murdered, all foreigners in Leningrad might have been politely asked at least to keep to their hotels for the time being.

The very reverse happened. I joined Kapitza at his invitation in a walk in a park. He did a lot of walking, and ate a lot of apples, in order to keep fit. His two bodyguards, described as protecting his safety, and no doubt intended to prevent any movement towards the Finnish frontier or kidnapping by foreign agents, followed at a respectful distance of about twenty yards. Presently we came to some woodsmen, felling one of the park trees. While axemen chopped at the base, others guided the fall with a rope tied to the upper part of the tree. They were having some difficulty, so Kapitza suggested that we should help to pull on the rope. He called to his bodyguards, who looked at each other, and then somewhat sheepishly came and also lent a hand. After the tree was safely down we resumed our walk and the guards dropped to their normal distance. However, as we approached a large villa they came up and followed us very closely and intently. We heard afterwards that Stalin and various other heads of state had come to Leningrad to look into the murder of Kirov, and we wondered whether some were staying in this villa, which was cordoned off and surrounded by soldiers.

I called on Pavlov in his flat in the block of academicians' flats in Leningrad, beside the river. He was then eighty-five years old, extremely alert, and very gracious. He said he did not speak any foreign language, but it became evident that he knew German pretty well. He, of course, knew of the official attitude to my visit. I handed A. V. Hill's letter to him. He said he would study it, and he expressed his admiration for Hill, Sherrington and the British physiologists, and then engaged in amiable social conversation. I noticed that the walls of his study were almost completely covered with pictures uniformly arranged. The point which struck me was the width of his interests. He was shrewd, and interested in ordinary things and people. His personality seemed to me to have much in common with Rutherford's.

My remarks to him were made in bad German. Pavlov helped out with the German for 'nightingale', when there was a difficulty in making a point clear.

When I went to Moscow two or three weeks later I was looked after by the Scientific Research Section of the Department of Heavy Industry. I was put up in a hotel not used by tourists but by engineers, and Central European Communists who had left Germany after the rise of Hitler. Many of these people were behaving badly; they seemed to expect that Moscow would provide them with exactly what they had been accustomed to in their original countries. I saw Russian officials, who were being disgustingly abused by them, return their behaviour with the utmost consideration.

VOKS presently took charge of my arrangements, and put me in the best suite on the first floor of the Hotel Metropole. This gave me space and opportunity for a large party, to which Hessen, Petrovsky, Rose Cohen, Tamm and other Moscow friends came.

Shortly after I had arrived in Moscow Kapitza also came to Moscow to see high officials. We met again, and he told me that he had come to terms with the Soviet authorities.

I called on Sokolnikov (who had recently returned to Moscow after his period as Soviet Ambassador in London) in his flat, and I described the attitude in England about Kapitza. I said I believed Kapitza might contribute more to science by working in England, where his Russian outlook was as valuable in widening that of the English as his scientific ideas. Sokolnikov asked me whether the English would be prepared to exchange Kapitza for Cockcroft.

Politics Change Scientific Circumstances 143

It was evident that our discussion was not reaching any useful end.

The subject was changed to the social relations of science, and Sokolnikov recommended me to read Mumford's *Technics and Civilization*. I noticed he had about a score of excellent foreign books bearing on the subject, and I asked him how he had obtained them. He told me they were from the library of the Soviet Foreign Office. I was impressed by this selection of references bearing on science and society, which I had not anticipated in a foreign office library, and wondered whether the library was as well stocked in other branches of specialist knowledge.

After talking to Sokolnikov, I was taken to meet Mme Sokolnikov, who already had a visitor. This was Malcolm Muggeridge, who was then in Moscow. He had a kind of lively sadness, as if he had recently discovered that existence was different from what he had expected.

I called again on D. A. Petrovsky, who had been transferred from the directorship of the Department of Higher Technical Education to be Administrative Director of the Soviet Geological Survey, a strategic post in connection with the planning of the development of the minerals of the U.S.S.R.

N. I. Bukharin had moved from the Scientific Research Section of the Department of Heavy Industry to the editorship of *Izvestia*. I called on him at night in his editorial office. He looked more at home in the *Izvestia* editorial chair than in the Department of Heavy Industry. There he had given the impression of being the perpetual revolutionary exile. He had had two able young scientific assistants, but his personal secretariat made a conspiratorial impression. It was clever and active, but one felt that science was only one of its concerns.

While I was talking to Bukharin in his *Izvestia* office, I could see the regard in which he was held by his staff. A succession of people called, and they all came in with a broad grin of pleasure at the prospect of seeing him. One of them was a cartoonist, who had drafted what appeared to be a very bad cartoon, but he was so pleased to see Bukharin that he lapsed into a permanent deep grin, and was unable to attend to what he said. Bukharin treated him kindly, suggesting that the drawing was not quite perfect.

The chief subjects of our conversation were Kapitza, and the literature of Marxist philosophy. With regard to Kapitza, he said politely, but very clearly, that the decision that he should remain in the Soviet Union was final, and there was nothing more to discuss. I asked him for recommendations of recent Marxist works for translation into English. When I got back to England I arranged for one of these to be translated by Ralph Fox. It was published in 1935, with the title *Marxism and Modern Thought*. Bukharin also presented me with an autographed copy of the U.S.S.R. Academy of Sciences' memorial volume of the 50th anniversary of the death of Karl Marx, published in 1933, as a source of his own views.

By the middle of January 1935 I had collected my notes for my book on *Soviet Science*. I was fortunate to have completed my programme, for it was becoming noticeably more difficult to visit Soviet scientific institutions. The security precautions were made much more strict after the rise of Hitler and the assassination of Kirov. They became more and more severe, and were not relaxed until many years later. This preserved the immediate interest of the information in *Soviet Science* much longer than is usual in that kind of book.

In 1934 Sir John Russell made his second visit to the U.S.S.R., and talked to me about the state of agricultural science there. His standard books on soil science had a much larger sale in Russian than in English. His Russian readers turned out in processions of hundreds, and met him with banners at railway stations. He said he had heard much more about world revolution on his first visit in 1930, and saw many pictures of Lenin. In 1934 he heard much more about the First Five Year Plan, and saw many more portraits of Stalin. Russell was a chemist by training, who had taken up agriculture. He looked more like an explorer than a scientist. He spent a great deal of time in the open, and had a bronzed complexion, darkish rough-looking hair, and a shaggy moustache. He was interested in the Le Play Society, an organization for the propagation of the ideas of the French engineer, who advocated the application of engineering and science to social problems.

Russell was keenly interested in the relations between science and society, but his ideas were different from those of the Marxists. (N. I. Vavilov had spent some time at Rothamstead, the English agricultural research station, which was a bond of interest between them.) I first saw Russell on the decks of the *Caronia* in 1924, and I did not at once perceive that he was a scientist. His scientific interests and sensitive disposition were not obvious at sight. His combination of qualities was rare and valuable. He attended to important things, and did not bother too much about administration, which annoyed some people.

During my visits to the U.S.S.R. in 1934 I got to know H. J. Muller, the American geneticist. I first met him at a dinner in Leningrad, where he had gone to direct research in the genetics department of the Academy of Sciences. Muller was not quite sure

Politics Change Scientific Circumstances 145

of his bearings in the U.S.S.R., and behaved with obvious caution; his American liberal radicalism was in fact not very compatible with Marxism. He was simple and straightforward, and his manner did not conceal his thoughts. We soon became friendly, and I was invited to his flat.

Muller was a short, pale, dark, thin man with a gentle and charming manner. His first wife was a blonde lady of medium height, who, I believe, had been a schoolmistress. After a while Muller went out of the room, and returned with his son an enormous, youth of fifteen, extremely fair, and bigger than himself. 'What do you think of him as a prospective Rugger blue?' Muller said, as an introduction. Shortly afterwards, Muller followed the transference of the Academy of Sciences to Moscow. We found ourselves in the same hotel, and as the Academy provided Muller with a car, he gave me some helpful lifts.

Muller's department contained twenty assistants, among them Offermann, the Argentinian geneticist, Kostov of Bulgaria, and Ruffel of the United States. Levit, the human geneticist, was already experiencing difficulties, through the controversies arising between the Mendelians and the Lysenkoites. Levit had data on seven hundred pairs of identical twins under scientific observation, and had obtained evidence that methods of teaching in infancy not only affected the acquisition of knowledge, but also the intelligence: good method could increase the intelligence.

I noted in my article on Soviet science that though there was a severe shortage of scientific directors, the new type of young Communist scientist was appearing. 'Will he succeed in simultaneously making scientific discoveries and adhering to the Communist party's political line?'

From Moscow I went to Dniepropetrovsk to see the new Physico-Technical Institute, created in relation to the big industrial developments in the region of the Dnieper Dam. Its staff included the able young scientists B. N. Finkelstein and G. Kurdumov, the former working in theoretical physics, and the latter on the X-ray structure of metals, on which he had already done notable work.

The weather was so cold that during the train journey the temperature fell to $-38°C$, and when I arrived in Dniepropetrovsk, it was still $-32°C$. The heat radiated from the boiler of the steam locomotive so quickly that the train was unable to proceed at more than about ten kilometres an hour. I arrived seven hours late, and found that the Dniepropetrovsk scientists had stayed up all night, so that they could come to the station to welcome me.

At this institute I saw Soviet science very much in the making; it was new, the men were young and able, and they achieved much. Kurdumov later became an Academician and a leading figure among Soviet experts on steel, which was to be so important in the Second World War.

CHAPTER FOURTEEN

1935
REALISM IN SCIENCE AND POLITICS

On 24 April 1935 the Soviet Embassy in London issued a statement that: 'As a result of the extraordinary development of the national economy of the U.S.S.R., the number of scientific workers available does not suffice and in these circumstances the Soviet Government has found it necessary to utilize for scientific activities within the country the services of Soviet scientists who have hitherto been working abroad.

'Dr Kapitza belongs to this category. He has been appointed director of a new institute for physical research under the Academy of Sciences. This institute has been specially founded for him by the Soviet Government, and large sums have been set aside for the building and its equipment under the directorship of Dr Kapitza, and in accordance with his desires and requirements. So far as his personal life is concerned, he is comfortably situated and receives a good remuneration.'

The Soviet Government's decision brought a variety of reactions in England. The most responsible of these in scientific circles were expressed by Rutherford and F. G. Hopkins, who was then President of the Royal Society.

Rutherford published a letter in *The Times* on 29 April, in which he described Kapitza's work in England, the lines of research he had conceived, and the facilities that had been provided for their prosecution. He said that 'while no one disputes that the Soviet authorities have a legal claim on Professor Kapitza's services, their sudden action in commandeering them without any previous warning has profoundly disturbed the University and the scientific world'.

He said that scientific men were watching with admiration the rapid progress of science in the U.S.S.R., 'but even under ideal conditions it would require much time to reconstruct in Russia the unique equipment specially constructed by Professor Kapitza in Cambridge. Science is international, and every scientist hopes that it will long remain so, and the facilities granted to Professor Kapitza in this country are a good example of this fact. Apart from the personal factor it may matter little in the long run whether the investigations that Kapitza had in view are ultimately made in Russia or in this country.' It was important to avoid waste of time and duplication of costly apparatus. 'But the personal factor is a vital one in creative work.' It was inevitable that Kapitza should have been greatly disturbed by the conflict of loyalties. He had heard that his health had been seriously impaired. 'Men of scientific originality and imagination like Kapitza require an atmosphere of complete mental tranquillity in which to do their creative work.' He feared that the Soviet Government, acting within its rights and no doubt with the best intentions, had not appreciated the psychological situation. It had given so many proofs of its interest in the development of science, that he hoped it would enable Kapitza to choose the environment in which he could most effectively utilize the special creative gifts with which he was endowed.

Hopkins supported Rutherford's letter with one published on 1 May. He mentioned the problem that had been raised for the Royal Society, a considerable portion of whose trust funds had been given for an object, the benefits accruing from which were now in abeyance. The 'country of Pavlov' was later in the year to hold an international congress of physiologists, and he hoped that the Soviet Government would 'further recognize the internationalism of science by allowing labours so well begun to continue where they can most rapidly progress'.

The letters of Rutherford and Hopkins were followed by one from H. E. Armstrong, who referred to himself as the senior Fellow of the Royal Society, and a member of the U.S.S.R. Academy of Sciences. He questioned their view of the loss of Kapitza. He said that according to his information, 'there is a feeling of relief, at least among our younger men'. He had protested a year or two ago 'against the importation by Manchester of a physical chemist from somewhere in the Balkans'. How was British science to progress if we did not employ those we train? 'At Cambridge and elsewhere men are hugging chains, eating their hearts out for lack of opportunity to develop their talent under practical conditions. Why, then, should we import foreign labour?' He questioned the value of Kapitza's work. In fact, he questioned whether 'relatively too much importance' was 'being attached to the doings of the atom-smashing brigade led by Lord Rutherford?'

'Do not our national needs demand work of a very different type —what is the use, for example, of atom-smashing when cattle are being slaughtered in thousands upon thousands because we know nothing of foot-and-mouth disease?'

Armstrong said that he doubted whether Ludwig Mond would have been in favour of his bequest being spent on a laboratory. His money had been virtually earmarked for the maintenance of the International Catalogue of Scientific Literature, which the Society had tragically allowed to lapse. Nor did he believe that Messel would have approved the utilization of his bequest to endow an academic professorship. He asked why Kapitza had been elected a Fellow of the Royal Society, when he was not a British citizen.

The Russians were setting out to order national development through systematic application of scientific method. 'Surely the Russians are right?' They are said to lack mechanical skill. 'If a paragon of mechanical ability be advertised as living abroad, they cannot logically do otherwise than recall him. Instead of leading a lotus life in Cambridge, he, too, may well be doing national work of a far higher order than even that involved in magnetizing atoms to destruction.'

The day after the Soviet Embassy's statement, I published an article on Kapitza in the *Manchester Guardian*. I referred to the facilities that had been provided for him, and to his projected researches on the combination of intense magnetic fields and very low temperatures. He had left for a summer vacation in Russia in 1934, from which he had not been able to return. 'From the purely scientific point of view it must seem a pity that his work should have been interrupted just when long, expensive, and elaborate preliminary work had been completed.

'Those who attach great importance to the principle of academic freedom will see in the incident of Kapitza's retention in Russia a serious infraction of that principle. It should be realized, however, that Kapitza is a loyal Soviet citizen, and at the present time Soviet Russia has need of her reserves of intellectual as well as material strength.

'Professor Kapitza has been working in Cambridge for fourteen years. The improvement in the relations between scientists in the Soviet Union and in Britain was chiefly due to him. Cambridge will feel the loss of his personality even more than his researches. He was the representative of another sort of culture, and he did much to broaden the minds of staid British students. In return, he benefited much from the steady, methodical help of prosaic British colleagues. It is questionable whether his abilities, which were of a peculiar type, could be as fruitful in any other environment as they were in Cambridge. The future will show whether that is so.

'As a loyal citizen in a period of national crisis it seems clear that when his Government has invited him Kapitza should work in Soviet Russia. But it also seems clear that the Soviet Government should have issued its invitation in a form that could not be misunderstood either by Kapitza or the rest of the world. The Soviet Government appear to have made a psychological mistake in this matter that may not be to the benefit of Kapitza's future researches or to Anglo–Soviet scientific relations.'

On 9 May 1935 Rutherford wrote to me:

My dear Crowther,

I thought your article in the Manchester Guardian an excellent one in all respects.

There is no further news to report about Kapitza, and unless I hear something definite within the next few weeks it will look as though they are going to adhere to their decision.

You may have seen the foolish article of old H. E. Armstrong in The Times. As you probably know, he is a recognized literary buffoon of the scientific world, but at the same time it is a pity that he cannot keep out of what he considers to be the limelight.

With kind regards,
Yours sincerely,
Rutherford

Negotiations on the future of Kapitza's apparatus and the Mond Laboratory were conducted between the University of Cambridge, the Department of Scientific and Industrial Research, the Royal Society, and the Soviet Government. The latter had started to build the Institute for Physical Problems for Kapitza, and offered to buy Kapitza's special generator for intense magnetic fields, which had been built at the expense of the D.S.I.R. It also offered to buy new apparatus and duplicates of old apparatus for transfer to the U.S.S.R. It offered to refund certain expenditures to the Royal Society, and agreed that any balance remaining should be used as a fund for future work in the Mond Laboratory. Rutherford and Cockcroft, who would conduct the Mond Laboratory in future, felt under a strong obligation not to pursue research on intense magnetic fields with Kapitza's larger generator, if Kapitza wished to pursue this line in Russia. Kapitza's large generator would probably be replaced by a large electromagnet for investigating the properties of atomic nuclei at very low temperatures. R. E. Peierls was to act as adviser on the future researches in the Mond Laboratory.

The effect of the changes was that Kapitza could continue his own researches in the U.S.S.R. at the expense of the Soviet Government, and the researches in the Mond Laboratory would be continued along rather different lines without financial loss. The net material result was that the Soviet Government made a considerable new grant for scientific research.

Kapitza's new Institute for Physical Problems had a fruitful history. Kapitza himself made major experimental contributions to very low-temperature physics. Landau became head of its theoretical department, and by his contributions became probably the leading authority in the world on the theory of the properties of matter at very low temperatures. In addition to this, Kapitza promoted much fundamental research on the engineering problem of the transmission of power by electromagnetic waves, one of the great developments which can be expected in the future.

His personal relations with the Soviet Government varied in cordiality from time to time, but he survived, with distinction. I had said in my article in April 1935 that the future would show whether his abilities would be as fruitful in the U.S.S.R. as in Cambridge. Certainly, in science, they were; but there was no substitute for the cultural influence he could exert as a Soviet citizen in Cambridge; that was a loss.

The growth of Nazi power, and then the outbreak of the Second World War, showed how proper it had been of the Soviet Government to insist that he should remain at home.

The return of Kapitza to the Soviet Union was conducted with good sense, and on the whole was a credit to all concerned. There was an almost complete absence of idealistic theorizing about academic freedom; a practical sense of scientific and political realities prevailed. It seemed to me yet another example of Rutherford's power of perceiving realities, and not being deflected from correct judgment by philosophic verbalisms. His actions were intuitive and nearly always correct, but sometimes did not coincide with his verbal expressions.

Pavlov attended the international conference of neurologists in London in the summer of 1935. I had a talk with him and found that, at the age of eighty-five, he still retained the most extraordinary vivacity. In his communication he described his work on the temperament of dogs, and how he had established that they possessed the same four fundamental temperaments noted in human beings by the Greeks. The vigorous demonstration of this was of great importance because it implied that experiments on dogs could provide significant information about the behaviour of human beings. The temperament of dogs and men could be explained in terms of stimulation and inhibition. The choleric type was deficient in inhibition: he belonged to it himself. In the phlegmatic type of stimulation and inhibition balance. The melancholic is over-inhibited and the least able to adapt himself to life. In the sanguine type, which is the most perfect temperament, there is great mobility of reaction and rapid adjustment to stimulation or inhibition.

On the basis of these four types of temperament, which are inborn, human beings have three more behaviour characteristics, the artistic, the thinking, and the ordinary, which combine with the others in differing degrees. The four basic temperaments came into action on direct stimuli from the external world, but the three specifically human types of behaviour arose from internal stimuli. They are concerned with signals of signals. Language and symbolism are based on this physiological mechanism. The artistic reacted more directly than the thinking type; thus it is physiologically more primitive, but, of course, the physiologist had nothing to say on the relative values of art and science.

Pavlov said that he had worked with dogs for sixty years, and five years previously he had begun to visit mental clinics to see whether his results had any bearing on mental disorder. He found that many could be paralleled in dogs. A choleric dog tends to have the same kind of higher nervous disorders as a choleric man. He said that the disorder of stereotyped movements, such as continuous combing of the hair was due to a defect in the stimulatory properties of the nervous system. Fixed ideas of the same type, instead of leading to the continuous movement of the hand, led to continuous, incompleted movement in the brain. He regarded paranoia as due to over-tension in the mechanism of stimulation and inhibition. It was usually due to delicacy in one part of the brain. This explained why paranoiacs often seem quite normal on all subjects except one, at the mention of which they become wildly excited.

On Freud, Pavlov remarked that he had many interesting ideas, but in his opinion he had the misfortune of not being a physiologist, and was therefore unable to express his ideas in a scientific, objective form.

Pavlov also described his experiments on producing higher nervous disorders in dogs. He had been able to produce in them parallels to claustrophobia, and the various forms of fear. He elucidated fear in terms of excessive inhibition.

The rise of science in the U.S.S.R. was a striking feature of the period, but that in the United States also began to make a new impression. In a general survey of recent scientific progress I noted that the Cavendish Laboratory was now becoming known to the public as well as to science, and I described how it had been directed by an incomparable sequence of four men of genius, Clerk Maxwell, Rayleigh, J. J. Thomson and Rutherford. However 'the most striking general advance in science in the last few years has been in the United States'. Until recently their vast expenditure on research had not produced comparable results, but the discovery of heavy hydrogen by Urey, of the positron by Anderson, of the cure for pernicious anaemia by Minot, Murphy and Whipple, and the contribution to genetics by T. H. Morgan and H. J. Muller had greatly increased the prestige of American science. For a long time the Americans had monopolized observational astronomy. I wrote that 'A review of science in the period from 1910 to 1935 shows that the frequency of the appearance of supreme ability remains fairly constant, but that there has been a large increase in opportunities for the moderately talented worker. There has been a large increase in fundamental research in America and Soviet Russia. In Britain the amount of scientific work of the very highest order remains much the same as before, but there is a large increase in the volume of useful research. An impartial judge would have had to admit that up to 1933 Germany on the whole made the greatest contribution to science, but that at present Britain shares the lead with the United States. The stimulation of science by the Great War has been the most remarkable feature of the period, but, unfortunately, the fabric of civilization was damaged so deeply by that catastrophe that it is doubtful whether the increase of science has made up for the loss of values in other directions and of confidence in the future of civilization. On the whole, the war destroyed more culture than it created, in spite of its stimulus to science.'

In England, research in physics had been very much concentrated in Cambridge and Manchester. One of the features of the time was the development of new centres, especially at Bristol. This was largely due to the talents of A. M. Tyndall, who had the vision to foresee lines of progress, the administrative skill to organize them, and the ability to persuade donors to provide the funds. He secured the appointment in particular of N. F. Mott as a young professor of theoretical physics. Mott started conferences on physics, rather on the lines of the informal conferences led by Niels Bohr in Copenhagen, in which frontier problems in physics were explored in a collective manner by a group of able men from all parts of the world. At Bristol, conferences on the physics of the solid state were started, in which English and foreign investigators collectively attacked the problems of the subject.

Among Tyndall's strokes of administrative scientific genius was his selection of C. F. Powell as a young research assistant. He called at the Cavendish Laboratory in search of a good man, and was recommended to see five young men. He saw them all, and then accidentally got into conversation with Powell, who was not on the list. 'This is the man for me,' said Tyndall to himself, and Powell came to Bristol, where he raised research in cosmic-ray physics to the highest world standards.

Tyndall was a modest, amiable man, rather slender, dark, and clean-shaven. He listened to people, and was persuasive. He was a type of administrative scientist whose value is being increasingly appreciated, as the administrative problems of science become more and more weighty.

My book on *British Scientists of the Nineteenth Century* was published in 1935. Shortly after the publication I happened to call on Rutherford. He asked me to sit down, and then said: 'Did I see a review of a book of yours?' I replied: 'It is possible.' He then said: 'A whole page in the *New Statesman*; you can't complain about that!' 'No,' I said, 'it was gratifying.'

The review was by David Garnett. In it he had spoken well of the book, but had disapproved of my comparison of Kelvin with Clerk Maxwell, to Kelvin's disadvantage. Who was I to award marks to these respective geniuses? In fact, my attitude reminded him of Chekhov's character, Dr Lvov, whose famous remark was: 'Do you think I am such a fool as to be unable to distinguish between good and evil?'

Evidently, the name had stuck in Rutherford's prodigious memory. 'By the way', he asked, 'who is this Dr Lvov?' I explained. 'Oh,' replied Rutherford, 'I thought he might have been one of those Communist fellows.'

While my book made Rutherford think of communism, it received a grudging review in the *Daily Worker*, on the ground that it was written by one insufficiently learned in Marxism. This drew a spirited protest, of which I was quite unaware, from Ralph Fox, who pointed out that if communists did not welcome intelligent books written by non-communists, however insufficient their knowledge of Marxism might be, they could scarcely expect to attract authors to their ranks. The book was still being bought steadily, if modestly, in 1968.

Rutherford used to note all kinds of things about the variety of people whom he knew, in spite of his immensely busy life. Each

Realism in Science and Politics

morning he used to receive a pile of correspondence from all parts of the world, and from scientists in many fields, besides physics. He used to sort out the papers on biology, chemistry and other subjects, and send them to those people in the University to whom he thought they might be of interest. It was a striking manifestation of his human qualities.

After the publication of *British Scientists of the Nineteenth Century*, in which Faraday is discussed, I mentioned that Faraday had declined to accept the presidency of the Royal Society. Rutherford threw his shoulders back and said: 'I suppose he didn't feel up to the job.'

I have not often applied for a job in my life, but on one of the rare occasions, I collected references from Rutherford, Joffe, Millikan and Heisenberg. Shortly afterwards, I went to report an address by Rutherford to a scientific society. I sat on the second row, in order to make sure of hearing what he said. After delivering his address to an audience of about four hundred, he stepped down from the rostrum to sit in the middle of the front row. He espied me, and immediately turned round on his seat, and in an enthusiastic stentorian whisper, which could be heard throughout the hall, said: 'I've just been giving you a reference for the . . . I'm surprised you think it worth while going in for, but I've done my best for you.'

Rutherford was a New Zealander, and did not belong to the English upper middle classes by birth. When he was a student in New Zealand, he became engaged to his landlady's daughter. Their marriage did not take place until Rutherford had guaranteed their future by historic achievements. Lady Rutherford was not personally interested in science, and took a detached view of his scientific prestige. In the midst of his most authoritative statements at breakfast, she would interrupt his eloquence with the interjection: 'Ernie, you've got some egg on your moustache.' F. P. Bowden, who was an Australasian, told me he once met a deaf old man near Malvern, who roared at him: 'So you come from Cambridge? Rutherford and I were at college together. We always said his landlady's daughter would get him, and she did.'

Rutherford once said to Niels Bohr: 'You may be a bit of a Bohr, but you've got a very charming wife.' He was such a great man that his remarks only endeared him the more to all concerned.

One of the interesting and significant events of 1935 was the unsuccessful rebellion in the Royal Society. The anniversary meetings of the Society were usually quiet and polite occasions, when the president delivered an address on the general aspects of science during the past year, and conferred the Society's medals on various scientists who had achieved special distinction.

This year, Sir Frederick Hopkins's five years of presidency ended, and the new president and officers were to be elected. The old council nominated the candidates, and its nominations were voted upon. For many years this vote had been formal, and the nominations confirmed more or less unanimously. But in 1935 the nominations were contested. A number of fellows, dissatisfied with the conduct of the Society, made other nominations for the presidency and chief officers, and a number of seats on the council. The old council had nominated Sir William Bragg for the presidency, while the opposition's candidate was Sir Almroth Wright.

This unusual meeting drew a full meeting of fellows, including Ramsay MacDonald and two ex-presidents, Sherrington and Rutherford. About half the total membership of the Society was present, and there was an air of amiable excitement. The nominees of the old council were elected by a majority of about 180 to 20. Sir Daniel Hall received about two hundred votes for a seat on the council, so that it was evident that both sides had voted for him, and that no other candidate had been supported by both sides.

A considerable number of younger fellows were dissatisfied with the way in which the Society had been conducted for a long time. They wanted to see it give a more positive lead on contemporary social problems which had a scientific aspect. They wanted to see the Royal Society doing more to help the world in its difficulties, and they would welcome changes in the Society's government that would make such action easier.

After the election one eminent scientist remarked to me: 'The National Government has been returned with an overwhelming majority.' I reported that the challenge indicated that changes in the government and attitude of the Royal Society would occur sooner or later. However, it subsequently turned out that the changes were to come later rather than sooner.

Sir Frederick Hopkins delivered his last anniversary address to the Society while the vote was being taken. His theme was the social responsibility of scientists. He referred to the efforts of the British Association for the Advancement of Science to deal with the question, and the criticism of these efforts by the press. He said that the importance of the question was clear, but it was not easy to see how scientists as a class should act in relation to it. Perhaps the scientist could perform his social duties only by taking care to be a good citizen as well as a good scientist, and that there was no action a scientist could take, as a scientist, to control the social effects of scientific discovery. He had noticed two attitudes to the problem; the optimism of General Smuts, who hoped and expected that a connection between science, religion and art would be discovered in the future; and the pessimism of the late Sir Alfred Ewing, who near the end of a long life as scientist, engineer and academic administrator became disillusioned with the results of the progress of science.

Hopkins said his sympathy was strongly with the optimists, and that men should strive to solve contemporary problems and not turn from them in despair. One task was to establish better relations between science and the public and government. Hitherto the Government had frequently sought advice through the Royal Society. He thought that the Society should no longer consider its role restricted to advice, but, on occasion when it had something definite to say, should take the initiative.

The oldest and most famous of British scientific societies felt the pangs of adaptation to new social and political requirements.

CHAPTER FIFTEEN

1936
SCIENTIFIC PERSPECTIVES MODIFIED BY HISTORICAL CONDITIONS

As the atmosphere of Soviet science changed after Hitler's acquisition of power, so that of science in Britain also changed, but in a different way. The Soviet change was in the direction of increased security precautions and military preparations, that in Britain of liberal political and intellectual protests. It became difficult for a foreigner to enter a Soviet scientific laboratory, and after 1936 virtually impossible for at least a decade. In Britain, Bohr and Rutherford were increasingly heard on subjects other than physics. The leaders of science reaffirmed liberal political and intellectual principles as the foundations of scientific advance, and took further discreet steps in helping scientists who were the victims of fascist tyranny.

Early in 1936 Rutherford took the chair for Bohr, who lectured on the humanistic aspects of science at the Warburg Institute for the Study of Primitive Art and Science, which had recently been moved from Hamburg to London, owing to the effects of Nazi politics on the conditions of scholarship in Germany. Bohr thought that the principle of complementarity, which he had used with such power in reconciling apparently irreconcilable principles in physics, might be applied fruitfully in other domains. He had derived the principle of complementarity from Heisenberg's principle of uncertainty, according to which the combined position and velocity of a particle cannot be ascertained beyond a certain limit of accuracy. The more accurately the position was known, the more uncertain knowledge of the velocity became, and vice versa. Thus these two fundamental aspects of any particle cannot be separated, and are in a sense complementary.

Bohr thought that similar reasoning might be invoked to reconcile the contradiction between vitalists and mechanists in biology. He argued that in order to learn the behaviour of a living organism it must be allowed some freedom. This freedom may confer on the organism the power of concealing some of its secrets. Thus the mutually exclusive principle is seen in operation when the biologist

Scientific Perspectives Modified by Historical Conditions 159

tries to make an experimental analysis of the organism's behaviour. Bohr pointed out that William James had remarked long ago that the difference between analysis in psychology from that in physics was that in psychology the very act of analysis, the fixing of attention on the idea, changed the idea. In fact, when analysis in physics is considered more deeply, it is found to have a similar quality.

Bohr contended that the principle of complementarity might help the understanding of national conflicts. He thought that the very power of understanding English culture might disqualify a man from understanding Chinese culture. Such a difficulty had, in Bohr's opinion, nothing to do with biological racial character.

Thus Bohr protested against the racial prejudice which was being exacerbated by the Nazis, and attempted to undermine it by philosophical argument. He suggested that race prejudices might be overcome in a way similar to that by which physicists had overcome their prejudices in favour of causality and pure objectivity.

Rutherford, from the chair, took another line, also characteristic of his personality. He said that the differences between people were only slight, and were due very largely to education and tradition. He would like to see for once a really expensive experiment in which all the races of the world were mixed together, and see what was obtained. Plenty of money seemed to be available for battleships, so he did not see why they could not try other expensive things.

It is frequently thought that Rutherford was prejudiced in favour of inexpensive experiments, but, as in a number of other misconceptions about him, this was not the case. He believed that it is always possible to conceive inexpensive experiments that are worth while, and was annoyed with scientists who became dependent on expensive apparatus, but this was not the same thing as disapproving of expensive experiments on principle. He condemned the lack of scientific imagination of those who could not think of anything worth while without expensive equipment. His concern was the source of the finance; I have heard it said that when he was offered large sums from America for the Cavendish Laboratory he refused them on the ground that if the British wanted research they could pay for it, and he would have been happy to accept any sum as long as it came from what he regarded as the proper sources.

While Bohr was wrestling with the intellectual problems of race prejudice, he also cautiously forecast in the same week the possibility of causing an atom to explode. He thought that the very energetic particles which had been detected in cosmic rays might have been ejected from exploding atoms. As accelerating apparatuses had already been made which could produce particles with one-hundredth the energy of cosmic-ray particles, he thought it probable that particles with cosmic-ray energy might presently be produced. As such particles appear to be the products of atomic explosions, they would, presumably, be able to cause atomic explosions.

Bohr invoked the knowledge which had recently been gained from experiments with neutrons to conceive a model of the atomic nucleus which might explain the process of explosion by an impinging particle. He compared the nucleus with a group of billiard balls lying on a circular table with low cushions. If an external ball is shot into the group, it starts a series of mutual collisions which cause the impinging ball to be captured. If the balls go on colliding, one ball may presently collect enough energy to jump over the cushion.

While this process envisaged the explosion of the atom it did not entail any considerable release of atomic energy, for the energy of the particle ejected from the exploding atom was supposed to be much the same as that of the impinging particle. But the conception of a model of an exploding atom was loosening the ancient prejudice in favour of the eternity of the atom. By this, and by his subsequent 'liquid-drop' model of the atomic nucleus, Bohr provided giant intellectual strides towards the coming release of atomic energy.

Bohr's very first physical researches as a student had been on the properties of water drops and jets, and he applied the mastery he had early acquired in this field to the apparently very different one of nuclear physics. He was a practitioner in the application of sets of ideas from one field into others in which, at first sight, they appeared quite foreign.

While the elders of science were struggling in this way with the problems raised by Nazism, the younger scientists displaced by it had to find jobs. Many of these at various times passed through London, and some called on me. Among these were Weisskopf and Delbruck. They visited London together, their boat train depositing them at Charing Cross Station. They walked up to a policeman, and one of them asked: 'Can you please tell us where we can get some good coffee?' The London Bobby looked them up and down, and then replied: 'On the Continent, sir.'

Weisskopf consulted me in 1936 on the form of an application for a readership in mathematics at the Imperial College. In spite of his brilliant qualifications he was not appointed. He went on to the United States, where he pursued a brilliant career. I was told that Heitler had applied for an appointment at the Imperial College. E. A. Milne was an expert assessor, and at the discussion of the merits of the candidates, spoke very strongly in his favour, but inadvertently referred to him as Hitler; Hitler had discovered this, and Hitler had done that, etc. Everyone was too polite to blink an eyelid.

Rutherford was Chairman of the Advisory Committee of the Department of Scientific and Industrial Research. He gave increasing attention to the impact of science on society. He gave the Norman Lockyer Lecture of the British Science Guild in 1936, on the subject of 'Science and Development'. He held that, on the whole, science had been overwhelmingly beneficial to the welfare of mankind. It was true that science might occasionally be used for ignoble ends, but the responsibility for this rested in the community rather than on the scientist and the fault was not his if the community failed to prevent the prostitution of science. He did not agree with those scientists who were depressed by such events, nor did he think that scientists could by their own actions control the wrong use of their discoveries. The precipitate introduction of new techniques into industry had, however, produced violent effects of a serious nature during the last few years. Some method of control of the rate of application of new ideas or inventions into industry was probably desirable in the public interest. In the present state of industry, when progress and change were rapid, it would be an advantage to the State to know the probable changes that were to be expected in industry before they were actually put into operation. He thought that the Government should set up a 'prevision committee' of an advisory nature. The function of the committee, which would be composed of representatives of business, industry and science, would be to form an estimate of the trend of industry as a whole, and the probable effects on the main industries of new ideas and inventions as they arose, and to advise whether any form of control was likely to prove necessary in the public interest.

Rutherford went on to say, in 1936, that a competent committee of this sort would no doubt have foreseen the coming competition between motor and railway transport, and have adjusted competing claims before they became acute.

Rutherford said that when he began research at Canterbury College in New Zealand in the 1890s, it was in a cellar, but quite a comfortable cellar. His first duty had been to fill Grove cells with nitric acid every morning, as they had not even got lead accumulators. When he arrived at the Cavendish Laboratory in 1895, which was certainly one of the best in the world, it possessed only one crude ammeter, one voltmeter, and one electrometer. He used to test whether the lead accumulators were charged by taking shocks through his fingers. The creation of the sensitive, practical and accurate modern instruments of which every student nowadays had dozens was largely due to Weston, an Englishman who had gone to the United States.

Townsend and Rutherford both utilized the single electrometer, an ancient instrument devised by Kelvin. They used it for the measurement for the electrification of gases, and Rutherford said that he believed this was the first important research ever done with a Kelvin electrometer.

He had seen the wonderful instrumental inventions of C. T. R. Wilson, culminating in his cloud chamber for revealing the behaviour of single atoms travelling at thousands of miles a second. He thought that the history of man would never contain an account of an instrument more unexpected and wonderful than this. The improvement of the vacuum pump had created the electric lamp and the possibility of atomic disintegration on a large scale. Forty years earlier the experimenter had a few small instruments; now he needed huge glass tubes, and magnets weighing up to a hundred tons. The importance of the inventor of new instruments and machines for the advancement of science should not be underestimated.

In 1929 and the six succeeding years I had made seven visits to the U.S.S.R. I felt it was necessary to start on a serious study of scientific and technical development in the United States, and also to apply the method I had pursued in *British Scientists of the Nineteenth Century* to the study of scientists of other nationalities. In order to apply this method it is necessary to have some knowledge of the general history of the country concerned. My knowledge of languages was not good enough for me to pursue this except in English, and as American history was written in English, it was immediately accessible to me. I therefore started work on *Famous American Men of Science*.

At the same time, visits to scientific institutions in the U.S.S.R. having become more difficult, and visits to German scientific institutions repulsive, I paid more attention to other countries. Owing to the conditions in Germany, the small countries such as Holland and Denmark were becoming increasingly important as meeting places for scientists.

I went to the Seventh International Congress on Refrigeration

Scientific Perspectives Modified by Historical Conditions 163

in Holland, and saw the Philips Laboratories at Eindhoven. After this, I went to Copenhagen to attend two conferences.

During the Refrigeration Congress, I visited Leiden, and in particular the Kamerlingh Onnes Laboratory for Low Temperature Physics, and the industrial reserach laboratory of the Philips electrical firm at Eindhoven. Though the two laboratories appeared to belong to quite different regions of research, they were historically very closely related.

The Leiden laboratory was an academic institution in the Leiden University. This was a State University supported by the Dutch Government, and its staff were civil servants. Consequently an influential professor of physics, who was himself a civil servant, was in a position to secure considerable funds for laboratory development. Kamerlingh Onnes had been such a personality, and he had used his power to create the first university research laboratory which made full use of modern industrial technique, both in equipment and organization. He included in his development of the laboratory a school for instrument makers as an essential part of the laboratory organization. The staff of this school served also as the chief mechanics to the research workers. As recognized teachers and civil servants these technicians had a higher status than elsewhere. They were not treated as mere workmen, and this had an influence on the quality of their work. Further, the school provided an ample supply of first-rate technicians. This enabled the Leiden investigators to make very large numbers of precise measurements with very skilfully made apparatus. They made the properties of many substances at temperatures near absolute zero as well known as at room temperatures, and they aimed at accumulating a large amount of exact data as a foundation for theoretical advance.

Kamerlingh Onnes's policy of laboratory development enabled him to overtake the British investigators in low-temperature physics, who had enjoyed a lead for fifty years since the days of Joule and Kelvin. Dewar had succeeded in liquefying hydrogen, but to his chagrin Onnes became the first to liquefy helium and open up the field of research at very low temperatures. He not only secured the lead in discovery, but provided a stream of scientists and technicians familiar with industrial methods, and hence useful in many parts of Dutch industry. The Leiden system greatly influenced the Russians in their planned development of scientific research.

The most interesting piece of experimental research done at Leiden at this time was the approach to 0.005 degrees above absolute zero. This was carried out by De Haas and Wiersma, following the theoretical suggestion of Langevin, Debye and Giauque, that as the arrangement of molecules in a body absorbs a certain amount of energy, a sudden rearrangement at a very low temperature might cause the absorption of nearly all the residual heat, and thus produce an extremely low temperature.

The low-temperature physicists were already thinking about the possibility of utilizing the phenomenon of super-conductivity for transmission of large electrical currents along very thin conductors, and the construction of powerful electrical machines in the space of a matchbox.

The Seventh Congress on Refrigeration held at The Hague illustrated the variety of application of this technique. Besides providing a meeting-ground for scientists, some of the most distinguished of whom were refugees, it also provided an interesting demonstration of the British activities in the field. Apart from the Dutch, the largest delegation was the British. It was distinguished by the work which had been done by Sir William Hardy and his colleagues, in their introduction of carbon dioxide atmospheres in refrigerated ships and warehouses, for importing apples and chilled meat. There was also a strong group from the new low-temperature school at Oxford, including Kurti, Mendelssohn and Simon.

The Congress covered a vast range of topics, from experiments with liquid helium to the manufacture of ice-cream, ice-hockey and skating rinks, and the application of refrigeration in the Dutch flower and bulb trade. The Soviet scientists sent a description of the freezing operations in the construction of the Moscow Underground through regions of mud and unstable earth. Paul Becquerel communicated a paper in which he suggested that as certain simple living organisms could survive at minus 27 degrees Centigrade, germs of life might survive in space. Their dispersal through space might have led to the population of planets.

A notable feature of this Conference was the long and excited discussions among the tea and coffee tables along the sunny beach of Scheveningen near The Hague, where scientists formerly separated by political events resumed scientific and personal exchanges.

If the low-temperature laboratory at Leiden was then unique in its application of industrial technique in an academic laboratory, the Philips industrial research laboratory at Eindhoven was unique among industrial laboratories for the quality of its research atmosphere, equal to that in Leiden University. It was then

Scientific Perspective Modified by Historical Conditions 165

probably the best industrial research laboratory in the world, superior to any in Germany or the United States and in advance of any in Britain.

The motive for the creation of the Philips Laboratory was, of course, industrial. During the First World War, the supply of special glass that they obtained from Germany for making their electric lamps was cut off. They had to set about making their own and solve the problems which this involved. This stimulation to solve their own problems led to the creation of their great research laboratory. They secured from Leiden Kamerlingh Onnes's assistant and collaborator, G. Holst, who was offered attractive conditions. He was not very sorry to leave Leiden, for Onnes had been a domineering chief. Much of the research leading to the discovery of super-conductivity had in fact been done by Holst under Onnes's direction, and Holst had actually been the first to observe the fundamental new phenomenon. But Onnes took the credit, as many heads of laboratories did in those days.

Philips were thus fortunate in securing an exceptionally able man for developing their new department of research. By 1936 it had a staff of 415, including forty physicists, twelve chemists, and fifty-three engineers. A feature of the organization was the virtual absence of administrative staff. A spacious and quiet building had been erected for them. The most important invention they had made up to that time was metal-and-glass seals. This enabled them to develop the sodium-vapour and mercury-vapour lamps that have since become much used for road lighting and other purposes. It also made possible the manufacture of large radio transmission valves consuming hundreds of horse-power.

While Philips were accomplishing research of the highest quality, their staff were making first-rate contributions in such subjects as the mathematical theory of radio circuits under van der Pol, whose study of the theory of oscillations led him to detect hitherto unrecognized types of oscillations in the heart; he made a model of the heart which manifested these oscillations.

Another unexpected line of work was on the synthesis of vitamin D. This had arisen out of research on the improvement of ultra-violet ray tubes for medical purposes.

The staff of the Philips factories at Eindhoven had already contracted from a maximum of 24,000 through rationalization and economic nationalism in other countries, and had created a difficult unemployment problem. Much of the town as well as the industry of Eindhoven belonged to the firm, and was managed in a paternalistic style. It was apparent that the firm attempted to think of everything and in fact did rather too much thinking on behalf of their workers.

From Holland I flew to Copenhagen to attend the two conferences in which Bohr took a leading part, one on theoretical physics and the other on the philosophy of science. About eighty physicists attended the former. The feature of the Bohr physics conferences was the study of theoretical problems in a collective and social spirit. Bohr's talent for inducing younger colleagues to make discoveries was even greater than his own tremendous direct contributions. He had a power of sympathetic intelligence which enabled him instantly to perceive the value of his colleagues' aims and efforts. Everyone felt that he or she was immediately understood, and theorists of the most diverse temperaments and nationalities worked together in amity.

The most interesting attender of these conferences was, for me, Heisenberg. He gave a theoretical explanation of the particles that appear in cosmic rays, which aroused great interest.

The physics conference followed the usual inspiring and well-marked lines of Bohr's conferences. The congress on the philosophy of science was different. To attend both in immediate succession, and see some of the chief scientists in the two different intellectual environments, was illuminating.

The philosophers met in a large room in Bohr's house. This was the celebrated mansion in the grounds of the Carlsberg Brewery which had been bequeathed to the nation as a house of honour for its most eminent contemporary scientist or scholar. Bohr was the present occupant. The room in which the philosophers met was decorated with a series of busts of eminent historical figures.

The roof of the room had a skylight. One day, wreaths appeared around the brows of the busts of the historical figures, and on another, the face of a young scientist could be seen from the philosophers' seats, peering down at them through the skylight. This was Paul Ehrenfest, the son of Professor and Mrs Ehrenfest of Leiden, both of whom were theoretical physicists, Mrs Ehrenfest being of Russian origin. Ehrenfest was subject to fits of depression, probably temperamental, but which he associated with difficulties in understanding the new quantum mechanics invented by the younger generation of his period.

Their son Paul was an attractive young man, who pursued research on cosmic rays at the high-altitude laboratory on the Jungfraujoch. He and his family were later on overwhelmed by tragedy. He was killed in an Alpine accident, and his father committed suicide.

Scientific Perspectives Modified by Historical Conditions

When I was in Leiden, Mrs Ehrenfest had invited me to stay in her house. She put me up in Paul's room, which was used as a guest-room when he was not there. A succession of visitors had been requested to write their signatures on the wall. I cast my eye down the list; it started with Einstein, and continued with almost equal distinction. Mrs Ehrenfest insisted that I should add mine.

The congress on the philosophy of science presented an extraordinary variety of ideas and personalities, compared with which the conference on theoretical physics appeared extraordinarily uniform, in spite of the wide range of personality among the physicists. It seemed that there was little unity of opinion among the philosophers, so that each wandered off along his own train of thought and lost touch with the others. The theoretical physicists, in contrast, were agreed about so much, that their differences in personality seemed much less, though intrinsically these were no doubt very wide.

In this year, the British Association met in Blackpool for the first time. Its organizers felt they were rather bold in having the courage to meet in the British metropolis of mass-entertainment, and the governors of the town felt they were enterprising in attempting to widen their appeal to include bodies of this kind.

The Association had become aware of the need to attend more to the relations between science and the public welfare. Sir Josiah Stamp devoted his Presidential Address to 'the Impact of Science on Society', and the Education Section had papers on 'The Cultural and Social Values of Science', but the Association did not proceed far in the direction of what was really needed, the expression of the collective opinion of scientists on the place of science in modern life. I wrote in my report for the *Manchester Guardian* that 'members of all sections should come together not only to explain but also to discuss how science bears on public welfare'.

Stamp, who was then the chairman of the L.M.S. Railway, naturally drew some of his data from transport. He discussed how the development of transport increased the mobility of the population, while trends in population growth and habits inhibited it. The percentage of elderly persons who did not wish to move was increasing, while unemployment payments decreased the tendency of unemployed workers to seek work in other regions where there was more employment.

Thus the fluidity of invention and the rigidity of society were increasing simultaneously. Stamp hoped that the increasing difficulties caused by innovation might be reduced by means of psychological research. But unfortunately, the expenditure for research in the social sciences was only about one-tenth that for the natural sciences. I commented that the disappointing feature of his address was that he did not suggest how this situation might be changed.

Stamp heard his views, which Rutherford subsequently approved, criticized by Hogben and others in the Education Section.

H. G. Wells had come to Blackpool to listen to this discussion, and to support the increasing demand for more study of the relations between science and society. He stayed in one of the large Blackpool hotels, which was quite unaccustomed to the kind of patrons it received on this occasion. I was staying in the same hotel, and witnessed a singular conflict of cultures between Wells and the industrious waiter who looked after him. The waiters and the hotel knew how to look after guests with a large appetite and thirst. Quantity, not quality, was their criterion, and Wells could not stomach the large quantities of unsubtle food that were placed before him. He regarded himself as a connoisseur of red wine, and refused to accept the bottle of red liquid which was produced. He presently saw that the hotel staff meant well and genuinely could not comprehend the difficulties raised by their fare.

The Blackpool meeting provided much interesting material in other directions, besides the impact of science on society. Jeans, Milne and Jeffreys spoke on the evolution of the solar system. Discussions on the structure of living matter, the shapes of animals or the architecture of life included contributions from H. J. Muller, Dorothy M. Wrinch, C. H. Waddington and Joseph Needham. The Blackpool Meeting provided much for the scientific journalist, and the *Manchester Guardian* gave me a great deal of space. It was during this meeting that Ritchie Calder remarked to me: 'We all envy you, J.G.!'

Calder himself did notable work at this meeting, propagating the idea of a World Association for the Advancement of Science. Some of its features were later realized by the founding of UNESCO.

Later in the year, one afternoon in October, as I sat in my flat in Russell Square with Ralph Fox, a letter arrived from Mr James B. Conant, President of Harvard University, which had been forwarded from my publisher. He had not been able to find my name in any of the usual works of reference. He had read my *British Scientists of the Nineteenth Century* with interest, and wished to make my acquaintance. He invited me to meet him at Brown's Hotel, and then have lunch with him at the Athenaeum.

He asked me whether I would be prepared to lecture at Harvard on some topic in the history of science. I said I would be very happy to do so and had in fact just finished a book on *Famous American Men of Science* which was due to be published soon. I could lecture on the substance of this before it was published. Conant suggested that perhaps I ought to spend some time in America before finally completing the manuscript, in order to obtain a more direct knowledge of the subject on which I was writing. I said I thought this would be a mistake, for if I studied the subject in America I would, no doubt, feel compelled to rewrite it entirely. It would be better, I thought, to leave the book as it stood, as a record of how certain famous American men of science appeared to a spectator on the other side of the Atlantic. Conant presently accepted my view, and it was agreed in principle that I should visit Harvard in the spring of 1937, and deliver six lectures on *The History of American Science*.

About a week after this I received a letter from Harold Laski, whom I had never met. He wrote that it would be a great pleasure to meet me. 'I hope you saw President Conant. After the picture I painted for him, I am hopeful that he will invite you to Harvard.' I lunched with Laski at the London School of Economics. We started with the two largest sherrys I had ever seen, and Laski discoursed in his famous manner, a mixture of generosity, genius and boasting. He told me how he had praised my work when consulted by Conant. Laski had been the hero of a celebrated struggle for toleration when he was on the staff of Harvard, and this had given him a special relationship with the University which henceforth took particular note of his advice.

During this lunch Laski told me that his struggle at Harvard had won him the friendship of Franklin Delano Roosevelt who had been concerned in his case as a Visitor of the University. Whenever he was in Washington he used to drop in to see his old friend 'Frank'. On one occasion, on entering the hall of the White House, Laski was transfixed by a portrait of Harding on one side, and of Coolidge on the other. He was shown upstairs into the President's study, and as he entered, the President asked: 'How are you, Harold?' 'Fine, Frank, but what are you doing with those portraits downstairs?' 'Ugh,' said the President, 'that's just to show how different it is when you get upstairs.'

I was told later that Conant had lent *British Scientists of the Nineteenth Century* to P. W. Bridgman, for his opinion on its physical competence. Apparently, Bridgman reported that he did not agree with my Marxism, but that I certainly knew my thermodynamics. I did not in fact know much thermodynamics, but I had followed very closely the magnificent review of this subject in Larmor's long obituary article on Kelvin, in the *Proceedings* of the Royal Society.

I was told, too, that during the summer of 1936 Conant had spent a holiday in Wales, and had read my reports in the *Manchester Guardian* of the Blackpool Meeting of the British Association. He had, in fact, gone to the beginning of the Blackpool Meeting, but found the arrangements in Blackpool so uncomfortable that he left almost immediately.

During the winter of 1936–37 I prepared six lectures based on *Famous American Men of Science*, for delivery at Harvard in the following spring. It was fortunate that this invitation to visit America came just at a time when I had felt the need for studying American science, and had devoted some attention to the subject.

During the autumn of 1936 the worsening political situation reflected itself in science in various ways. One which made a particular impression on me was the search for substitute materials in order to secure economic self-sufficiency.

In this year also, P. E. Cleator's book on *Rockets Through Space* came to me for review. Cleator was then President of the British Interplanetary Society. He explained that the interplanetarians hoped to devise a rocket whose crew could control the jet of gases so that it is gradually accelerated against the earth's gravity, until its speed reaches seven miles a second, and it leaves the earth. They calculated that such a rocket driven by a mixture of oxygen and hydrogen would weigh about 50,000 tons, and would cost £20 million. If atomic energy became available, the whole arrangement would be less clumsy.

I commented that while there was no prospect that the hopes of the interplanetarians would be achieved soon, they were not impossible. 'If humanity can prevent itself from destroying science, science may carry it through the universe some day.'

CHAPTER SIXTEEN
1937
HARVARD AND PARIS

EARLY IN THE spring of this year I went to America to deliver my lectures at Harvard University. Mr Conant, who was a young President, then forty-four years old, had entered upon his appointment four years before, with the intention of stimulating the development of Harvard in various directions. He invited a number of Europeans to lecture there, among whom was J. D. Cockcroft. I heard later that he was offered a chair of electrical engineering, but he did not wish to leave Cambridge.

Cockcroft told me he had travelled third class on the ocean liners, and had found it quite all right, so I decided to do the same. I travelled steerage in the French liner *Paris*, but it was a rather insanitary old ship, and the food was not good. I arrived in New York with a sore throat, and took a taxi straight to the apartment of my publisher, W. W. Norton. He and his wife very kindly looked after me.

Norton had a considerable acquaintance among scientists, and he invited Professor and Mrs L. C. Dunn and Dr and Mrs Peyton Rous to dinner to meet me. Dunn and Rous were critical of J. B. S. Haldane and other British biologists of his generation; they thought that they had not published very much original research.

Norton had read a review of Carrel's *Man the Unknown*, which I had written. This led to a discussion of Carrel. Dunn was very sharply critical of him. It was said that Carrel would look in certain directions, knowing that persons in distant parts of the U.S.A. lived there, and would then read their thoughts. Norton, Dunn and Peyton Rous were all in favour of the Spanish Republican Government, and Norton and Dunn were members of a committee in its support.

Peyton Rous invited me to visit the Rockefeller Institute for Medical Research after I had returned from Harvard. He gave a lunch for me in the Institute, with Gasser, then Director of the Institute, Landsteiner, the famous discoverer of blood-groups, and A. E. Cohn, the biochemist, among the guests. Rous said he had decided not to invite Carrel, after our remarks on him earlier.

Gasser worked on electrical impulses in nerve and similar fields. He was a man with a strange physique; tall, thin, and with a very high-pitched voice, presumably due to hormone abnormality. He was a good if not great scientist, and his limited range of human feelings had, presumably, been regarded as a positive qualification in a director—such a person might be more than usually objective in making judgments of the comparative values of lines of work, and decisions on which were to be promoted. One could see the positive element in this, but it was doubtful whether it more than counterbalanced the absence of other ordinary qualities which facilitate social relations.

Landsteiner was typically Austrian; his hair and moustache were dark, and his face pale. He looked very much the Viennese intellectual in exile, quiet, polite and resigned.

After lunch I was shown round the laboratories of Gasser and Peyton Rous. I had the impression that Gasser depended on instrumentation more than the British school led by A. V. Hill. When Peyton Rous took me to his laboratory, and described his celebrated work on cancer produced in rabbits by a virus I noticed a change in personality as he spoke of his research. I have seen this in some other scientists, the most striking being that of Frederic Joliot. Both Rous and Joliot seemed to me to grow in weight when speaking of their research. Both were exceptionally charming and amusing in ordinary social exchanges, as hosts or guests. But in the laboratory another man appeared: all amusement had disappeared, and an authoritative earnestness became evident. I suppose that in such cases, an extremely strong system of conditioned reflexes about research has become established, and when this system is stimulated by the appropriate conditions, it takes charge of the personality. A deeper and more serious side of life is suddenly revealed.

Cohn told me that my introduction to *British Scientists of the Nineteenth Century* contained the clearest expression of the belief that the success of discovery depended on the adaptation of psychological traits in the scientist to the characteristics of the development of science at the time.

The subject of my six lectures was *The History of American Science*. They dealt with the relation of Benjamin Franklin's Discoveries to the Conditions of his Time; The Influence of Science on the American Constitution; Joseph Henry, and the Smithsonian Institution; Josiah Willard Gibbs; Thomas Alva Edison, and the Rationalization of Invention; The Significance of the History of Science and a Programme for Research.

Harvard and Paris

Conant had asked me to include a lecture on the last subject, no doubt with the aim of contributing ideas for the extension of Harvard studies in the history of science. The policy for these studies had not yet been satisfactorily worked out or organized. George Sarton had long been working in Harvard, but he had then no appropriate position there. He was supported from Washington by Carnegie funds, and, as he was not on the Harvard professorial staff, did not have adequate accommodation. He was working in the Widener Library, in an improvised room made by shutting off the end of a corridor. Sarton sat in this fastness, surrounded by barricades of thousands of books.

My impression was that the Harvard authorities were not quite happy about Sarton's conception of the history of science; it was thought to be too antiquarian, and not bearing sufficiently on the current life of science.

After I had drawn up the synopsis in London for my sixth lecture, I sent it to Harold Laski for his comment. 'That sounds fine—for a start,' he wrote. The theme of the lecture was that the problems of society cannot be solved rationally without a thorough understanding of the role of science in modern civilization, and that this was impossible without a knowledge of the history of science. I subsequently expanded the theme into my book on *The Social Relations of Science*, which was first published in America in 1941.

The other five lectures contained the substance of *Famous American Men of Science*, published later in 1937. There was an audience of about a hundred at each lecture. I felt that the first, on Benjamin Franklin, was warmly appreciated. I admired Franklin, and enjoyed writing on his great contributions to science, and to America, and I also had what I believed were one or two new things to say about him.

My second, on Science and the American Constitution was, I thought, the most original. I had examined the attitudes of the chief founders of the Constitution to science, and found that they were both positive and varied. Incidentally, I came across a number of coruscating remarks, especially by John Adams, on the influence of bankers and party caucuses and the dominating role of corruption in American politics.

My remaining lectures were well attended, and members of the audience continued to come to me afterwards, and make amiable comments. But what I had said in my second lecture was not appreciated by all. At my lecture on Edison, I noticed a tall, elderly gentleman in the middle of the front row, wearing dark glasses, and accompanied by a solicitous lady; he evidently enjoyed what I said. I had referred to Edison's power of choosing and engaging able men, and that these included A. E. Kennelly, the famous discoverer of the Kennelly–Heaviside layer. After the lecture I was asked whether I would like to meet Kennelly, who was the old gentleman I had observed. I was delighted at the prospect, but felt nervous at the idea of having had an old colleague of Edison's in the audience.

Kennelly, who was now almost blind, was charming. He told me that he had listened with interest and agreement to what I had said. I had treated science in a new way, dealing with it as a social activity. He had been with Edison six years. He also told me of a technical error in my account of the 3-wire system of electrical distribution. I was able to put this correction in the proofs of my forthcoming book.

Kennelly asked me to lunch, but I had no free date, so he asked me to call in his laboratory. He repeated his remarks about the importance of the study of the social significance of science, in which I had engaged. He mentioned that he had been born in India, and the population of that country had increased by 150,000,000 during his own lifetime; this was a disaster. Man must understand what he is doing when he applies science. He wished the American Middle West could be near to Europe, just to frighten sense into the Middle Westerners. He was strongly in favour of increasing armaments, because nothing else, no other burden, seemed to bring men to their senses.

When I arrived at Boston I had been met by the President's Secretary, Stephen H. Stackpole, a pleasant, quiet and efficient man. He was the first of the highly efficient American presidents' secretaries with whom I became acquainted. I had not met anyone like them in the British academic world, in which elderly office boys, and sometimes office girls, without adequate intellectual training or experience, often managed the business of the heads of universities for them.

Stackpole made the arrangements for the programme of invitations and visits most agreeably. About two hours after I had arrived, I was taken by the President to dine with the Fellows' Society, where I was introduced to the former president, Lowell, A. N. Whitehead, the philosopher, L. J. Henderson, the physiologist, G. D. Birkhoff, the mathematician, and others. Henderson sat me down on a sofa, and told me that the Fellows' Society was supposed to combine some of the features of the All Souls and Trinity common-rooms; it had been started about five years before. There were some eight senior and twenty-four junior fellows. Henderson had asked A. V. Hill why Cambridge had produced so many good physiologists, and Hill had said that it was due to the Cambridge system of 'prize fellowships' so they had introduced the system in Harvard.

At dinner I sat between Haskins, the medieval historian, and the physicist Fisk, who told me he was glad I was not J. A. Crowther, as he did not like his books. Poor J. A.! As I later learned, he had had a raw deal from the scientific world, for, on balance, he made a distinguished contribution to biological radiology, and his textbooks had been widely used.

Henderson was chairman at the dinner, and mixed the salad ceremoniously. I was told that he had chosen the wines, and insisted on including a bad Californian port; I should fill my glass and leave it, as a protest.

In the course of conversation on the history of science and invention, I was reminded of A. P. Usher's excellent *History of Mechanical Inventions*. Haskins invited me to lunch to meet Usher, and Henderson asked me to dine at his home.

Then Conant gave a lunch at his house for me to meet Harlow Shapley, D. A. Little, the Secretary of the University, Wild, the Director of the News Office, Greene, the Director of the Harvard Tercentenary, and some others, to discuss the problem of how the University might present to the general public the results of its scholarly work. Shapley had himself started life as a newspaper reporter. He was keenly interested in science journalism, and was a leading figure in *Science Service*, which had been founded to improve the provision of scientific information to the press.

The provision of academic news at Harvard had not been well organized during Lowell's presidency. Lowell had disliked the press, and he had had a very uncomfortable time with it over a controversy on the restriction of the number of Jewish students. Wild, who had been on the *Chicago Daily News*, had been brought to Harvard to improve the press arrangements.

The Harvard authorities were also concerned with journalism from another perspective. A large sum had been bequeathed to them by a Wisconsin magnate. Unfortunately, after the will had been contested only about an eighth of the original sum was to come to the University. The problem of how to make the best use of this was engaging their attention.

I talked on my views and experiences in scientific journalism. Conant mentioned that Rutherford had said that he had absolute confidence in me as a scientific journalist. Among my remarks I repeated my ideas about the influence of ball-games on experimental science, especially in particle physics. Shapley commented that he could tell from the way a student played ping-pong whether he would make a good observational astronomer.

Haskins gave me lunch at the Signet Club to meet Usher. Two young men were also there; they were the only men introduced to me at Harvard who were not on the staff. One was David Rockefeller, and the other F. D. Roosevelt Junior. I noted Rockefeller's aplomb, and F.D.R. Jr's nervousness. Apparently they had been invited because it was thought that our conversation on science might be good for them.

On the next day I was entertained to dinner by L. J. Henderson. He lived in a wooden house in the Colonial style. He liked talking about Barcroft and A. V. Hill. I had never been able to understand why Henderson's book on *The Fitness of the Environment* enjoyed such prestige. It seemed to me obvious that the environment must fit the organisms it had produced. I found that Henderson's political views were very conservative, and I concluded that this may have increased the acceptability of his scientific ideas.

I had half an hour's talk with E. B. Wilson, the authority on Gibbs. I learned from him that Gibbs appears to have worked on a half-salary at Yale, in order to be free from routine, and devote himself to research.

My next engagement was with D. A. Little, who was in charge of the appointment bureau for Harvard men. He invited me to lunch in the Faculty Club with Sarton. Sarton showed me the accommodation he then had in the Widener Library. He had no natural light to work under. He gave me a signed copy of his book on the study of the history of science. He subsequently entertained me in his home.

Sarton told me he had met Nicholas 'Miraculous' Butler. When he arrived in New York as a refugee from Belgium in 1915, he had an introduction from Sir William Ramsay. He was ushered into Butler's presence and before he had found his bearings, he discovered himself standing on the mat outside the door again.

Sarton had been a pupil of Henri Poincaré, and regarded him as his master. When he published the first volume of *Isis*, he issued it with a portrait of Poincaré as the frontispiece. As he had not thought it necessary to give it a legend, he was surprised to find that many persons supposed that it was himself.

Sarton's house in Belgium was near the frontier. After the

invasion in the First World War it was occupied by the German General Staff. Sarton had buried some of his MSS. in the garden before leaving.

Sarton said that Lowell had been completely uninterested in the history of science. It appeared to me that Conant believed that the history of science would be advanced best by persuading professional historians to specialize in it. He did not think that ordinary successful scientists would care to leave their own researches in physics, chemistry etc. to engage in the history of science.

The Sartons, with their daughter May, who was going to stay with Conrad Aitken at Rye in England to embark on a novel, were a civilized family. Mrs Sarton, who was English by birth, had acquired a quiet French charm of manner.

I went to the Harvard College Observatory, where Shapley had invited me to tea. I was put in charge of Cecilia Payne, who told me about her book on *Variable Stars*. There were about thirty astronomers, sitting around in a large circle. They included Miss Cannon, Mentzel, and H. N. Russell. After information had been given of a new white dwarf discovered by Miss Cannon, and a new comet, Mentzel talked on synthetic optical glass.

Shapley asked me about the position of women scientists in England. Good women research workers were then having considerable difficulty in obtaining appropriate positions and support in Britain. In contrast, at Harvard College Observatory, there were several eminent women astronomers, especially Miss Cannon, Miss Leavitt and Miss Payne. This problem in the social relations of science started a lively discussion of the social responsibility of scientists.

The weightiest scientific attendance at my lectures appeared to be at the one on Gibbs. It included Bridgman, Van Vleck, Birkhoff and Bainbridge among the physicists. The apocryphal story that Gibbs always insisted on mixing the salad, on the ground that he was a better authority on the equilibrium of heterogeneous substances, drew an audible laugh; I wondered whether this had anything to do with L. J. Henderson. Professor Baxter, the Master of Winthrop House, told me that he had been impressed by my description of Yale. His father had been at Yale, and so had his grandfather, and from what he had heard from them, life at Yale was then exactly as I had described it. I had had the benefit of reading that remarkable book, *Four Years at Yale. By a Graduate of '69*, published in 1871; I had accidentally found a copy in the London Library.

A. P. Usher gave me lunch, and introduced me to his gifted pupil Robert K. Merton, who had already published his excellent paper on *Puritanism, Pietism and Science*. Merton struck me as the most able of all the coming men whom I met at Harvard. He seemed to have a subtler cultural sense than Harvard men in general. This was all the more conspicuous because he wore an old and rather bucolic suit, the air of which contrasted with the fineness of his mind. He was like a highly talented young man from a country town, whose intellect had carried him far beyond the local environment from which he had emerged. I do not know anything about Merton's origins: this was merely the impression he made on me in 1937.

In my view, his early work on the social interpretation of the history of science belonged to a higher and more important region of intellectual endeavour than the stream of American sociology. I think it a pity that he presently devoted his main effort to this field.

The new perspectives which were opening in the history of science were very promising.

In the autumn of 1937 Charles Singer invited me to stay with him in Cornwall. He told me that the treatment of the history of science in its social context had made his own work out of date, and that if he were starting on that subject again, he would do so from that point of view.

In December 1937 he wrote to me that he had read the manuscript of one of my lectures, I think it was the one at Harvard on the history of science, with the utmost interest: 'In my opinion, you have an entirely new method of approach and your treatment of the subject is most convincing.'

Singer reviewed my book on Francis Bacon, which was published in 1960, shortly before he died on 10 June of that year. His review appeared in the *British Medical Journal* on 16 July. I am proud to have had his attention in the final days of his life, and that one of my books should have been the subject of possibly his last writing on the history of science.

I was much interested in Usher's remarks on the origin of his own studies in the history of invention. He had been led to this from his study of economic development. He had developed a *Gestalt* theory of how inventions are made. I had the impression that Usher was rather isolated at Harvard, and that his work had not received sufficient appreciation. He had a rather rough and emphatic mode of expression, which had perhaps been developed in order to resist opposition; this, in turn, reinforced opposition. But it seemed to me that Usher's work was more valuable than some which was at that time more fashionable.

Besides being entertained by Professor Murdock, the Master, I was also invited to dine in the hall at Leverett House. I was placed between Murdock and Salvemini, who was amiable in a nineteenth-century Garibaldian way. Murdock told me that he had recently been staying with W. S. Adams at All Souls in Oxford, and had found him remarkably reactionary in his opinions. I was surprised by this, for W. S. Adams had seemed to me one of the more liberal of the Fellows of All Souls.

On this occasion, as at most of the others during this Harvard visit, Howard Mumford Jones was present. He struck me as one of the most valuable and attractive persons in the University. As a professor of American literature, he had a deep and sympathetic understanding of American culture. He combined this with a sense of humour, and an interest in the relations between science and literature.

The difference between the dining traditions at Harvard and at the ancient English universities was notable. The college or house system for accommodating students had only recently been introduced at Harvard. The undergraduates did not sit at long tables in the English medieval manner, but at individual tables for four, covered with cloth, as in a teashop. Many of them read newspapers during the meal. The atmosphere was a forerunner of what came to Europe thirty years later. I noticed, too, that the undergraduates did not look particularly prosperous.

Cecilia Payne and her husband Dr Gaposchkin had me to lunch at their home. They had a new house with a large garden several miles from Cambridge. It was very comfortable and light, and Gaposchkin had given it a Russian atmosphere. He was then about thirty-eight. He had originally been a fisherman in the Black Sea, and was blown by a tempest into Turkey. He wandered from there to Berlin, where he gradually acquired some education and began learning science. He slept in the barracks for Russian refugees and worked at odd jobs, such as harvesting in the fields. With the rise of Nazism he went to the United States. At Harvard his scientific position was much less than his wife's. He had the obstinacy and spirit of a Russian peasant, and he complained of the absence of soul in America.

Cecilia Payne was originally an Englishwoman of partly German descent. She studied botany at Cambridge, England, and turned to astronomy. She went up to Shapley after one of his lectures in

England, and asked for a job, which he promptly gave her. She spoke highly of Shapley's business ability.

The Harvard authorities arranged for me to visit the Massachusetts Institute of Technology to meet K. T. Compton, Vannevar Bush and Van de Graaff. K. T. Compton was a handsome man, one of the finest examples of what might be called a liberal-conservative American executive. Like Conant, he had an extremely good young private secretary, again far superior to any corresponding ones I had met in Britain. I noticed that the two young private secretaries seemed to have a very good mutual understanding of their own.

I had heard that negotiations had been proceeding for J. B. S. Haldane to be appointed to a chair at M.I.T., and I asked Compton whether anything had been fixed. He told me that Haldane had written saying he could not come until the Spanish Civil War was over, as he was collecting information on gas warfare.

Compton told me of some of the developments at M.I.T. They had founded a department of biological engineering, and had built wind-tunnels for aerodynamical research; they had also organized meetings between business managers and young trainees for executive work. As an example of one kind of help they gave to industrialists, he instanced the problem of a cement manufacturer who could not find a satisfactory name for his product. All of these activities were of a kind which became prominent in European academic institutions thirty years later.

The lean, keen visage and figure of Vannevar Bush were impressive. He was very shrewd, and also kind. I have been told that he was known in some circles as the Texan fox. We admired the achievements of Cockcroft together, and then I was shown his famous calculating machines. His biggest one was smaller than the improved version of it which had been made by D. R. Hartree at Manchester in England.

Before I left Harvard I was asked to record my lecture on the history of science. After it was finished, a few stretches of tape were run off, and I heard my own voice from a recording for the first time. I thought it was much more cultivated than I had expected, but too prosy and deliberate, and I sounded as if I were an amiable uncle of about fifty. (I was then thirty-seven.)

Before I had come to the United States, my literary agent, David Higham, had arranged for me to discuss the project of a lecture tour with the American lecture agent, Colston Leigh. This was fixed for 1938. My Harvard hosts kindly arranged that I should be invited to lunch at the Harvard Club in Boston, as it was thought that this might be useful in connection with the 1938 tour. I was also put up as a temporary member of the Harvard Club in New York, through Mr. Walter Lippman, who was at the time an Overseer of Harvard University, besides being a 'newspaper writer', as he described himself.

I was most grateful to Conant and to America for this reception, which was of a kind which, I believe, had never been given before to a scientific journalist. It helped to raise the status of my profession as I learned from the keen interest in it of my American science writer colleagues. They had been hoping that this kind of recognition might have come to one of them; but this was not so easy, for the selection among American science writers would have been more invidious than that of a foreigner. Also, the American conception of science writing was not quite the same as mine of scientific journalism.

The American science writers gave me a very friendly reception, and elected me an associate member of their National Association. I believe about ten years passed before another European scientific writer was elected to their associateship.

My visit to Harvard was very interesting and encouraging, and I greatly appreciated the honour it gave me; but almost from the first, I received two strong impressions. One was the difference in ideology between educated Americans and educated English people. This came out at once in the attitude at that time to Nazi Germany. I had taken it for granted that the primary present consideration was the defeat and destruction of the Nazis, and I was therefore very much surprised to find that there were highly educated Americans who did not take this view. They were already thinking about the more distant future of Germany, which they evidently regarded as the chief seat of power in Europe. This, in their opinion, being so, it seemed that determined efforts were being made to keep in close touch with Germany, whatever its form of government.

Visits to Germany, and relations with science in that country were to be continued, in spite of the Nazification. Political power was being placed above principle. Perception of this attitude extinguished permanently any wish to settle in America.

I wondered how this attitude, which later proved to be of particular interest in relation to what happened in occupied Germany after the Second World War, was to be explained. One factor in it, I concluded at the time, was the indebtedness of American science and culture to the German academic tradition. Up to the First

World War, American scientists and scholars had generally gone to Germany to pursue post-graduate research.

An aspect of this heritage which impressed me at Harvard, and in other American intellectual quarters, was the regard for Max Weber. Scholars who had been so deeply influenced by Weber could scarcely write Germany off, whatever she did, and indeed what Weber himself had done. He ended his days as an outright nationalist, on the excuse of the Versailles Treaty, contending that as Germany would never receive justice, there was no alternative to a completely nationalistic policy. His critics could say that, ideologically, he had cleared the way for the Nazis, of whom, through his works, he ought to have been one of the most formidable posthumous opponents. The influence of Weber on American thought appeared to me to portend profound intellectual troubles in the future.

The substratum of Weberism would increase the difficulty of coming to an intellectual understanding with America. This intellectual obstacle was more serious than the obvious superficialities of American life.

Externally, Harvard appeared comparatively new and uniform. Many of the buildings were recent, and consequently reflected the contemporary period. The social atmosphere was more uniformly middle class than in Oxford or Cambridge. Social competition was simpler and more clear-cut. Harvard appeared to be more completely a product of a capitalist society than Oxford or Cambridge, which drew their traditions from a pre-capitalist era. There was much energy, ability and innovation, and more efficient management, but it seemed to me that it would be even more difficult for a socialist to survive there than in the ancient English universities.

I left Harvard with mixed feelings. On the one hand, it was a large and vigorous intellectual institution, on the other, the intellectual life seemed more brittle and insecure than at comparable institutions in England. The controlling forces of American society naturally reflected the principles of business, management, and efficiency in a more powerful and less complex manner than in England; consequently, these criteria were apt to be more directly applied in intellectual matters. The continuous concern with academic freedom at Harvard was in itself an expression of the existence of pressure on it, and the fear that it might be curtailed.

While in New York on this occasion, I met Bernard J. Stern, Henry Hart, and other leading supporters of *Science and Society*.

Hart had reviewed *British Scientists of the Nineteenth Century*, which had been published in the U.S. under the title *Men of Science*. He was interested in the relation between science and religion, as exemplified in the life of Faraday. He had just come from Spain, where he had spent three weeks, reporting for one of the New York newspapers. He had been in Bilbao, and described how he went into a hotel there, and had noticed an English youth and girl, with fresh faces against a sombre background. The young people came to speak to him and his colleagues, when they heard them speaking English. Miss Jessica Mitford shyly asked if he would inquire in a hotel in Bayonne for letters to Mrs Romilly, as she had had to pose as Esmond Romilly's wife, which she later became. Esmond Romilly asked Hart later whether it would be cricket to write in an article that he had been in a village in no-man's-land in a valley, whose sides were in the hands of opposite forces, when he had not been there. It was very dangerous in the village.

By a happy chance, Gropius arrived in New York, on his way to Harvard, just three hours before my boat left for Europe. I went to meet him on the pier; I warned him that architecture was not very modern in Boston, but the people were very hospitable.

An hour or two later, I was on the *Normandie*, returning to Europe.

After my return to England, the trend of science in Europe drew me to France. The revival of science there had become striking. Several factors had contributed to this. The rise of Nazism threatened France, and stimulated action for the defence of the nation. A wide range of opinion in the political field agreed to act together in a Popular Front, under the leadership of M. Blum. In the new atmosphere of vigour, M. Blum appointed Irène Joliot-Curie as Under Secretary of State for Scientific Research, the first occasion on which a scientist, as such, was 'taken into the counsels of the French nation'. The appointment invoked the fame of the Curie's, as well as bringing into a position of political authority a scientist who was also a woman.

French science, which had held its own up to the beginning of the First World War, had suffered a severe decline after it. This was attributed to the disastrous destruction of young Frenchmen of talent in the war, and to the ossified condition of French higher education in science. Many of the older science professors not only did not understand the new scientific theories, but were actively opposed to them. The University of Paris had acquired an excessive pre-eminence, which had attracted too much ability from the provinces, and damped down research in them; this in turn damaged the intellectual breeding-grounds for Paris.

As the degree of the development of science had become one of the chief measures of the strength and civilization of a modern state, perspicacious Frenchmen became concerned. A Council of Scientific Research was formed in 1933, to encourage the cultivation of science, because 'disinterested researches in pure science are the source of all progress in human powers' and 'apart from motives of idealism and prestige it is of practical importance to discover those capable of scientific research'.

The economic crisis of 1935 had prompted the amalgamation of the various State funds for science into one unit, the National Fund for Scientific Research. The Blum Government invigorated and transformed a movement for improving French science that had already started.

Another inspiring factor was the emergence of new leaders among the younger scientists who had survived the First World War. Foremost among these was Frederic Joliot, who, with his wife Irène Curie, had in 1934 discovered artificial radioactivity. This restored France to the front rank of nuclear research, and was the biggest single scientific contribution to the revival of French scientific morale. Joliot belonged to the post-war generation, and was free from the inhibitions afflicting so many of the older French scientists. The combination of Joliot and his wife was a formidable addition to the leadership of French science.

Irène Joliot-Curie gave her special prestige to the Under Secretaryship for Scientific Research for a short time, and was then succeeded by Jean Perrin, who, with Paul Langevin, had done most among the older French scientists to sustain French science.

These appointments were of special interest to the British observer, for both Irène Joliot-Curie and Jean Perrin were Nobel laureates; no British Nobel laureate had as yet been placed in political charge of the British State organization of scientific research.

Perrin had appointed Pierre Biquard scientific secretary of the department. I had met Biquard some years before in Kapitza's house at Cambridge. Kapitza had been in touch with Langevin on problems in magnetism, and Biquard had been one of Langevin's distinguished pupils, carrying out important researches on the effect of ultrasonic waves on the diffraction of light, a field in which Peter Debye also worked. But besides being one of Langevin's pupils, he was the closest and oldest friend of Joliot. Consequently, when I visited Paris in April 1937 I was able through Biquard to

Harvard and Paris

meet and interview both Jean Perrin and Frederic Joliot-Curie, and at about the same time I became acquainted with Pierre Auger, eminent in atomic and cosmic-ray physics.

Perrin explained to me that the motive of cultural idealism had a large influence in the new Government's science policy. He considered that the dignity of the human spirit demanded the cultivation of pure research, and that science provided the only means for the liberation of humanity from the restrictive conditions of nature.

Other French scientists expressed a similar view to me. The French tradition of intellectual freedom and culture was aggressively alive, and expressed in the Department of Scientific Research. One of the first aims was the construction of a large astrophysical laboratory in the Provençal Alps. Another was to provide adequate equipment for the Joliot-Curies.

I was introduced to Joliot in his laboratory at the Collège de France, where he had recently been appointed professor. After my first interview I wrote that 'Professor Joliot has a remarkable personality besides being a great physicist. He is probably the most important experimental physicist of the younger generation, as he combines a power of leadership with the highest scientific ability. He is small and dark and possesses the most characteristic French "élan".'

Joliot's laboratory at the Collège de France was rather cramped. It was a small site, but on this were eight floors. I wondered why the site could not have been extended. Apparently this was because the Collège was extremely careful of preserving its great tradition, and any large change might have indirectly affected this.

The Collège had no connection with the University. It awarded no diplomas, and professors required no formal qualifications except the ability to promote free culture. They might lecture on any subject, and anyone could come in to listen. The audiences were mixed, and in cold weather some persons came in from the street to keep warm and doze through lectures on abstract subjects. The Collège had a number of features in common with the Royal Institution, but with wider facilities for members of the public.

As he described his plans, Joliot was accompanied by several of his chief colleagues, in particular, by Halban, Pontecorvo and Kowarski, whom I met for the first time. They stood around Joliot's desk, as he himself stood in front of it explaining his aims and programme of research, and his plans for development. He described the new high-tension laboratory which was being built for nuclear research at Ivry and he sketched a plan of it for me.

I was struck by the international character of his chief colleagues.

Savel and Namhias were French; Halban, Austrian; Pontecorvo, Italian; Feldenkrais, Palestinian; Kowarski, Russian; and Zlotovsky, Polish. Joliot remarked on the difficulty of establishing good international relations, especially among literary students who were apt to be very nationalistic, whereas the scientists were internationalist. Good international relations existed in science and these could be a factor in the improvement of international relations in general. He had noticed that men of all nationalities when they have worked together in the laboratory tend to preserve contact after they return to their own countries.

He said that the system of the Cité Universitaire in Paris, where students of the same nationality lived together in the same hostel, had unfortunately increased nationalist sentiment. I was told that according to French law no foreigner could be appointed a professor in a French university. This prevented the appointment to chairs of talented foreign students and eminent scientific refugees, in particular from Germany.

I believe that Joliot's views on the internationalism of science contributed at least as much as any other individual factor to the subsequent formation of the World Federation of Scientific Workers, and the Pugwash movement. Other people had similar ideas, but Joliot had more strength in promoting action towards their realization. He became the leading spirit and chairman of the executive committee of the International Congress of Science, held later in the year in connection with the Palais de la Découverte.

Besides the emphasis on the cultural values of science, the other point in the new French science policy which particularly struck me was the inadequate financial provision. It was only about one-third that in Britain, which was already too small. The French Department of Scientific Research spent £160,000 in 1936, while the British figure was about £500,000 in the same year; both figures seem negligible compared with the corresponding expenditures in 1968.

The French department was, however, organized more like the British Medical Research Council than the Department of Scientific and Industrial Research. It had been laid down that leaders of the administration must continue their own research and teaching. Their appointment to the administrative service was temporary and part-time, for periods of five years. This system was adopted in Britain in 1967, about thirty years later. It differed from that of the old Department of Scientific and Industrial Research, in which the chief administrative appointments were permanent, as in the Civil Service.

Harvard and Paris

Joliot told me in this interview that the development of scientific research should start with the discovery and encouragement of men rather than the building of institutions. He was in favour of building small installations, and exploring all methods, in order to discover which to adopt for laboratories on the largest scale.

I wrote accounts of my interviews with Perrin and Joliot for the *Manchester Guardian*. I heard later that Joliot had approved their content, with the exception that he did not consider himself small in physique; among Frenchmen he did not feel that he was below the average height, though he was not quite as tall as the average Englishman.

I went to France again in June 1937 to visit the World Exposition. Three of the most striking exhibits were a gigantic photographic enlargement of Mr Neville Chamberlain fishing, in the British Pavilion; and on the top of a tower over the Soviet Pavilion, a huge shining statue in alloy metal. I learned twelve years later that this had been erected under the supervision of S. P. Tambovtzev, with whom I had been concerned in 1929-30, with regard to technical education in the U.S.S.R. The Spanish Republican Pavilion was distinguished by Picasso's great painting of Guernica. They had a pool of shining Spanish mercury, in which coins for the Republican cause were thrown, and floated.

As part of the World Exposition, a remarkable Palais de la Découverte, or Palace of Discovery, had been designed and organized under Perrin's direction. It was to remain as a permanent institution after the Exposition had ended. Its aim was to explain science to the public, not by concessions to the marvellous but by the determined effort of a large number of scientists to make the exhibits simple, yet interesting and instructive. Perrin told me that it had been conceived in talks with Henri de Jouvenel, who had a wide acquaintance with French science, art and literature. It had taken two years to carry out the idea. It was not a museum, but a living organization, which would show new inventions and discoveries daily. It was to promote relations between laboratories, and the applications of discoveries; parties of students and workers were to visit it. The character would be truly French, but give appropriate illustrations of foreign work. Paul Valéry and others were among his intimate friends, and under their guidance, the exhibits had been made with exquisite modern taste.

Many of the exhibits had been newly thought out by good scientists, and contained novel ways of demonstrating scientific principles and theories. The presentation was designed by artists of equal ability, and the choice of topics was made with similar discrimination. Owing to the strict scientific and artistic control, there was an admirable absence of exaggeration of particular French contribution to science.

The International Congress of Science in connection with the Palace of Discovery was held in the autumn of 1937. Its conception and execution was largely due to Joliot and the younger French scientists. His executive committee included Auger, Francis Perrin, Ephrussi, Teissier and Wurmser.

Under Joliot's influence especially, the aim of the congress was to strengthen the international bonds of science and culture. It was hoped that it would be the first of a series of international scientific conferences of a general character. A splendid series of lectures was arranged, all of which were given by scientists from other countries. The French scientists took part only in the discussions. Altogether, the total membership of the congress amounted to 1,500.

The British lecturers included Blackett, Cockcroft, J. D. Bernal, W. L. Bragg, Joseph Needham, C. H. Waddington and J. B. S. Haldane; from other countries there were Niels Bohr, Raman, Debye, Lemaître, Bothe, van der Pol, Northrop, Ruzicka, R. Kuhn, Timofeeff-Ressovsky, H. J. Muller and others.

Bohr spoke on his evaporation theory of nuclear disintegration. Under certain conditions, parts of the nucleus just, as it were, dried off into space. He also mentioned that analogies between the structure of the nucleus and the structure of carbon compounds had come as a great surprise.

In 1936, while attending Bohr's conference at Copenhagen, a conversation he was having with me was suddenly interrupted, and Bohr had been diverted in the middle of a sentence. At the Paris conference in 1937, as soon as he saw me he came to me and started by completing the sentence he had not been able to finish in the previous year; an illustration of his remarkable memory.

Raman gave a picturesque address on ultrasonics and the experiments of Debye, Biquard and others, in which a liquid is made to act as an optical diffraction grating by ultrasonic vibrations. He described them as making sound waves visible. Lemaître had already been using Bush's computer at M.I.T. for calculating the effect of the earth's magnetic field on cosmic rays, and Northrop spoke impressively on crystalline enzymes. H. J. Muller on mutation caused by radiation was followed by F. Holveck, who discussed the effect of radiations on bacilli. He described how radiations could be directed on to a single cilium, or hair, on a small cell, so that the cell spun round. Holveck was murdered by the Nazis during the occupation of France.

The French scientists had started a revival, of vision and quality. They had advanced many ideas of the future, but events presently proved that their efforts had come too late to be effective. Scientific idealism, even of the finest kind, was insufficient if it was not combined with an adequate economic, technological and military development.

Nevertheless I returned to England full of hope. I thought the French scientists at the Department of Scientific Research had a better grasp of the principles of scientific policy than their British colleagues, though they were much weaker on the economic and technological side. I knew nothing at the time of the brilliant British secret advances in military science.

A few days after I arrived in London, a fearful misfortune occurred. On October 19th Rutherford died, entirely unexpectedly. When I heard the news, I wept more than I did when my own father died.

He was to have attended the annual meeting of the Indian Science Association, and had already written his address. He had an operation for a slight hernia, which was normally safe and quick, in the expectation that he would be perfectly fit when he reached India; apparently, however, his internal organs did not recover from the anaesthetic.

Earlier in the year, when my book on *Famous American Men of Science* was published, I had sent him a copy. He had helped me with securing a photograph from the Cavendish Laboratory museum of the model of a Gibbs thermodynamic surface, which Clerk Maxwell had made.

He wrote, on 10 May 1937:

Dear Crowther,

I was very pleased to receive your last book 'Famous American Men of Science', but have only had sufficient time so far to read bits of it here and there on Franklin, Willard Gibbs and Henry. I can appreciate that you have had to do a lot of reading to give a picture of their scientific work and opinions.

I was particularly interested in reading the story of Henry, which I did not know well, and his work on electromagnetic induction. I have not yet found time to read your account of Edison. I think I told you once I met him in his old age in War time, and was not at all impressed by his mental outlook or knowledge of Science. I did, however, see something of his enthusiasm and interest in everything he was doing.

I am sure your book will have a good sale, and it will

undoubtedly be read with great interest by scientific men. The thermodynamic surface reproduces well from the photograph.

I hope you enjoyed your stay in Harvard and that your lectures went off well.

Yours sincerely,
Rutherford.

This was the last letter I received from him.

Rutherford had been my most eminent encourager. I had intended to ask him, after I had returned from my forthcoming lecture tour in the United States in 1938, for permission to write a biography of him. His death removed the possibility of carrying this out in the way I had in mind.

For me, as for many other people, Rutherford was irreplaceable. Facing 1938 and the future without him was a numbing prospect. I spent a good deal of time gazing out of the window of my top-floor flat at the trees in Russell Square, reflecting on the scientific scene, so suddenly and apparently unnecessarily deprived of its greatest figure.

CHAPTER SEVENTEEN
1938
NEAR THE END OF AN EPOCH

I LEFT FOR America again in the last days of 1937, in order to carry out my lecture tour, and make a number of scientific visits. Before starting on the tour, I went to Indianapolis to attend the Meeting of the American Association for the Advancement of Science. I found it bigger and rather different from the British Association. It was more like a group of simultaneous meetings by specialist scientific societies. There were fewer general receptions, and scientists working in diverse fields seemed less likely to meet each other.

The most striking event at this meeting was the passing of a general resolution in favour of inviting the co-operation with 'its prototype', the British Association, and with all other scientific associations with similar aims throughout the world, 'not only in advancing science but also in promoting peace among nations and intellectual freedom', so that science could continue to advance and spread more abundantly the benefits to mankind.

In the American press there were references to the Blackpool Meeting of the British Association in 1936 as already historic on account of this development, which owed much to Ritchie Calder's initiative.

I was invited to attend the gathering of the American science writers who were covering this meeting. They were very friendly, and much interested in my visit to Harvard in the previous year; they still remembered my article on the neutron in 1932. Their association had well-defined aims. They belonged to, and acted in, the American journalistic tradition. Among them were David Dietz of the Scripps–Howard newspapers based on Cleveland, and William L. Laurence of the *New York Times*. Laurence was of Russian origin. When a boy of about fifteen, he had become involved in the Revolution of 1905, and had had to leave the country. In the United States he became an American citizen and a journalist, covering crime, sport, and the usual topics. He saved up money and studied at Harvard where he was awarded a degree in philosophy.

Laurence struck me as the most interesting of the American

science writers. He had an imaginative appreciation of the significance of discoveries, which was more important than details of technical exposition. He understood my political views, though he did not agree with them. He said to me more than ten years later that he admired my freedom from disillusion, though he was unable to share it. During the Second World War he was more sympathetic to General Groves than many atomic scientists in the atomic bomb project. Groves, for all his limitations, had immense managerial energy. I suppose the Americans might easily have had a more liberal manager who was less capable.

I felt able to understand Laurence's regard for Groves, though I profoundly disapproved of Groves's political line. The same problem had arisen in England; was the Prime Minister, Churchill, to have a more liberal and perhaps less capable scientific adviser than Lord Cherwell? In my opinion, it was better that Churchill had Cherwell, than a scientist of less influence and scientific ability.

Laurence led the reporting of science for the *New York Times*, which also had Waldemar Kaempffert as Science Editor. Kaempffert edited a magazine page on science. I was impressed by the handsome library of scientific literature which he had in his office; this looked into a dark chasm in the *Times* building. Kaempffert had been the first curator of a new museum of science and invention in Chicago, before he joined the *New York Times*. I found his style somewhat flatly expository and less interesting than Lawrence's. Having two such prominent science writers on the paper must have occasionally led to delicate situations.

Indianapolis in 1938 had a provincial air. The most noticeable item in the city's diet appeared to be roast turkey, which had a grey appearance and an insipid taste.

I went to Pittsburg to see the research laboratories of the Westinghouse Company, which had recently launched an interesting plan for assisting fundamental research, under the direction of E. U. Condon. He showed me a van de Graaff apparatus for nuclear research, enclosed in a large steel pear-shaped chamber about fifty feet high, which could be filled with air at a pressure of 64 lb. per square inch. The air under pressure improved the insulation, and enabled higher voltages to be obtained, without increasing the overall size of the apparatus. It preceded any pressurized van de Graaff apparatus in England, in spite of the priority of Cavendish Laboratory research in atomic disintegrations.

I saw evidence in several directions that American science was progressing swiftly. This was all the more noticeable in contrast with the decline in Europe after 1933, owing to the Nazification of

Germany and Austria. Up to about 1920 the growth of science in America had been comparatively slow. This was due to several causes, such as the absorption of energy in settling a new country, the strength of the classical tradition in the older American universities, which had been founded to educate preachers, and the almost universal custom of making post-graduate studies in Germany. American scientists could not achieve sufficient intellectual self-confidence under these conditions.

The First World War hastened the process of emancipation. Post-graduate study in Germany was halted abruptly. This was followed by a number of first-rate American discoveries. Before 1923 Americans had received only three Nobel Prizes; after that year the rate increased rapidly.

One of the most striking American discoveries had been that of heavy hydrogen by H. C. Urey in 1931. I called to see him at Columbia University in New York, and he related to me his experience during this period of American scientific development. He was born in Indiana, where his grandparents were the first settlers. He graduated in zoology, and during the First World War engaged in industrial chemistry. He was awarded a fellowship in 1923 which enabled him to study in Niels Bohr's Institute of Theoretical Physics at Copenhagen. There he received a deep impression of the subtlety of European scientific thought. When he visited Europe again fifteen years later he noted a remarkable relative change. American laboratories were, by that time, generally better equipped than European, and Americans were beginning to hold their own in ideas. I reported that 'as this change is proceeding rapidly, American world-leadership in science may soon become pronounced'.

I observed that native American achievement was being assisted by the better use of the large number of eminent scientists who had recently arrived from Europe. Among these was Stern, whom I had called to see in the University of Pittsburg. The most impressive development was at Princeton, where the Institute for Advanced Study had been founded in 1930. I went to call on its director, Dr Abraham Flexner, a former Secretary of the Rockefeller Education Board. Flexner had persuaded Mr Bamberger, a wealthy store-owner in Newark, New Jersey, to found the Institute. Within eight years its school of mathematics had become the most active in the world, and Einstein had been persuaded to join the staff. The Institute had as yet no buildings of its own, and the members of its staff were given hospitality in Princeton University.

On the day before I arrived in Dr Flexner's tiny office in a commercial building in the town Hitler had occupied Austria. Flexner was excited and enraged by this event, and it was long before he got round to talking about the Institute. He had recently seen Eden and other British statesmen, and had pleaded with them not to allow the foreseeable dreadful event to occur. He had been given confident and soothing assurances that he need not worry. I do not remember to have heard anyone so effectively criticized as Flexner criticized the British statesmen. His discourse was heartfelt, penetrating, and well expressed, an illustration of how knowledge is the father of eloquence. Flexner had become famous for his searching criticism of American universities; he was now speaking on the yet higher subject of European and human civilization, and he excelled himself. As a European, I felt that his discourse was a sign of the increasing transfer of intellectual and cultural values from Europe to America.

The mathematicians met in Fine Hall, the mathematical building of the University. Oswald Veblen was its director. His remarkable personality enabled him to create the conditions in which the new school of mathematics could grow and flourish. As I have already mentioned, I had met him twenty years before, in 1918. This charming man of Norwegian descent had contributed much to the unifying of a diverse body of scholars. He took me over Fine Hall, where, besides Einstein, Hermann Weyl, J. von Neuman, Infeld and others worked. The Institute for Advanced Studies already had a staff of twenty-nine scholars, ranging over economics, finance, international relations, archaeology and classical studies, besides mathematics. One of Einstein's sayings was carved over a fireplace in the lounge of Fine Hall: 'Der Herr Gott ist raffiniert, aber boshaft ist er nicht,' which Americans have translated as 'God is slick, but he ain't mean', in his construction of nature; the scientist can always have hope of solving his problem, if he has chosen a genuine one, however difficult.

My primary reason for calling on Veblen was not, however, mathematical, but to ask him about his famous uncle, Thorstein Veblen, in whose intellectual and social background I was interested. Oswald Veblen told me that he believed his uncle was influenced in his views by his origin and social background. He came of Norwegian immigrant stock and had grown up on the American frontier, then still near the Middle West, in a Scandinavian agricultural community, which was egalitarian and socialist in outlook. Thorstein Veblen felt himself to be detached from both Europe and capitalist America, which gave him his peculiarly objective view of social affairs.

After lunching with him, Veblen took me across the Princeton campus to Fine Hall. On the way he suddenly remarked, 'There's Einstein, you must meet him.' We sprinted gently up to him, and Veblen introduced me as 'The Scientific Correspondent of the *Manchester Guardian*'. Einstein bowed and said, 'The *Manchester Guardian* is the greatest newspaper in the world.' That was all he said, and he then walked on. After I had returned to London, I was rung up by one of the stalwarts of the *Manchester Guardian* in its London Office, who asked whether I had heard rumours about Einstein. 'What rumours?' I asked. 'Well, there is a rumour in Fleet Street that he has gone mad.' 'If that is so,' I said, 'it is unfortunate for the *Manchester Guardian*.' 'What do you mean?' my interlocutor asked. 'I met him a short time ago, and he told me "the *Manchester Guardian* is the greatest newspaper in the world".' 'Oh,' said the old *Manchester Guardian* man solemnly, 'it is evident that our information is incorrect.'

The developments at Princeton especially indicated that the United States was developing talent, and providing means, on a rapidly growing scale. It had in fact accumulated the men and resources which made her enormous scientific effort in the Second World War possible.

My lectures, including two in Canada, were delivered to a variety of audiences; university departments, cultural societies, women's clubs, and business men's clubs. The subjects covered the biographies of scientists, aspects of the social relations of science, and science in the U.S.S.R.

Henry Sigerist engaged me to speak in the Welch Institute for the History of Medicine and Science at Johns Hopkins University. I was taken for lunch in a club in Baltimore, in which H. L. Mencken was the leading personality. James Bone of the *Manchester Guardian* was friendly with him, and members of the *Baltimore Sun*, which syndicated the *Guardian*'s articles, were also there. Sigerist expressed radical opinions in an unstrained way, which was striking in his environment. He combined charm of personality with wide learning, and an agreeable, clear style.

In Chicago I was engaged to speak in the school of modern art, of which Moholy-Nagy had recently been appointed director. I spoke for an hour and a half on American scientists; the audience was the most appreciative and understanding that I had in America. I stayed with the Moholy-Nagys, who were striving hard to consolidate the school and their position.

In 1968 there were numerous appreciations of Moholy-Nagy on the fiftieth anniversary of the founding of the Bauhaus. Many of these were written by authors who had not met Moholy. They did not, in my opinion, pay enough attention to Moholy's human and practical qualities. One of the most remarkable was his shrewdness. He made innovations in art and photography and painted abstract pictures of patent artistic merit. Some other artists may have done as much in these fields, but Moholy could persuade owners of large stores and businesses to make use of his art. At Chicago he invited wealthy personages to dinner, to try to secure their support. Moholy was wonderfully skilful and ingenious at handling these situations. I thought his talent in getting the new art utilized was far superior to his philosophizing on it. His gifts were artistic and practical, not intellectual.

In my view, the most difficult part of the Bauhaus achievement was their success in getting their innovations adopted. They solved the problem of the relation between artistic innovation and ordinary life, and thereby showed an understanding and mastery of both. This was vastly more important than achievement in art by itself, or in ordinary life by itself.

My other lecture in Chicago was to the Women's Club. This was in a million-dollar building. There was a lunch before the lecture, and I had to stand beside the chief officers, while hundreds of middle-aged and elderly ladies queued up in a long line to shake hands with me; nine out of ten wore black dresses. I was asked about the state of health of a number of British personalities, starting with that of James Ramsay MacDonald, recently Prime Minister. The only time in my life when I had ever met him was in a railway compartment in a train in York railway station; I had nothing useful to report. The occasion reminded me of the story about Koteliansky, the translator of Chekhov into English. He went to a London reception, and on entering it, whispered his name to the announcer, who roared out in a tremendous voice: 'The Right Honourable J. Ramsay MacDonald.' Koteliansky, utterly flustered, squeaked out as loudly as he could: 'No, no, the very opposite: Mr Koteliansky.'

The incident that I remember about this lecture occurred in the lift. As we slowly ascended, one lady said to another: 'Did you listen to the President's fire-side chat last night?' 'Yes,' was the reply, 'I do believe our President is going mad.' The first lady then said, as we waited for the lift door to open on the office floor: 'My dear, have you only just found that out?'

In Chicago I called on R. M. Hutchins, the President of the University. He was then a tall, handsome man of thirty-eight years. He had severely criticized the superficiality of higher learning in the United States, and had espoused the teaching of classical and medieval philosophy as an antidote. I found him a masterful personality in conversation, combining earnestness with joy of combat. He seemed to watch out of the corner of his eye how one was taking his strictures. He made shrewd and provoking criticisms of the cultural limitation of scientists, and demonstrated unanswerably that educational changes were necessary. He appeared to agree with the principles of the Oxford tradition, but interpreted them in a direct American way very different from the indirectness of the Oxford spirit. There seemed to be a roguishness in his attitude. After he had spoken at length, I suggested that another way of remedying the cultural limitations of scientists was to teach them more about the history and social relations of science. He listened, but I never saw any future effect on those lines.

The Women's Club at Springfield, Massachusetts, engaged me to speak on the Curies, in the middle of the morning; I concluded that the ladies must have finished all their domestic chores by about 9 a.m., if they were prepared to listen to a discourse on such a subject at such an hour. After the lecture, which was in a cinema, I was taken for a drive round Springfield. I asked my hostess about the various buildings, of which she appeared to know nothing. To each question, she said: 'Now, let me see, what can that be?' and then called to her chauffeur, who immediately gave the answer. After several of these inquiries, we came to a succession of buildings several miles long; I asked what they were, and again she consulted her chauffeur. 'That', he said, 'is the United States Arsenal.'

I was taken to lunch with the committee of about a score of ladies, and one or two silent men, no doubt to chaperone me. I tried to make conversation, telling such stories as I knew, including Harold Laski's about President Roosevelt and the portraits of Harding and Coolidge. This story was received with a long silence, after which the lean, tough, humorous lady who was in the chair, looked at me with a penetrating glint in her eye and said: 'This is Coolidge country here, you know.'

At Ottawa I had been engaged to address the Women's Club of Canada. When I arrived at the hotel for the reception the club secretary was looking out for me in the lobby. As I approached, she gave me a penetrating look, and her first words were: 'You do look different from your photograph.' I concluded that I had been engaged on that, and the first impression of the real thing was disappointing.

In the Middle West I had to speak at Louisville, and at a college in Missouri. On my way there I called at St Louis, to visit the physics department of Washington University, to hear about some controversial experiments that had been done there. I was interested to see the city where T. S. Eliot was born. I did not notice much 'waste land', but a great deal of waste river; the Mississippi there is straight and sluggish, and its water has a disagreeable yellowish-brown colour. The main part of the city was on the west side of the river, whose front bore several railway lines; on the opposite bank was another series of railway lines. At first glance, St Louis appeared to consist of two sets of freight railway lines with a broad river between. No wonder T. S. Eliot had found his way to England.

I journeyed on to the college in Missouri where I had been engaged to lecture. There I had to speak with the president and staff assembled on the platform behind me, and two thousand students in the audience in front. The president, who was in the chair, introduced me, and then called on me to speak on a subject other than the one for which I had been engaged. As I walked from my chair to the rostrum, I wondered what to do. I decided to speak on the subject the chairman had announced: it was not very difficult to do, because it was one on which I had spoken in other places. This lecture went down rather well, which I attributed in part to the stimulation caused by the mistake.

The fee from the lectures enabled me to live in the United States for about three months, during which I took a small flat in Madison Avenue, New York. I found that living there was then not much more expensive than in London. It was quiet, and within a few minutes' walk of Central Park. Life in New York could be as peaceful as in England, free from the racket and strain commonly associated with that city.

Karl Darrow arranged for me to visit the Bell Telephone Laboratories. These were an outstanding example of industrial research. I was particularly concerned with the question of the stimulus which industrial research might give to research in fundamental physics. The laboratories were at that time in factory-like quarters in the southern end of Manhattan Island, in New York City. They had nearly one million square feet of floor, and had a staff of 4,200, about 2,000 of which were engineers and scientists. Such was this industrial laboratory's development, already in 1938.

The problems of telephony directed attention to many aspects of nature. The one which impressed me most then, was the research on sound, stimulated by the efforts to improve the

Near the End of an Epoch

efficiency of the telephone as an instrument for dealing with the peculiarities of speech and hearing. Harvey Fletcher and his colleagues had made a searching investigation of this problem, and had thereby revivified the science of sound, in which academic laboratories had rather lost interest.

Another line of research which had led to outstanding results was in solid-state physics. It was necessary to know as much as possible of the physics of metals, and the other materials used in telephony. As part of this general line of research Davisson and Germer were examining the reflection of electrons from a single crystal of nickel. They had observed in 1921 that the reflected beam was not uniform, but split into a bundle of rays; in fact, it was diffracted. When de Broglie proposed his wave-theory of matter in 1924, it was seen that Davisson and Germer's observation provided an experimental proof of it. A little later G. P. Thomson gave another elegant experimental proof. For their work, Davisson and Thomson shared the Nobel Prize for physics in 1937.

On the research development side, the Bell investigators had already produced the special iron alloys on which very high speed telegraphy and very long distance telephony depend. They could already claim that perfectly clear conversation through a cable 15,000 miles long was possible.

They had also progressed with the development of the co-axial cable, which enables hundreds of messages to be transmitted simultaneously. They pointed out that it would be very suitable for wired television, and would greatly reduce the number of separate wires required for telephony.

Thus, developments which were to have major social effects in Britain twenty years later, were already under way in the Bell laboratories in 1938. The political atmosphere in the Bell Laboratories was distinctly Republican.

In Karl Darrow's company in the faculty club of Columbia University, I first met I. I. Rabi. He had recently made an elegant extension of the work of Stern on the polarization of electrons, and he showed me the fine apparatus he had devised for his experiments in his laboratory at Columbia. On the same occasion, I met Dean Pegram. Pegram and Rabi were to become very influential during the Second World War and after, and it was particularly valuable to be acquainted with their personalities.

The view of science I had obtained during the previous ten years had enabled me to see the beginning of the emergence of the U.S.S.R. and the U.S.A. as super-powers, and I feared for the future of Britain. In 1938 I wrote an article with the headings:

Scientific Research: Britain Lags: Other Countries Compared: More Grants Needed. I began with the observation that 'the Munich negotiations on the dismemberment of Czecho-Slovakia have shown in a striking way how the relative power of Great Britain is decreasing in Europe'. I said that evidence for this tendency was coming from various directions, including 'the relative development of science in the leading countries of the world'. As science had become the chief positive characteristic of civilization, the degree of its cultivation and utilization provided a measure of a country's cultural, industrial and military strength. I estimated, as well as I could, the relative expenditures on science, and concluded that 'the relative Russian and American expenditures on research are respectively ten and five times as great as the English'.

The disparity between British and American academic institutions was rapidly growing. The growth in the Massachusetts Institute of Technology, for example, was particularly significant. It was a type of technological university for which there was then no adequate parallel in Britain. It combined first-rate fundamental research with equally good technological research. 'They will soon overtake British research laboratories, where they have not done so already, unless a far bolder and more generous policy towards research is adopted here.'

While the relative British decline caused deep foreboding, the cultural and academic side of British science was flourishing brilliantly. This was evident at the British Association Meeting of 1938 which was held in Cambridge. It proved to be the last general public manifestation of a period of British scientific development which was to end with the outbreak of the Second World War in the following year. As Cambridge was then the most famous centre of science in the then-existing British Empire, and still, perhaps, in the whole world, meetings there had always attracted exceptionally distinguished participators. The fourth Lord Rayleigh was President of the Association, and C. G. Darwin of the physics section. Blackett, W. L. Bragg and Oliphant were among the speakers on physics. J. W. Cook spoke on the synthesis of cancer-producing compounds, and A. R. Todd on that of vitamin B_1. R. F. Harrod was president of the economics section, Gordon Childe of the anthropological, and R. G. Stapledon of the agricultural.

A remarkably large number of the leading British scientists participated in this Meeting, and the foreign visitors were eminent. It provided a splendid illustration for the general public of the quality and vigour of British academic science. But the atmosphere was still that of the educated and ruling upper classes; full evening dress and dinner jackets were worn by nearly everybody at the chief social functions. It was still the science of the British Empire.

But besides the excellent scientific papers there was also new activity during the Meeting. The American Association for the Advancement of Science sent a strong delegation to confer with the British Association on the problems of science and society and discuss what action might be taken with regard to them. Were the contemporary social upheavals due to the advance of science and technology? Do scientists create modern society, or does modern society create them? Is freedom of thought necessary for the progress of science? If it is, how can scientists help to preserve it, and so protect science and themselves?

Besides the discussions between the British and the American Associations, there was a similar movement in the International Council of Scientific Unions.

In response to these developments the British Association decided at the Cambridge Meeting to set up a new division for the Social and International Relations, under the chairmanship of Sir Richard Gregory. As it dealt with problems common to all sciences, it was not just an extra specialist group like the Association's existing sections, but a division in which scientists of every kind would have an equal interest.

The extensive discussions leading to this development drew many distinguished people to Cambridge, and inspired numerous private meetings. J. D. Bernal spoke on the social function of science, the subject of his famous book, which was to be published in the following year. H. G. Wells and Boyd Orr were in his audience. Sir Frederick Hopkins took the chair at a conversazione organized by the Association of Scientific Workers, of which he was then President. Bernal also spoke on this occasion, and Kaempffert of the *New York Times* described how German palaeontologists had had to conceal new discoveries in the Near East, because they ran counter to Nazi theories.

The Cambridge Anti-War Group held a packed meeting in the rooms of Trinity College to discuss science and war. Accounts of the Group's work on air-raid precautions were given. Gordon Childe expressed himself in favour of the British Government's peace-at-any-price policy. Bernal said that the scientist must inevitably have influence in war because of the nature of his work, and he might use this influence in the direction of peace. The

Society for Intellectual Liberty held a crowded meeting in St John's College, at which the indefatigable Bernal also spoke.

The Cambridge Meeting of the British Association gave an impression of great intellectual vigour and considerable sense of social and moral responsibility. British science certainly did not appear craven, however unsatisfactorily it was integrated into the national life.

I was staying in the University Arms Hotel, which has a lounge at the end of a short corridor from the entrance. H. G. Wells, C. P. Snow and others assembled in the lounge, and looked as if they were engaged in uncomfortable conversations. Sitting in the lounge, one had a good view of members of the Association entering the hotel and walking down the corridor towards the lounge. Lord Stamp, who had been President at Blackpool, had given some offence by accepting an invitation to Hitler's 1938 Nuremberg rally. He returned from the rally in the middle of the Meeting, and I happened to see him enter the hotel and stride down the corridor in the masterful Nuremberg style; he almost looked as if he was about to raise his right arm.

However, brilliant as the Meeting was, the attention of some was distracted from time to time by the famous marathon test match at the Oval between England and Australia. This was the match in which Hutton scored 364; his innings lasted over three days, and from 200 onwards some of us went out to buy newspapers every couple of hours or so. He went up to 250, then 300, and 350; would he reach 400? Would he pass Maclaren's 424? Would he exceed Bradman's record score? He achieved the test record for his day, but not that for all first-class cricket. The performances of the scientists were absorbing, but they did not altogether divert attention from those of Stamp and Hutton; there were other sides to life.

The British–American co-operation on the social relations of science was followed up by the attendance of Sir Richard Gregory at the American Association's 1938–39 Meeting at Richmond, Virginia. In the course of his address, on the subject of religion and science, Gregory said that 'scientists cannot stand aside and watch' fascist attacks on human progress. They had a duty to assist in the establishment of a rational and harmonious order. The view that the sole function of science is the discovery and study of natural facts and principles without regard to the social implications of science could no longer be maintained. Birkhoff, who was President of the American Meeting, said in reference to Gregory's address, that he hoped the collaboration between the two Associations would lead to the further unification of the whole scientific world.

The political disasters of 1938 stimulated the scientists to take further stock of their social and moral position.

Finally, in the late autumn of 1938 came the news from Germany of the discovery by Hahn and Strassmann of the uranium phenomenon, later named fission by Frisch. It appeared that the uranium atom could be made to divide, with the release of a great deal of atomic energy. This major discovery immediately seized the attention of all the leading physicists in the world working in that field.

The fact that the discovery was indeed published, placed it within the science of the epoch that was ending: in the following year other major discoveries to which it led were already being kept secret.

CHAPTER EIGHTEEN

1939-40
FRUSTRATION

AFTER 1938 THE scientific scene changed rapidly. A great deal was happening, but scientific attention was no longer fully focused on science: there was a preoccupation with the political and military situation which interfered with the reporting of scientific events. If they had obvious military significance, scientists generally became chary of talking about them. Consequently, it became more difficult to write articles based on good scientific information and judgment.

The situation did not appear to the scientific journalist quite in these terms at the time; all that he noticed was that the information which was forthcoming was less satisfactory than it used to be. He knew that an increasing number of scientists were placing themselves at the service of the military, but he had no knowledge of what they were doing or whether they were doing anything at all.

The absence of obvious scientific military preparations was depressing to the civilian observer. Its effect was to create pessimism on the outcome of a war with Nazi Germany. I could see no hope of the defeat of Nazi Germany by a nation led by Neville Chamberlain. I felt merely that I would prefer to be killed rather than submit to Nazism. So far as I could see, British industry was too effete to produce armaments that could compare with Hitler's, and I knew nothing of the secret research in military science then in progress.

British society was not, however, as decadent as it seemed. It could still react effectively to the situation. The concrete terms in which it was doing this in science were secret, and unknown to me. But the reaction was expressed also in theoretical forms, which were publishable. The most striking of these was the publication on 9 January 1939 of J. D. Bernal's book on *The Social Function of Science*. I received it for review for the *Manchester Guardian*, and arranged that it should be noticed in a turn-over article, that is, more than a column long, starting on the news page and continuing overleaf.

The book was in this way treated as important news, as well as a distinguished contribution to the literature of science. Bernal's book contained many of the ideas about science which were to become better known during the next thirty years. He treated science as a function of society, an integral part of it, which acted upon, and was acted upon, by the rest. Besides clearly recognizing and definitively depicting science as a social function, Bernal conceived scientific organization and development on a scale which was new in Britain. He proposed the multiplying of organization and expenditure by factors of ten or more, which seemed fanciful at the time. Besides this major step in the scale of thinking about science, Bernal dealt with numerous particular problems. For example, he drew attention to the conflict between literary and scientific cultures, which received so much attention thirty years later. Thus, Bernal had produced a programme for the new scientific epoch that was just beginning.

A few months later, I reviewed a short book on *Science and Civilization*, by a young physicist named A. C. B. Lovell. This was Sir Bernard Lovell's first book, written before the war, and before he became a radioastronomer. I was happy to write that it was 'an admirable introduction to a new subject'.

Meanwhile, the world's nuclear physicists were exploring the discovery of uranium fission with intense effort. In the first few days of January 1939, Fermi and Dunning in New York and the Joliot-Curies in Paris swiftly demonstrated the effect, and Meitner and Frisch gave it its correct interpretation.

Joliot-Curie discovered and published evidence in the spring of 1939 that a chain reaction might be possible. I wrote in June that 'the discovery by several investigators in different countries that atoms of uranium explode when penetrated by neutrons has brought the control of atomic energy for human purposes nearer. F. Joliot, of Paris, has shown that when a uranium atom is exploded by an incoming neutron, three or four fresh neutrons will explode four more uranium atoms, and that the sixteen neutrons from their explosions will explode sixteen more uranium atoms, and so on. In this way a chain of exploding atoms in a lump of uranium might lead to an enormous output of energy in a short time. The arrangement seems simple, and some of the greatest authorities believe that almost inexhaustible sources of atomic energy for scientific an industrial purposes may ultimately be released by it. But no practicable quantity of energy has as yet been obtained. The end of the world is not at hand, but the large-scale release of atomic energy is now a subject of serious research and discussion.'

With this paragraph, the kind of scientific journalism that I had practised since 1926 began to come to an end. It had been based on first-hand knowledge and discussion of publishable material. Henceforth, such material became only partially available. The writer on science in the press became dependent on such information as was released to him by official sources. He had to write in the knowledge that some, or most, of the essential information on any scientific subject might be withheld from him.

Writing on science in this situation is a different art from that which I had been propagating and practising. It is a very necessary and important art, but of another kind. When an institution decides to release only such information as it thinks fit, it engages in a technique of public relations, in which information is subordinate to relations. The interpretation of such knowledge as is released becomes a detective rather than a socially expository operation.

I only slowly recognized the profound change in the nature of writing on science for the press, which gathered momentum early in 1939. Its presence manifested itself to me rather in the shape of inadequate information. In those branches of science less immediately connected with possible applications, the change came less quickly. For example, in astronomy it was possible to deal adequately with H. A. Bethe's explanation of the origin of solar and stellar energy. He had suggested a plausible mechanism by which atoms of hydrogen could be transmuted into atoms of helium in the interior of stars, with the release of large amounts of nuclear energy. He received a Nobel Prize for this, and other important work, thirty years later.

In general, however, this kind of reporting was at an end. It continued to be possible only in those subjects, such as food and agriculture, which had no direct bearing on armaments. The British Association's new division for the Social and International Relations of Science held its first meeting in Reading in March 1939, and devoted its discussions to the food value of milk, and related topics. At its first London meeting in May, it discussed the relations between science and society. The historian Sir Ernest Barker contended that the first aim of science was to proceed with a single regard for truth and discovery, and that it was the business of society, through its leaders, to control the social changes brought about by the impact of science. The social duty of the scientist was to contribute his technique, to use his special training in assisting this second aim as well as the first.

However, the British Association's new division, though starting useful work, approached controversial matters in so gingerly a manner that it was slow in directing attention to those that were most important: the economic, political, and military implications of science.

The British Association met in its customary way at Dundee on 31 August. The crisis had had remarkably little effect on its programme or attendance; three thousand members were present. The President was Sir Albert Seward, who prefaced his address with brief but very earnest remarks on the desire of scientific workers in all parts of the world for the establishment of conditions which would enable them to work together in harmony as members of one great brotherhood. Then he spoke on his special subject of the deduction of the conditions of the Western Isles of Scotland in past geological ages from their deposits of fossil plants.

The Association proceeded with its programme, which, as usual, contained some very interesting items. One of the most striking was by Dr Zworykin of the Radio Corporation of America, to whom the modern scientific development of television perhaps owes more than to any other individual. However, on the third day, after the news of the invasion of Poland, it was decided to cancel the remainder of the Meeting. Sir Richard Gregory, who was Chairman of the division for the Social and International Relations of Science, was elected the next President of the Association; a very important election, for it had the effect of making him the leader of the Association's affairs for the six years of the war.

Among the numerous addresses which were cancelled was one by a young Oxford economist, who had been given the last place on the last day of the original programme, to deliver a paper to the Economics Section on *Exports and Imports in the Trade Cycle*, by John Harold Wilson. Six years later he was to emerge as a junior minister in Mr Attlee's first administration in 1945, and twenty-nine years later, in 1968, he was, as Prime Minister, still attending to the same topic.

Sandbags had already been piled in front of various buildings in Dundee. We returned to London and other places in calm excitement, with a single and open-minded attention to the situation, for which the Government had provided no adequate leadership or preparation. The moral and ethical considerations of the division for Social and International Relations were insufficient for many scientists who desired to undertake some more concrete and practical action. They agitated for this in various ways, through their own societies, both formal and informal.

The Ministry of Labour had started compiling a Central

Register of persons with professional and other special qualifications, who might be called upon in the event of war. Scientists pressed for a scientific section in it, and in March 1939 the Royal Society was asked to co-operate by compiling the part of the Register dealing with people normally engaged in, or qualified to undertake, scientific research. A. V. Hill, who was then the senior Secretary of the Royal Society, supervised the launching of this register of scientists.

Like many others I had grave doubts of the political morality of the Chamberlain Government; I felt very uneasy at being involved in a war under its leadership; but I did not agree with the pacifists who opposed war on principle. As the situation worsened, I reflected that, at any rate, there never could be a stronger case for war than opposition to the Nazis. I saw A. V. Hill, under whom I had worked in the First World War, and asked to be enrolled on the Register of Scientists, which was duly done.

Meanwhile, I wrote on the proper use of science and scientists in the war. I described the Register of scientists, which was being compiled. This would facilitate their engagement in tasks of immediate national importance, and utilize their talents and specialities efficiently. It would minimize the waste of genius that occurred in the First World War, through the enlistment of scientists in ordinary military duties for which no special knowledge was required; the memory of H. G. J. Moseley was still in the minds of many. He had been the most brilliant of Rutherford's experimentalist pupils. He had already made major discoveries when he enlisted in the army. The American physicist R. A. Millikan had said that in 'a research which is destined to rank as one of the most brilliant in the history of science, a young man, twenty-six years old, threw open the windows through which we can glimpse the sub-atomic world with a definiteness and certainty never dreamt of before. Had the European War had no other result than the snuffing out of this young life, that alone would make it one of the most hideous and most irreplaceable crimes in history'. Rutherford had said of his great protégé, 'our regret for the untimely end of Moseley is all the more poingnant that we can but recognize that his services would have been far more useful to his country in one of the numerous fields of scientific inquiry rendered necessary by the war than by exposure to the chances of a Turkish bullet'.

In August, a fortnight before the Second World War began, I drew together some of the chief points being made by the new movement for the study of the social and international relations of science, with the heading *Where Britian Lags*. I wrote: 'The news

Frustration

nowadays is unpleasant reading for Englishmen. New concessions are made to competitors nearly every day. Tariffs are raised against us, or Fascism is established in friendly democratic countries, or we are forced to depart from the traditional principles of British justice. Some may not be worried by these events so long as their comfort is not immediately disturbed, but this attitude cannot satisfy those who care for the future. The defeat of Britain would mean much more than personal servitude, for it would be accompanied by the cessation of the British contribution to civilization. The tradition that produced Newton and Darwin would be ended. This raises two reflections. Is everything possible being done to protect and strengthen science in Britain, and is the best use being made of our heritage of science for the solution of our present difficulties?'

I pointed out that the United States was spending at least nine times as much on research as Britain, and I quoted the striking statistics in Lovell's *Science and Civilization*, in which he showed that Britain was eighth among the leading countries of the world in the proportion of her population receiving a university education; even Italy had a higher proportion. I gave instances of the British failure to exploit British inventions, and in particular the purchase by the Cavendish Laboratory of a Cockcroft and Walton apparatus, originally invented in the Laboratory, from a Dutch engineering firm which had had the enterprise to solve the problem of marketing the first of the atom-smashers. Incidentally, in the 1960s, the Cockcroft and Walton atomic disintegrator was the type still most widely used in the laboratories of the world.

I mentioned, too, the drift of British scientists to the United States. R. P. Linstead, the discoverer of the new dyes, of which monastral-blue was the example, had just accepted a chair at Harvard University where, no doubt, his facilities would be larger. I argued that it was absurd that such men should drift away. Fortunately, Linstead did not become lost to Britain; he subsequently became Rector of the Imperial College of Science and Technology.

The war continued for one year and eight months before any official use was made of my services. As the months went by, the frustration among those scientists who perceived the major importance of science policy became increasingly intense. The Government had had the wisdom not to send scientists directly into the armed forces, but it had as yet little positive policy on their general utilization.

In the autumn of 1939 I transferred from the London Office of the Oxford University Press to the Clarendon Press, its headquarters in Oxford. I completed my book on *The Social Relations of Science*, which I had begun in 1938. It was published by Macmillan, one of whose advisers was Sir Richard Gregory. The member of the firm whom I saw in connection with it was Mr Harold Macmillan who was back in the family office after resigning from the Chamberlain administration. Owing to the war conditions, Macmillan had it printed in the United States. My old school friend, G. L. Brayshaw, then on the staff of the *New Yorker*, kindly corrected the proofs for me. It took two years to get it through the press and it appeared in 1941. The book quickly went into a second impression, which was sold out in 1943. Then a difficulty arose. The American end of the firm did not want to reprint it again, so it was agreed that the plates should be sent to England, and the third impression run off here. However, the plates contained zinc, and it was against the U.S. regulations for zinc to be exported at that time. I could probably have had it issued as a paperback in England, but I objected to that, as I thought it ought to be available in durable form. This was a mistake, for the effect was that the book remained out of print all through the crucial early years of the post-war period, and was not republished until 1967.

The early period of the war was very frustrating for the younger scientists. Many of them were burning to do something, but could find no adequate line of work. E. V. Appleton, who was then Secretary of the Department of Scientific and Industrial Research, told me in October 1939 that on his advice the Government had decided to preserve, as far as possible, all important groups of research workers in the country, in order to retain the efficiency of industry, and keep the country ready for industrial competition after the war. Accordingly, very few key men had been taken from the research laboratories of the industrial concerns. Fewer research units had been left in the universities. The branches of science in which there was a serious shortage of men were in meteorology and electrical communications. A. P. M. Fleming had been asked to advise on special courses for training men in these subjects.

Appleton said that in spite of his heavy duties, he did a little research every night, and was still working on some papers. He found the quiet of the blackout evenings distinctly helpful to study.

One of the most important developments in this frustrating period was the emergence of Solly Zuckerman's dining club, the Tots and Quots, as a creative influence in thought and action on science policy. During the decade before the beginning of the war it had been a stimulating society for exchanging scientific ideas and news, and deepening scientific friendships. The sociable side of Zuckerman's gifts had perhaps been uppermost. The crisis now turned it into a serious instrument for forming ideas on science policy and explaining them to influential people. As the war extended, Cherwell, H. G. Wells, the Free French, the Joliot-Curies, Sir Archibald Clark Kerr, Herbert Morrison, Admiral Nimitz, J. B. Conant, and many others were entertained by the Tots and Quots under Zuckerman's leadership, and enabled to meet a score or more of the then younger scientists, so that they had an opportunity of hearing their views under informal conditions. No one but Zuckerman was able to conduct this society adequately. When he later was away on war work, no one else was able to carry on really effectively in his place. It was a striking demonstration of the unique nature of his combination of social and scientific gifts.

The kind of relieving and inspiring discussion which took place is illustrated by my notes and recollection of a dinner on 23 November 1939, at which Bernal spoke on the disorganization of science. All that we had feared before the war would happen, had happened, and was in fact worse. Those traditionally in charge had no influence, though they had prevented anyone else from doing anything, on the ground that they were busybodies. Funds for science were liable to end at any moment. This was a mistake from the point of view of those running the war. The Government research departments formed vested interests. The situation in the services was lamentable. There was obstruction of research on things which should have been done years ago. He had heard three years ago of an important device; research on it started two weeks after the war started. We were not getting any effective utilization of science, and in addition to this, academic science was being stopped. The threat of absence of funds stopped research. This was happening because the Government and its servants were unfitted to use science. No plan for science could be expected from them. No use of the knowledge of de-lousing gained in the First World War had been made in the evacuation of lousy children.

The leaders of the Royal Society were upset at being left out in the cold. There was a complete lack of initiative in Government scientists and in the societies. This would have to come from the scientists themselves. The stream of students was being reduced. If there were no new first year students, there would be no new scientists. What should be done to keep science going?

C. H. Waddington remarked that as yet biology was not regarded as a war science. Biologists were therefore inclined to go on with their pure research.

R. F. Harrod said that economic policy was just as muddled. There were few good economic statisticians, but after the war had gone on for a few weeks those in Oxford wanted to go to man guns, because no suitable war work had been found for them. He started research on War Economics, with the help of the Rockefeller Foundation, to give scope to these men.

These and many other points were made, as the Tots and Quots at their frequent dinners discussed the situation and made clearer to one another the policy problems that faced scientists, and the scientific problems that faced the nation. Their activities contributed a great deal to forming a corpus of intelligent scientific opinion on these crucial problems, which spread to wider circles, and began to affect public opinion.

In the last months of 1939 the Tots and Quots were very active in focusing ideas on the place of science in the war. Julian Huxley produced a paper on the organization of scientific research in Britain, in which he emphasized the need for more executive power, planning and co-ordination in Government science. He thought a Ministry of Science would be too ambitious and impractical at that time; but a Scientific Council should be set up, with direct access to the Cabinet through an appropriate minister, so that its findings should become definitely known. It should not be described in its title as Advisory, thus helping to ensure that its findings would not merely be of passing interest, but would be regarded as matters on which positive action must be taken.

At a Tots and Quots dinner on 2 December 1939 W. K. Slater remarked on the confusion then existing in the county agricultural advisory services. The scientific advisers controlled by the county agricultural committees were taken over by the county war executive committees, none of whom were scientists. Consequently, scientific advisers in the counties were being left high and dry. They did not know what Government policy was, and were playing golf to take their minds off the situation.

Under the influence of Stapledon, the Ministry had advised ploughing up grassland, but when the war started they wanted corn; this was not the same as re-sowing grassland, according to Stapledon's principles. Since the First World War the farmers had been encouraged to breed special pigs and poultry. These depended on imported food. Maize and barley were out of stock. With the adoption of convoying of ships, the volume of imported fodder had been reduced to about one-third the peacetime figure. During the

Frustration

last few weeks pigs had been slaughtered all over the country. Old laying birds were being killed, and young laying birds were being sold to farmers for scratching round ricks.

Cattle were decontrolled when pasture area was reduced, so prices soared, and large numbers of cattle were being slaughtered. He thought a science committee was needed to advise the Government. It should be on industrial lines; the I.C.I. had a research committee which advised the board. Direct scientific representation to the Cabinet was needed.

As the situation became increasingly desperate, the Tots and Quots came in contact, through Zuckerman, with others, not closely in touch with science, who were equally desperate. Among these was the publisher Allen Lane, who had successfully launched his *Penguin* and *Pelican* sixpenny paperbacks, against the general opinion of the British publishing world. He wanted to do something, and so did the Tots and Quots.

It was agreed that Zuckerman should edit a book on *Science in War*, to be written by members of the Tots and Quots. The book was written, and published in tens of thousands of copies, within one month. It was a co-operative effort. I contributed especially to Chapter II. The chief figure in its production was Zuckerman. He was still working in the laboratories of the Zoological Society, and I went there to assist him in correcting proofs. He was in his shirt-sleeves, and he worked on the proofs with an energy and speed which I have never seen equalled. This extremely determined, quick, strong worker was the obverse of the sociable scientist who had previously been familiar to many of us. We now saw the tough hand beneath the charming manner.

When the book came out, I helped it along with a review in the *Manchester Guardian*, in July 1940. I observed that: 'It would be ridiculous to be defeated because of a refusal to use existing scientific knowledge. One of the greatest and most productive tasks that could be done at this moment would be to ensure that full use is made of science, and this book is a gallant contribution to that aim.'

In February 1940 several leading French scientists visited England in order to strengthen scientific co-operation during the war. They included Langevin, Auger and others, and were entertained by the Royal Society. Auger had recently been requested by the French Government to organize a Department of Scientific Information. He had quickly launched a well-conceived scheme. Some of the members spoke to the Society on French scientific preparations for the war. One of them, not Langevin or Auger, spoke at length in a manner which indicated even to people like myself, who were not in touch with military science, that he was quite out of touch with the realities of the situation. Tizard happened to be sitting immediately in front of me; I heard him groan and murmur: 'Good Gawd, we shall lose the war.'

The anxiety among the unofficial younger scientists was perhaps even greater. Zuckerman got in touch with the French diplomats in London, and it was arranged that the naval officer Métadier should attend a Tots and Quots dinner to discuss what could be done to promote and strengthen relations between French and British scientists. A number of tentative suggestions on the problem were made by members of the Tots and Quots. It was suggested that one of their members might go to Paris to discuss the matter with French scientists. When Métadier was asked to speak, he said that one of the needs was a better mutual knowledge of the science and scientists of the two countries. He then produced from his pocket one of the pamphlets issued by Hermann in their *Actualités Scientifiques*; it was the French translation of the chapter on Humphry Davy from my book on *British Scientists of the Nineteenth Century*. He observed that it was the kind of book which helped the French to acquire a better understanding of British scientists. One of the things that might be done to improve French and British scientific relations was to have more books like that. Zuckerman said to Métadier that the author of the book in question was down at the other end of the table. He was astonished, and became even more complimentary.

Zuckerman then said that after this, it was obvious that I ought to go to Paris on behalf of the Tots and Quots, and explore what could be done to strengthen Anglo-French scientific relations. The possibility of founding an Anglo-French Society of Sciences might be investigated.

On 8 April 1940, I flew to Paris with Métadier. We took off at Heston for Le Bourget in a large unheated plane, on a bumpy flight, as it was necessary to keep near the ground. Over the Channel the weather was fine and still; the spring sky and calm sea looked very peaceful. It was impossible to imagine from the appearances that the war had already gone on for more than half a year.

After arriving in Paris we went straight to the restaurant Capoulard for lunch. Presently, I heard a voice saying: 'Who have we here?' It was Freymann, the publisher of Mexican origin, who ran the Hermann firm, and had published the French translations of *British Scientists of the Nineteenth Century*.

Frustration

Métadier, who belonged to the political Right, was anxious to interest M. Louis de Broglie in the proposed new society, and possibly secure him as president of the French side of it. He was friendly with Destouches, a protégé of De Broglie, who worked at the mathematical Institut Poincaré. Métadier and Destouches introduced me to de Broglie, a shy, very highly-bred aristocrat. He agreed to give support to our proposed society; but there was no mention of the possibility of his becoming the president of the French side. I heard afterwards that he disliked the sea, and would not be inclined to undertake anything that involved crossing the Channel.

I now had a considerable discussion with Métadier, and we agreed that we would try to persuade the following scientists to be officers of the French side of the society: President, Joliot; Committee, Destouches, F. Perrin, Monnier, Rougier, Ephrussi, Lyot, Auger, Bruhat, Valiron; Vice-Presidents, Langevin, L. de Broglie, J. Perrin, Lapique, Brillouin, E. Cartan, Roussy.

Métadier was very anxious that we should secure the support of M. Morize of the French Ministry of Information. I was not clear about his reasons for this, and was uncertain how the interview would go. In the meantime, I went to the Institut de Biologie for lunch. Duclaux, Auger and others came in, including the Joliot-Curies. When Joliot saw me, he exclaimed 'Quelle surprise!' I asked Joliot for aid and support for our proposed society, and whether he would accept the presidency of the French side; he immediately agreed. I then asked him if he would accompany me at the interview with Morize later in the afternoon, and again he agreed at once. He telephoned Morize, and asked me to meet him at his laboratory an hour later. He took me to the Hotel Continental, and went to Morize's anteroom, which was just being entered by Métadier.

When we were shown into Morize's room he got up and welcomed me at the door, recalling that we had met at a cocktail party at Howard Mumford Jones's at Harvard. He asked us to sit down, and invited me to explain our project. I did this, and then he asked Joliot for his views. Joliot put the case for the Anglo-French Society of Scientists brilliantly, and Morize was much impressed. It was agreed that a document on our views should be prepared, and laid before the Ministers of Information and Education.

We then went to Auger's office, where Joliot immediately dictated his conception of the project to Auger, who got all the points down with great speed and accuracy, and presently expressed them in a statement. I was impressed by Joliot's instinctive gift for leadership, and Auger's extraordinary intellectual efficiency. He was a brilliant example of the Normalien training.

Next day, I went to an excellent small restaurant off the Boulevard Raspail that I had used on former visits to Paris. It was as full as ever, the cuisine as good and reasonable, and I recognized several of the regulars. Afterwards I went to the Dome, formerly the rendezvous of many intellectuals, for a coffee. I was the only person on the whole of its terrace.

I stayed in Paris with the Augers, whose home reflected many of the finest qualities of French civilization. During my stay, the news of the invasion of Norway by the Nazis came through. I reflected on the deceitful calm I had seen when flying over the Channel a few days earlier. Auger was very agitated by the news; he said to me that it was inconceivable that the British Empire would be defeated.

I was invited to dinner by Henri Laugier, the Secretary of the Council for Research in Pure Science. He gave it in a restaurant near the Hotel de Ville. The staff was elderly, and it was evidently a place accustomed to entertaining ministers. There were about a dozen people there when I arrived. Laugier introduced me to several of his guests, and I chatted with Langevin. One of the gentlemen was rather tall, about forty, with a round, slightly flabby face, darkish, with a small moustache. He wore a lightish suit, and, I think, spats. He had an English A.R.P. Badge on his lapel. Laugier introduced him as 'le Comte de . . .' I did not catch the name. I was puzzled by the badge, and imagined that he might be a French diplomat normally in London, where he undertook voluntary A.R.P. duties.

When dinner was served, Laugier took me to the place of honour in the middle of the table, putting this gentleman opposite me. Joliot was on my left side, and Jean Perrin on my right. Langevin was on the right of the gentleman opposite me. I presently noticed that my *vis-à-vis* spoke English fluently, so I engaged him in conversation. He spoke most enthusiastically about the Edinburgh professor of pharmacology, A. J. Clark, and said that C. G. Darwin had a most marvellous brain, being able to solve problems in a moment in his head, which had baffled their proponents for weeks.

I asked him if he was in France for a long time, and he replied 'for the duration'. I then realized that he was the Earl of Suffolk, the British Government's Scientific Liaison Officer in Paris. He later enabled Halban and Kowarski, Joliot's colleagues, to escape from France. They brought the latest results of the crucial French research on chain reactions in uranium, and the whole French stock of heavy water which was then the largest in Western

Frustration

Europe. Suffolk also brought out stocks of industrial diamonds and valuable machine tools. Their ship drew out of Bordeaux under fire from the enemy but reached England safely with its precious cargo. Halban and Kowarski continued their experiments in the Cavendish Laboratory.

Suffolk had a passion for dangerous activities. After his return he was engaged in bomb-disposal work, and terrified colleagues by attacking bombs with a hammer and chisel. He was killed in 1941 in a very dangerous operation in which a bomb exploded, and was awarded the George Cross posthumously.

At the Paris dinner he told me his family had done nothing for six hundred years. When he was a student at Edinburgh he was older than the average, and his friends were rather among the staff than the other students. He was invited to some staff function, and was the only student there. The Principal of the University, Sir Thomas Holland, had apparently snubbed him for being out of place. Suffolk then said that he wrote to Holland that he would retire from the university unless he received a public apology, and he claimed that he got it.

He said that his mother had been American. He went on to express the opinion that the Americans should be taught their place in the world, after the war. England and France controlled one-third of the world, so they could force the Americans to submit to tariffs; this would punish them for their present attitude.

During the meal Joliot told me that three personages had consulted him as to who should be the next Cavendish professor in succession to Rutherford. He had suggested Blackett. One of the personages said to him: 'But isn't he rather left?'

Suffolk presently startled me with the remark that he was entirely concerned with applied science. Of course, our projected society would be concerned with pure science. Our spheres were absolutely distinct. I remarked that of course our concerns were absolutely distinct, but complementary. He said he would do anything he could to assist us. I thanked him warmly, and asked where he might be found. He said at the Ritz.

Obstruction from Suffolk could have been troublesome. It was evident that the French, and especially Laugier, had squared him beforehand, so that there could be no possible clash between the pure and applied departments in France, as represented by their Secretaries, Laugier and Longchambon.

Francis Perrin remarked how glad he would be to help Anglo-French scientific relations, as he felt that England was his second country.

I returned to London on 13 April, and went to Oxford three days later to see Zuckerman, who fully approved. He was joined by Bernal and Waddington. We agreed that Dirac should be invited to be president of the British side. Zuckerman reported that J. B. S. Haldane was opposed to the whole project.

Two days later I went to Cambridge to see Dirac. I explained the project, and asked him whether he would be prepared to accept the presidency of the British side. After pondering on it, he said he would accept the presidency for one year. Sir Alfred Egerton, Professor A. V. Hill and Sir Henry Tizard accepted vice-presidencies of the British group, and Bernal, Blackett, Cockcroft, Darlington, Waddington and Zuckerman formed its executive committee. I became its secretary. The French group, with Joliot as president, had Langevin and Laugier as vice-presidents, and Irène Curie, Denivelle, Ephrussi, Lwoff, Francis Perrin, Rapkine, Wurmser and Wyart its executive. Auger was its secretary.

At the time of the fall of France, many French scientists fled from the country before they could be captured by the Nazis. Among the first to arrive were Laugier and Longchambon. Zuckerman was out of London, so I had to help on behalf of the Tots and Quots, and they came to my flat. Laugier was utterly cast down, but Longchambon seemed to be already finding reasons for going back to the occupied country. I telephoned Cockcroft, who came in the evening to see them; it was an intense and concentrated moment.

The great figure that emerged at this time was Louis Rapkine. I had been acquainted with him, but had not hitherto known him well. By 1939, when he was thirty-five, he had settled in Paris and become a French citizen. He was already a biochemist of standing, noted for his enthusiasm for science as well as his good work. At the outbreak of the war he had been mobilized, and sent to London to the office of the French coal-importing organization, where he had to compile statistics. When France was invaded Rapkine set himself to save French scientific workers from the enemy.

He compiled a list of fifty-seven, whom he endeavoured to rescue. Without any authority other than his own character, he secured the support of the highest scientific and political personalities in London and Washington. He organized the escape of about forty leading scientific workers and their families from France to the United States. He obtained posts for them, and even got their hospital bills paid. He became known among his friends and admirers as The Prophet, leading his scientists out of captivity. He brought them back to London after the Allied

Frustration

invasion of Europe in preparation for return to their liberated country.

Rapkine died prematurely in 1949 of lung-cancer, probably due to excessive smoking. It was a disastrous, and, it seems, unnecessary loss. He had been born in White Russia in 1904, the son of a Jewish tailor. His father and the family migrated to Paris in 1911, and after a difficult year went to Montreal, where he worked in a factory. Louis grew up as a Canadian citizen. He gained education by scholarships, and was an excellent athlete. He attended McGill University for three years, supporting himself by work during the summer vacations.

When he was twenty he went to France to pursue scientific research. He worked under Caullery and Fauré-Fremiet, and under Wurmser in the Institute of Physico-Chemical Biology. He supported himself by hawking in the countryside and selling boots at country fairs; he was so poor that at times he was starving, and was reduced to eating glucose from laboratory bottles. In spite of this he did good work on cell physiology, and in 1926 was awarded a Rockefeller fellowship. He was presently led by his researches to the study of enzymes, and made observations on the role of sulphydryl groups, at which Sir Frederick Gowland Hopkins had independently arrived. When Hopkins heard of the younger man's work, he withdrew, so that Rapkine could be assured of priority.

Rapkine had become naturalized as a Frenchman just before the outbreak of the war. He combined the fresh patriotism of his new citizenship and the enterprise of his Canadian background with his extraordinary personal gifts of moral strength, charm, intelligence, kindness, generosity and selflessness, besides practical and scientific ability. Many of those who knew him well regarded him as the finest man of his generation.

The French mathematician S. Mandelbrojt, who had been one of Rapkine's group, suggested that there should be a Rapkine Memorial Fund. This was founded, following a meeting at the Society for Visiting Scientists in March 1949, and since then, an annual Rapkine Memorial Lecture has been delivered in London on the interests to which Rapkine devoted his life.

One of the first things I heard from Rapkine when I began to collaborate with him closely was that he had observed from the trends of the graphs of the coal imports into France that the leading business people must have known about the coming capitulation well before it happened.

There were of course many other French activities in London touching on science. In the summer of 1940 I was invited to attend a meeting at which André Labarthe announced his intention of launching his cultural review to maintain French science, literature and art in a free environment. Raymond Aron spoke on its aims and programme.

At the end of the summer, J. J. Thomson died. With his departure, the great achievements of the Cavendish Laboratory at the end of the nineteenth and the beginning of the twentieth centuries became historical memories. An immense memorial service was held for him in Westminster Abbey.

Then, in the autumn of 1940, the formation of the Scientific Advisory Committee to the Lord President of the Council was announced. Its members consisted of three officers of the Royal Society, Sir William Bragg, Professor A. V. Hill and Professor A. C. G. Egerton, and three senior scientific civil servants, Dr E. V. Appleton, Sir E. Mellanby and Sir E. Butler. The chairman was Lord Hankey. In a leader in the *Manchester Guardian* I described the new committee as 'a move towards a more liberal use of science and scientists'. I suggested that 'the final aim might be to have a statesman in the Cabinet who could personally speak on the general scientific aspects of the chief problems of policy. We are engaged in a scientific war and live in a scientific age and it is strange that there should not be a direct spokesman of science in the ultimate council of the nation.'

Much was happening, but not much that I could write about; my income became very small. However, in the autumn of 1940, the Nazis extended the air war, especially by bombing London. This provided fresh scope for articles. The brilliant success of the Royal Air Force in repelling the Nazi attacks had not been foreseen by the general public, nor had it originally been more than a hope to those most intimately concerned in it.

I wrote about such information as was obtainable; for example, on the careful physiological and psychological selection of pilots. I mentioned the surprising information that from the medical point of view, the young men who were shooting about the sky in fighters were following a sedentary occupation, so they did not require to be muscular. Because of lack of exercise in the planes they needed a good deal on the ground. The pilot was rarely called upon to make a severe physical effort, but was exposed to severe threats of danger. Emotional stability was the most necessary quality, and the dashing, excitable type was by no means the best.

The first important new military science on which it was possible to write publicly was the research by Bernal, Zuckerman and others on the effects of bombing on structures and on people. It was discovered that bombing was enormously less destructive than had been supposed.

Zuckerman investigated the effects of the blast from bombs on animals, and showed that they were mainly due to damage by the pressure of the blast on the lungs. If the lungs were protected, the animal could stand a great deal of blast. The pigeon, for example, which has thick muscles over its chest, was astonishingly resistant. If a man wore a jacket of thick sponge rubber, he also could stand far more than had been believed. The head was remarkably resistant.

This work was of great practical importance for the design of protective equipment, and also of great psychological importance, in showing that protection against bomb blast was far less hopeless than had been assumed.

Bernal started the systematic study of the effects of Nazi bombing on British cities. He succeeded in measuring with reasonable accuracy the destructive effect of bombing. From this he was able to conceive and start a new science of the air offensive. It became possible to calculate what kind and how much air bombing would be necessary to achieve various objects. It provided the scientific basis for the planning of the invasion of Europe. This may well have been the most important military scientific work done during the Second World War.

Bernal gave a lecture on the physics of bombing at the Royal Institution at the end of 1940, which I reported, but the full development and implications of his work could not be written about until the end of the war.

Among other aspects of bombing on which one could write were the effects of moonlight on air raids. There was keen public interest and apprehension in the increase of bombing on moonlight nights. I looked into the meteorology of night-flying, and wrote that the best combination of moonlight and absence of cloud for mass night raiding over a small area, such as a city, is rare; it was not very probable that it would occur on more than one or two nights every month.

Another topic of immediate interest was Trueta's plaster-cast method of treating wounds. It had been found effective in the Spanish civil war, and simplified the surgical treatment required for the kind of wounds caused by bombing. I still knew nothing about radar.

Sir William Bragg delivered his last anniversary address to the Royal Society. In it he said that the young scientists of the day were doing something new. Besides adding to knowledge of the facts of nature, they were beginning to consider a new problem, the relation of those facts to society and to the government of nations. After long expectation, the Scientific Advisory Committee under Lord Hankey had been appointed. Its new and significant feature was its close and direct association with the Cabinet. Hitherto, scientists had been employed on useful items in the machinery of government. The new committee was not part of an executive body, or hampered by traditions and set habits. It had the time to consider the whole field of scientific knowledge and its possible influence on practice.

On an evening in 1940 the chief guest of the Tots and Quots was H. G. Wells. They met in a Soho restaurant. The menu consisted of Hors-d'œuvres Variés; Sauté de Boeuf Bourguignonne; Soufflé au Vanille, Sce Chocolate. The discussion was on the reconstruction of the world after the war. While it went on, a considerable air raid developed overhead. It was held that nutrition should be planned for all Europe, to be carried out after the war. Certain perishables should be semi-processed, so that they could already be stored for use after the war. Dairy stock should be built up abroad. Seven million tons of maize could be held in the Argentine, and so forth. There was a concentration on practical proposals. Tom Harrisson remarked on the decline of interest in ideology.

Meanwhile, a stick of bombs was dropped across Soho. The building was shaken, and pieces of plaster floated downwards in the succeeding silence.

H. G. Wells looked at the floating plaster ruminatively, and murmured a remark about 'posthumous nourishment'. He regarded himself as a connoisseur of red wine, and ours that evening had scarcely been the choicest. He looked at his glass, then stared hard at the ceiling, and announced: 'At the last I could say I was drinking good old Empire.'

CHAPTER NINETEEN

1941-42
SCIENCE FROM THE ADMINISTRATIVE ASPECT

WHEN THE YEAR 1941 began I still had not been called to do national work and I continued to write, when I could, on scientific aspects of the war. These included such matters as the political orientation of scientists towards the war, the scientific management of the nation's diet, the training of new and retraining of old engineering workers, and the beginning of extensive war co-operation between the U.K. and U.S.

In 1940 Joseph Needham toured the United States, lecturing in numerous scientific institutions throughout the country during a period of five months. He gave me a copy of a detailed report he had made on his experience. It contained many of the ideas which made his subsequent diplomatic work in science so distinguished.

The first article I published in 1941 was based on Needham's material. He had found the Americans very well disposed to the English, but affected by Nazi propaganda. Like the British, they were extremely ignorant of what the Nazis had written on science. He found that quotations from their works had a stunning effect on American audiences. He suggested that the Ministry of Information ought to prepare a critical review of these works.

He had found that nearly all Americans had believed after the collapse of France that the war would be over in a few days. The achievements of the Royal Air Force came as a complete surprise, and were passionately admired. He noted that the Americans, as one would expect, had a complete confidence in their own technical powers, and had no doubt of the outcome of any technical competition, such as modern war, in which they took part. American scientists were only too anxious to see more of their British colleagues. He thought that a less timid British information policy would be welcomed. It was more important that the Americans should see people they wanted to see, rather than that these should be, in the opinion of the British authorities, paragons of discretion.

Needham had shown himself a masterly scientific-political propagandist.

While personal endeavours to promote better understanding

between scientists were not as yet always welcomed, official policy had begun to move towards international co-operation. In March 1941 Dr. J. B. Conant visited England as President Roosevelt's scientific emissary. The Royal Society gathered what was probably the largest assembly of distinguished scientists since the beginning of the war to meet him.

With regard to the development of opinion on the war (which had not yet led to the assault on Pearl Harbor), Conant thought that academicians had shown up rather well. They had in general appreciated the seriousness of the threat of Hitler's power and philosophy, especially his philosophy, earlier than others. American academicians were determined more than ever to do what they could to help.

He had come on President Roosevelt's authority to exchange scientific information on instruments of war. The U.S. National Defence Council was mobilizing the scientists of the United States. He regarded the defeat of Hitler's philosophy as the ultimate aim, and the scientists of America intended to bring to the British and their fighting forces 'sufficient aid'.

At about the same time, the Rockefeller Foundation and medical schools in the U.S. and Canada launched a scheme by which British medical students could go to those countries to complete their medical training. The strain on the British medical services by the war, and the evacuation and bombing of hospitals had made the provision of medical training very difficult. The *Manchester Guardian* in a leader described the scheme as 'timely, generous and profitable'.

Such was the attitude of young British medical men at that time, that it had been foreseen that many of the most suitable candidates would not wish to leave the country. The Minister of Health had, accordingly, made it clear that effective study in America was the best national service that many of these men could perform. Nothing was more important than the strengthening of ties between the British and American peoples in medicine and science, as well as in other spheres, and young men could make a unique contribution because the future of their countries would ultimately be in their care. It was hoped that the plan might be widened to include students in science and other branches of learning. After the war the movement should develop into an exchange, so that large numbers of American students could come to Britain, while British students would go to America. The movement might grow into a system akin to generalized Rhodes scholarships.

The need to increase the output of munitions created a demand for more working engineers. Men and women who had formerly been in distributive and luxury trades had to be trained in draughtsmanship, fitting, instrument-making, machine-operating, sheet-metal-working, and welding. Training centres were set up, but there was difficulty in persuading workers to come forward for retraining, in spite of improved working conditions, such as spaciousness and cleanliness, working in groups, more informality and smoking at the bench, and good canteen facilities. Attempts were made to attract more workers through films, but these were not very effective, for eminent actors were engaged to play the chief roles, instead of people who had actually changed their occupations through the training system. A good deal of valuable experience in retraining of workers was acquired, which could have been used to more effect, not only during the war, but a quarter of a century later.

The British effort in food and agriculture was one of the more satisfactory developments to write about. Bomb-scarred London was supplied with vegetables of a freshness and quality hitherto not generally available. This was due to the planned production, in the Fen district and elsewhere, of vegetables which had formerly been imported from Holland, France, the Channel Islands and other countries.

The vast stores of canned food attracted attention to its keeping qualities. Two tins of food, which had been canned in 1824 for Sir Edward Parry's second voyage for the discovery of the North-West Passage to India, were taken from the maritime museums and opened, and the quality of their contents examined. They were still edible, if unattractive, after more than a century. The meat-fat, after all that time, retained about half of its original vitamin D.

I asked the Editor of the *Manchester Guardian*, W. P. Crozier, whether we could run a campaign for the better use of science in the war. He was keenly in favour of the aim, but said that paper rationing made it impossible: the size of the paper had been cut to the point that 'we now have only one important article a day'. That was the one on the leader page, and might deal with any topic. In spite of this, Crozier devoted continuous personal attention to scientific matters.

There were occasionally civil scientific events which were so notable that they merited attention. One of these was the death of J. J. Thomson in August 1940. I wrote a long obituary article and a short leader on Thomson. With regard to the latter, Crozier wrote to me that 'We are giving tomorrow your excellent Short

Leader on J. J. Thomson. This is the second contribution of the kind we have given from you within the last few weeks. It leads me to ask you whether there are any other scientists now growing old about whom we ought to have Short Leaders on hand. You will be able to remember those that you have already done, and I would be grateful if you would send along suggestions, say, for half a dozen others that we should add to our collection.'

This was a spur to one of my chief occupations as a scientific writer in 1940-41, the compiling of biographical notes and drafting obituaries and commentaries on contemporary scientists, most of whom I had met.

There was still space for reviews of exceptional books. G. H. Hardy's *A Mathematician's Apology* was sent to me, and I had great pleasure in writing about my revered former examiner. Hardy was one of the best writers of English, expressing his acute ideas in an exquisitely clear style. He said that he knew of no first-rate mathematician who had abandoned mathematics and attained first-rate distinction in any other field, but that if he had had to adopt another profession, the one in which he was confident he might have gained distinction was journalism. I was encouraged by this expression of opinion, for it agreed with my own conception of scientific journalism as a proper activity in its own right, capable of providing scope for the finest gifts.

Hardy made another point with which, however, I did not agree. He attempted to prove that the finest mathematics was of little interest, except to mathematicians, quoting the theorem of Pythagoras that the square root of two is an irrational number. Pythagoras deduced from this that there is a fundamental irrationality in the universe, and therefore in God who had made it. This attitude had profound social and political consequences, through its influence on Plato and others.

Meanwhile, as I wrote articles, obituaries and reviews, the bombs fell around in Bloomsbury. In the spring of 1941 Blackett occasionally called to see me. He was, of course, in the forefront of war research. My own lack of official work was all the more conspicuous in comparison with the activities of my scientific friends. Blackett remarked that something ought to be done about it.

A month or two later, in May 1941, I received a letter from Professor B. Ifor Evans, who was then director of the educational activities of the British Council, informing me that the Council had formed a Science Committee under the chairmanship of Sir William Bragg, and of which Blackett, Cockcroft and others were members. Would I be willing to discuss the possibility of my being appointed Secretary of the Committee, which was to work within his department? I presumed that the approach to me was at Blackett's suggestion.

I saw Evans, and agreed to accept the post on condition that it would not involve more than a four-day week, which would leave me free to continue to do what I could for the *Manchester Guardian* and the Clarendon Press. Since 1930 I have been primarily a self-employed person, however much work I may have done at various periods for particular institutions.

I believe one of the most serious snares for a writer is full-time engagement in such work as editing, publishing, academic work or administration. Some people have enough energy to be able to do a full-time responsible job in one of these and also write; they are few, and most of them damage themselves and their work by too ruthless conduct of their lives. Practical experience of these activities is, however, invaluable for a writer as long as they do not absorb all his attention or become his main and permanent job.

In the pressure and excitement of the first years of the war I did not formulate this to myself with such conscious clarity. The defeat of the Nazis, which was at first far from certain, was the overwhelming consideration for everyone not afflicted with diseased ideas or personality.

Though my engagement at the British Council was part-time, thus indicating my conception of it, I was soon doing at least a full-time job. It was intended that the Science Committee should function as a whole with regard to advice on general scientific matters, and for particular branches it should set up panels; for example, one on Engineering and another on Medicine.

Members of the Science Committee provided chairmen for these panels. In the case of Engineering, the Chairman was Sir William Larke, the Director of the Iron and Steel Confederation, and of Medicine, Sir Edward Mellanby, Secretary of the Medical Research Council.

When I arrived at the British Council, I was provided with a desk and a chair. That was almost the whole of the science administrative organization, apart from a medical literature and information unit under Dr Howard N. Jones, which had been taken over from another government department.

The British Council had been set up in 1934, especially with the aim of countering anti-British cultural propaganda in South America. The outbreak of the war in 1939 greatly increased the scope and urgency of its work. It was extended, and incorporated by a Royal Charter in 1940, which defined its object as 'the promotion of a wider knowledge of the United Kingdom and the English language abroad, and the development of closer cultural relations between the United Kingdom and other nations'.

The founders of the British Council, in the very characteristic manner of the ruling classes of those days, had naturally identified culture with language, literature and arts. It was not that they were prepared, after reflection, to deny that science was a part of culture; it was merely that it had scarcely occurred to them to include it. With the outbreak of the war, however, foreign peoples became acutely interested in British technique: they wanted to know whether the British had the scientific and technical competence to win the war, as well as whether they still had an impressive literary and artistic civilization. Evidently, if they could be supplied with accounts of British scientific achievements, and could hear British scientists lecture and teach in their countries, they might form a more hopeful opinion of British survival. The satisfaction of the new interest abroad in British science and technology would strengthen scientific and technical exchanges, which would lead to increased trade, through a better knowledge of what Britain had to offer in the more advanced scientific and technical industries.

These ideas were vaguely in the air, but had not been precisely formulated. The Science Committee was a collection of eminent and very busy persons who were available for advice, but originally had no very clear views of what they were supposed to do, or of what I was supposed to do. They were full of goodwill, but left me to get on with it, so I did.

I thought about the elements of a programme which could be discussed with Sir William Bragg and laid before the Science Committee as an agenda, together with points that cropped up from other departments of the Council, such as recommendations of scientific books and periodicals, and scientists as lecturers and teachers abroad. It became evident that short, simple, clear pamphlets on British scientists and engineers and their contributions to science and engineering were required, so I drew up a comprehensive list of titles for a series of pamphlets to deal with *Science in Britain*. This was received with some alarm, for it outlined a considerable publishing venture, which might, among other considerations, be regarded with disfavour by the publishing trade.

I suggested the starting of a brief, simple monthly science news sheet, with the title *Monthly Science News*, which could be distributed abroad in various languages.

Science from the Administrative Aspect

As these and other activities extended, I needed secretarial and administrative assistance, and a departmental organization came into existence. This presently became established as the Science Department.

My first administrative assistant was Miss B. M. H. Tripp. She was on the staff of the Institute of Petroleum, and had helped in the editing of the *Science of Petroleum*. She was interested in Russian, and after the invasion of the U.S.S.R. she called to see me to ask whether there was any opening that would enable her to do work of more immediate national importance, especially in the field of Anglo-Russian relations. I was very glad to secure her assistance. She was my administrative deputy for three years until she was appointed to a post at the British Embassy in Moscow; she did a great deal of the administrative work of building up the department. In May 1941 there had been virtually nothing; by 1946 the department had a staff of about fifty, and an annual estimate exceeding £50,000.

My work was made easier and much more satisfying by the change of atmosphere which took place shortly after I joined the British Council. Hitler's invasion of the U.S.S.R. a few weeks later led to an improvement in Anglo-Soviet relations, and in the attitude to socialist views generally.

The British Council had originally been created to a large extent by Lord Lloyd, who was an admirer of Mussolini. Its staff still contained more than one who had formerly been an open admirer of Mussolini, and a considerable number not very far removed from that, while there were extremely few in sight who might conceivably have been socialists.

The emergence of the U.S.S.R. as an ally meant that socialist thoughts could no longer be openly treated with contempt. They had at least to be regarded as 'interesting', which made things easier. My knowledge of Soviet science and scientists was suddenly transformed from a liability into an asset.

Working with Sir William Bragg was an invaluable experience and a profound pleasure. I had not known him closely before. He appeared to me to be a person of extraordinary natural wisdom, by which I mean he was able to make correct judgments on the facts before him, irrespective of his general set of ideas. He was widely regarded as a very charming man in a conventional way, and no doubt he regarded himself as conventional; I found him free from prejudice, and astute in affairs. Two incidents particularly impressed me.

I was called to appear before a meeting of the Scientific Advisory

Committee to the War Cabinet, with Lord Hankey in the chair. Sir William was a member, and most skilfully and energetically managed the various questions which were put to me, so that the Committee made all the reactions to my answers which he and I had previously envisaged and discussed. He was seventy-eight years old, but he behaved with extraordinary mental agility and polite insistence. He must have spent considerable time and effort before the meeting in persuading certain members of the Committee to come to the correct conclusions.

The other incident which particularly impressed me, and endeared him, was in connection with the Athenaeum. We had been discussing some business one morning at the Royal Institution, and it had become rather late. He said: 'Come and have lunch at the Athenaeum.' When he rang up for a table, he was told there was none available. He was really annoyed; it was the only time that I ever saw him angry. He said sharply that it had become a place of business, where the people for which it was originally intended were being crowded out. I suggested he should lunch with me at the Café Royal. During the war it had retrieved a little of its former character. Under the threat of air raids, and the occasional whistle and bump of bombs, patrons and staff got to know each other. There was less formality, and sometimes unusually good food; I suppose they were more concerned with keeping going than with single-minded devotion to profit.

Bragg became quite excited at the idea of lunch in the Café Royal. He told me that it was many years since he had been there. We sat at one of the marble-topped tables, and he chatted and looked around in a very lively fashion. It was certainly different from the axe-grinding down in the Pall Mall.

As one of my British Council projects for spreading a knowledge of British contributions to science I organized a number of broadcast addresses by British scientists on the field of research in which they had been engaged. Sir William Bragg acted as the introducer of the talks, and entered into their preparation with extraordinary liveliness and zest. He and his eminent son attended rehearsals in the B.B.C., in which they discussed problems of the presentation of scientific ideas and particular points. Sir William, at the age of seventy-nine, had the understanding, alacrity and enthusiasm of a born editor. He introduced his son Sir Lawrence Bragg's broadcast on X-ray metallurgy; it was the first and only time that two Nobel Laureates, father and son, appeared together in a broadcast.

Sir William died only three weeks after the delivery of this broadcast, and the *Manchester Guardian* published the obituary article and short leader I had written some time before. Sir William's death was sad, especially while his mental powers were still so active, though at seventy-nine it was not to be unexpected. I was glad that I had had a part in securing the joint recording of himself and his son, probably the last recorded words of his life.

Other recorded addresses in this series included Sir Robert Robinson on 'Atoms and Molecules', and 'Making Substances to Order'; Blackett on 'Cosmic Rays', Cockcroft on 'Shattering the Atom', and a discussion between J. D. Bernal and Sir William Bragg on 'The Origin of Life'.

Sir Robert Robinson talked about the subject of his addresses with me before he made the recordings. I was deeply impressed by the power and quality of his imagination. He seemed to be seeing the atoms and molecules as they moved in space, bending, straightening, and catching each other by their ends.

These addresses were collected and issued in the pamphlet *Science Lifts the Veil*.

I had Sir William's advice and help for nearly a year before he died in March 1942. Though he became physically rather frail, he retained a remarkable mental enthusiasm up to the last weeks of his life. He approached everything I discussed with him in a positive spirit, whatever conclusions we came to about it. By the time of his death the main lines of the science activities and organization in the British Council had become established, and confirmed as a science department under my direction. The science publications started, and the inquiries from other departments for scientific advice and information increased.

Hitler's invasion of the U.S.S.R. led to a widening and deepening of the Council's scientific activities. The U.S.S.R. became an ally, with whom scientific exchanges became officially desirable. This affected the social and political perspective of a great range of scientific matters, in addition to scientific exchanges in themselves. In my office I made arrangements for dealing with scientific exchanges with the U.S.S.R., which were carried out by Miss Tripp, who became secretary of the Anglo-Soviet Scientific Collaboration Committee.

That was something, though we were not able to go very far, owing to the persistence of deep political attitudes. In the autumn of 1941, during the days of the threat to Moscow, I received a letter from Joffe, which had taken only five days, at that desperate time, to reach me from Moscow. He informed me that a committee for scientific exchanges had been formed with him as chairman, and including Kapitza and Sobolev; I was invited to conduct the London end of the exchanges. I told the appropriate 'official quarters' of Joffe's letter, but I never heard anything more of the proposal. Yet while official bodies were unable to countenance such approaches as this, activities in other directions widened.

In July 1941 the Committee of the British Association's Division for the Social and International Relations of Science decided to organize a conference on *Science and World Order*, arising out of a suggestion by Sir Richard Gregory. The Committee appointed a Programme Sub-committee, consisting of the general officers of the Association, Ritchie Calder, S. Chapman, and C. H. Desch, A. V. Hill, D. P. Riley and me. It had been intended to hold an opening day's meeting in London, and two further days of meetings elsewhere, but it soon became clear that the conference was expanding into a demonstration of the unity of science in the anti-Nazi world. The Royal Institution, still under the direction of Sir William Bragg, placed its famous lecture theatre at the disposal of the Conference.

Foreign governments were informed of the Conference, and invited to send representatives to it. There were at the time many governments-in-exile in London. Consequently, scientists of at least twenty-two nationalities appeared at the meeting. The chairmen of the main sessions were chosen to reflect this. Sir Richard Gregory, the President of the British Association, gave the opening and concluding addresses. Lord Samuel presided over the session on 'Science and Government', Mr John Winant over 'Science and Human Needs', Mr I. M. Maisky over 'Science and World Planning', Dr Benes over 'Science and Technological Advance', Mr Wellington Koo over 'Science and Post-War Relief', and H. G. Wells over 'Science and the World Mind'.

Besides taking part in organizing and speaking at the Conference, I wrote two articles on it for the *Manchester Guardian*. Among the features to which I drew attention was the fact that such statesmen as Benes and Maisky, Wellington Koo and Winant attended the sessions not once but several times, and delivered carefully prepared addresses. It was a sharp break with the bad old traditions; it was evidence that statesmen were now seriously concerned with the scientific aspects of world affairs.

The variety of speakers and subjects at the Conference was fascinating. The session in which I spoke was presided over by H. G. Wells. He had been called to order in a previous meeting for exceeding his time, which had put him into a petulant temper,

Science from the Administrative Aspect

for he believed that he had been cut off in the midst of solving one of the problems of the human race. After this he was determined that no one else should have a second of extra time.

When he took the chair at the session on 'Science and the World Mind', he ostentatiously pulled out his watch and placed it conspicuously in front of him. Then he stared at it almost constantly. The first speaker was Lancelot Hogben. When Hogben reached the very last second of the allotment of time for his address, he began slowly to sink from view, like Minerva disappearing beneath the waves, dragged down by Pluto. Wells had in fact got hold of him by the seat of his trousers, and was gradually but firmly pulling him down. Hogben tried to recede as gracefully as possible.

I had to speak next, and seeing what had happened, I cut two or three hundred words from my paper, to be sure that I ended well within my time. I spoke on 'The Education of the Public', with regard to science and its social significance. In it I showed a graph of the awards of the Nobel science prizes, and the significance of the decline of awards to French scientists which began about 1914, and the sharp upward trend of awards to Americans beginning in 1933.

A most interesting contrast in the conduct of sessions was seen between Maisky and Wells. The chairmen had been provided with a push-button for operating a red light by which speakers could be warned of the approach of the end of their time. The push-button lay on the bench some way from Maisky's right hand, which kept stealing out like a cautious mouse, and suddenly pausing and withdrawing, and then starting again. One thought one noticed in it an unconscious demonstration of the caution of the very careful diplomat, who reflected many times about what he was doing before he came to a decision.

At a preliminary stage Wells gave a small dinner party in his house at Regent's Park, attended by Benes, Maisky, Koo, Winant, Gregory, Ritchie Calder and myself. Wells's interest and action did much to consolidate the idea of the conference, and secure influential support for it.

I was drawn in as a channel through which support from the British Council might be secured. The Council gave a luncheon at the Savoy Hotel to British and foreign representatives assembled for the Conference. Mr Anthony Eden attended, and addressed the Company.

I was asked for advice on arranging the places at the luncheon tables. I put H. G. Wells about four or five places from Eden, and Negrin, the last President of Republican Spain, opposite Wells.

I noticed that during lunch Wells and Negrin appeared to be getting on famously. But I heard afterwards that consternation had arisen in some quarters when it had been learned that Dr Negrin had been placed only four or five places from Eden.

The British Association commissioned O. J. R. Howarth, D. P. Riley and myself to write an account of the Conference, to be issued as a Penguin Special with the title *Science and World Order*; it appeared in 1942.

The British Council's offices in other countries found that science was a new and useful medium for the extension of their activities, which had hitherto been largely concerned with teaching English and the promotion of literary and artistic exchanges.

As scientific exchanges grew, offices abroad needed scientists on their staffs to deal with the increasing volume of business concerning visiting scientific professors and lecturers, the arrangements for science students to come to Britain, the supply of scientific equipment, and presently engineering equipment, for teaching and research purposes.

At first, the science officers appointed abroad were regarded as specialist assistants not concerned with general policy. The heads of offices abroad were usually men drawn from the classes which at that time supplied diplomats. As the war went on, new situations arose in which new combinations of qualities were required. The Japanese attacks on the Americans and the British Commonwealth enormously increased the immediate political and military importance of China. It was realized that after the war there might be a rapid development of the huge Chinese market, which would be of great benefit to British trade.

In February 1942 I took part in a joint meeting of the British Council and government departments concerned with British cultural representation in China. About a score of names were put forward for consideration; they ranged from bishops to female explorers, and were patently unsuitable. A glum and desultory discussion went on for a long time, as cultural activities in China were considered important.

When it became evident that there was no general agreement, and discussion had virtually died into empty silence, I suggested that Dr Joseph Needham might be a suitable person. I heard someone murmur: 'Who is Dr Joseph Needham?' I explained Needham's eminence as a biochemist, who had long been interested in China, and in the Chinese students of biochemistry who came to work under him at Cambridge.

A kind of cautious interest in Needham began to arise in the

Science from the Administrative Aspect 235

meeting, and presently a young man ejaculated something like: 'Oh, of course, if it is *possible* to have someone like *Needham*, that would be *wonderful*.' When it became clear that Needham might indeed become a possibility, various people spoke up for him, and the meeting ended with a recommendation of him, and another.

Needham was approached and appointed to lead a British Cultural Mission to China, and an office was set up in my department to deal with scientific exchanges with China.

After Sir William Bragg's death in March 1942 the British Council invited his successor as President of the Royal Society and as Director in the Royal Institution, to succeed him as Chairman of the Science Committee. This was Sir Henry Dale.

I made my first call on Sir Henry as Chairman of the Science Committee while he was still at the National Institute for Medical Research. He told me that if he had been Chairman of the Science Committee at the time, he would have intervened with regard to Needham's mission. He said that the British Council had been referred to as 'an upstart body', and he indicated that the scientific activities of the British Council ought to come under more control from certain other scientific quarters.

Criticism reached such a pitch that efforts were made to recall Needham, but after about a year the agitation subsided, owing to his own brilliant work, and the refusal of one government department to accept interference from another.

But the general objective remained. Sir Henry subsequently told me that before he retired from the presidency of the Royal Society, he intended to secure a reorganization of the Council's Science Committee on what he regarded as sound principles. I gathered that in essence this meant that the British Council would not be able to initiate any important scientific activity of which the Royal Society did not approve.

Thus, from the spring of 1942, a conflict on policy arose: was an office operating on a Foreign Office vote to be controlled ultimately by a private body? This fundamental problem was with me for the next four years, until I left the British Council in 1946.

Meanwhile, on many questions of ordinary scientific business in which this problem did not arise, Sir Henry was very helpful.

CHAPTER TWENTY

1941-42
INDIVIDUAL WORK AND INSTITUTIONAL INFLUENCE

ENGAGEMENT IN THE administration of scientific activities, while it absorbed a great deal of interest and energy, also stimulated demands as an exponent of science. *The Social Relations of Science*, which had been completed in the autumn of 1939, appeared at last in New York in 1941.

The book received considerable attention. The subject had been reformulated only recently, and there was not much literature on it. I received copies of ninety-five reviews and notices. The reviewers included Sir Richard Gregory, J. B. S. Haldane, F. A. Hayek, C. E. Lucas, P. B. Medawar, Naomi Mitchison, Olaf Stapledon, Barbara Wootton and M. Polanyi.

After reading the book, C. H. Desch sent me an interesting letter, in October 1941, containing a copy of his pamphlet entitled *The Social Function of Science*. He wrote: 'You may perhaps be interested in the enclosed, which I printed for private circulation ten years ago. The title was not as hackneyed then as it has since become. You will find it old-fashioned as I am not in the least a Marxist, but I believe that the views which some of us derived from Comte still have some value. The pamphlet was entirely without effect; I sent out nearly 500 copies, and received five replies.'

After I joined the British Council I was invited occasionally to broadcast, especially on science in the U.S.S.R., and to participate in 'brains trusts'. In one of these, in which C. E. M. Joad was also taking part, a question arose about the atom. Joad asked me aggressively whether atoms exist. I treated it as a nonsense question, and refused to be provoked. Joad lost his temper and began to make silly remarks that destroyed the point of his position. He was desperately anxious to monopolize the discussion.

In the middle of the war Henry Martin, Editor-in-Chief of the Press Association, became keenly interested in the organization of Aid to the U.S.S.R., and I met him through this. Thereafter, I sometimes received as many as one hundred press-cuttings if I made a speech anywhere. But presently, though I continued to speak in the same manner, the number of cuttings dropped to a

Individual Work and Institutional Influence

few, as before the war. These experiences impressed on me the influence of institutional connections on publicity.

Shortly after I joined the Council, two of the elder and most well-intentioned leaders of British science independently suggested that they should put me up for membership of the Athenaeum; this was another example of the institutional effect. I thanked each of them, and declined.

The planning of science, and the revision of traditional views on the independence and freedom of the individual investigator which it implied, became the object of a reaction which began in the spring of 1941, and was presently organized in the Society for Freedom in Science. The views of the proponents of planning were attacked, including the chapters in the *Social Relations of Science* in which I had commented on the practical and moral inadequacy of the concept of absolute freedom in science. My strictures on the behaviour of Galileo, and approval of Bacon's conception of science and its social relations were also attacked. The leading part in the restatement of the traditional view of the freedom of science and the scientist was undertaken by M. Polanyi. A number of scientists publicly supported his views.

Sir Henry Dale expressed similar views in his first Anniversary Address as President of the Royal Society, on 1 December 1941. He concluded it with the observation: 'If science should become entangled in controversial politics, through the over-eagerness of its advocates and champions to invoke the sanction of science, or to claim its potentialities, in support of any special political doctrine, then indeed I believe that the threat to its freedom might become a real danger. Let there be no misunderstanding of my meaning. I am not abusing the privilege of this Chair by using "controversial" as an epithet, to be applied to political opinions which I do not happen to share. I see danger if the name of science, or the very cause of its freedom, should become involved as a battle cry in a campaign on behalf of any political system, whether its opponents would describe it as revolutionary or reactionary. If science were allowed thus to be used as a weapon of political pressure, it would be impossible to protect science itself eventually from the pressure of sectional politics. If that should happen the dangers are, I believe, beyond dispute—the danger, for example, that fundamental researches, having no immediately practical appeal, would be allowed to fall into arrears through relative neglect; or the danger that the rigid standards of true science would be relaxed, by allowing the convenience of results for policy or for propaganda to enter into the assessment of their validity as evidence. This Society, with its firm and unbroken tradition of complete aloofness from political controversy, may still find it an important part of its function, to keep watch, and, if necessary, to stand without compromise, for the right and the duty of science to seek the truth for its own sake, in complete freedom from any kind of extraneous influence. I hope, indeed, that there will never be need thus to invoke our tradition, to protect the freedom and the integrity of science from the enthusiasm and the advocacy of any of its friends.'

Sir Henry's views were fundamentally in conflict with mine, and consequently co-operation between us became increasingly difficult. He took his own position to its logical conclusion in 1948, by resigning his Honorary Membership of the Academy of Sciences of the U.S.S.R.

All scientists were to be free, but only in the way that conformed with Sir Henry's own principles.

As I became increasingly involved with the science work for the British Council, I had less opportunity for writing, but the *Manchester Guardian* published eighteen contributions from me during 1942.

The first of these was an account of the Association of Scientific Workers' Conference on Science and the War Effort in January. It came a few weeks after Sir Henry Dale's remarks to the Royal Society on the freedom of science. This was one of the several very stimulating wartime conferences organized by the Association. I noted that 'Like all serious intellectual gatherings at this time, its sessions were packed by large audiences, and many speakers had interesting points to make direct from experience.' As the Association was largely run by its younger scientists, so its views threw light on those of the leaders of tomorrow. Its President was then R. A. Watson-Watt, who had just become recognized as the chief inventor of radiolocation, and one of the saviours of his country.

Two of the events about which I wrote in 1942 were the tercentenaries of Isaac Newton's birth, and of Pascal's invention of his calculating machine.

Another event which I reported in 1942, was the suspension of the office of Secretary in the Zoological Society, for the duration of the war. The Secretary was Julian Huxley, and the decision amounted to his departure from the office. It occurred after dramatic debates. Among the supporters of one side, which turned out to be rather the larger numerically, were enthusiastic big-game hunters and various kinds of amateur animal-lovers, while among

Individual Work and Institutional Influence 239

those of the other side were doctors, scientists, and writers, such as Lord Horder and H. G. Wells. The substance of the dispute was whether amateur or professional scientific interests were to take precedence in the Society. The disputants became involved in irrelevant personalities and irrational arguments.

Julian Huxley's departure from the Zoo was one of the oddest events that happened in the world of science in my time. He bore a name which, through his grandfather's work, and his own, stood beside Darwin's. He was an unsurpassed exponent of biology and, in my opinion, an unsurpassed broadcaster. He felt the blow severely, and was made quite ill.

Besides such matters as these, there was also some new scientific information to report. One item was progress in research on desalination, much in the news in 1968. In 1942 interest in it was acute, owing to the perishing of torpedoed sailors from thirst. Today, the desalination of sea water on a large scale promises to be the most practicable method of meeting the increasing demand for fresh water by the world population.

The Association of Scientific Workers organized a Conference on Scientists of the United Nations and the War Effort. Sir Richard Gregory, Sir John Russell, and Professor Cassin were prominent in it. The aim was to increase the solidarity of all scientists opposed to the enemy, and urge the best use of them in the war effort, whether they were of British or other nationality.

In these war-efforts there was participation in united efforts by men of every kind in science, from the senior leader of research to the junior technician. The unity was in marked contrast to the position a quarter of a century later. The research scientists have tended to look more and more to their specialist bodies, while the technicians have concentrated on economism, such as a too exclusive attention to higher wages.

CHAPTER TWENTY-ONE

1942-46
SOCIAL NEEDS OF SCIENTISTS

As THE WAR extended, scientists of many countries came to London, for a variety of reasons. Besides those who escaped from the occupied countries, there were scientists attached to government missions as scientific advisers on military, medical, industrial and other matters. By 1942 this heterogeneous body of scientists temporarily in London had grown to a considerable number.

At the beginning of the war, the scientists, like others, were concerned only with survival; after two years, however, their activities began to acquire a degree of permanence, and a need for organization became increasingly felt. These scientists required looking after, besides being enabled to carry out their official duties.

The British Council had a department for helping visitors from abroad. It provided clubs and other amenities, and was conducted with outstanding skill and success by Miss Nancy Parkinson. One day in 1942 she sought my help in looking after visiting scientists. Many of the scientists in London in 1942 had serious personal problems. They had difficulties in finding accommodation and jobs, and were worried about relatives in occupied countries.

This inquiry by Miss Parkinson led me to start the foundation of the Society for Visiting Scientists, which became established in 1944, after two years of informal operation.

In 1943, as Secretary of the Conference of Allied Ministers of Education, Miss Pakinson consulted me on the re-equipment of schools and universities in devastated countries with scientific apparatus and books. This was the beginning of the movement that led to the incorporation of scientific activities in UNESCO.

With regard to the first of these inquiries, the scientific societies could not cope with such problems in the volume in which they now occurred. At the same time, they had a long tradition of receiving and helping visiting scientists in more specialist matters. My office undertook a good deal of personal and informal work, but it seemed to me that a club not unlike those of the Fondation Universitaire in Brussels and the Harnack House in Berlin was required. The scientific societies had a traditional concern with contacts between scientists, but had neither the means nor the staff to provide the extended social attention that was now required.

I therefore asked the Chairman of the British Council to write to the President of the Royal Society, suggesting that the Royal Society might form a subsidiary society in which visiting scientists could meet, and that a joint committee with the British Council be set up to regulate the arrangements of such a society, if formed.

The President replied that the Royal Society was unable to accept the proposal, but expressed willingness to offer limited help in achieving the object in mind. It appointed Professor F. G. Donnan to co-operate with the British Council. This was an excellent suggestion, for Donnan was one of the most sociable figures in British science. He drew up a memorandum of proposals in consultation with A. V. Hill, one of the Secretaries of the Royal Society.

Donnan brought the memorandum for discussion with Miss Parkinson and myself. A skeleton plan was worked out for a society which should have an office, preferably near Burlington House and the British Council, which should be at the service of all visiting scientists, whether members of the society or not. There should be occasional scientific and social meetings, open to all full members and to guests invited by the Society's officers.

It was proposed that Professor Donnan should be President. I should be Secretary, and, if possible, Dr Louis Rapkine Assistant Secretary. The Executive Committee should be substantially the same as the Exploratory Committee, whose members included British, Dominion, American and other Allied scientists.

I suggested that Rapkine, who was then in New York, might be brought in to run the new Society because, in spite of his unique efforts in saving French scientists, he had no adequate means of support. His qualities for the kind of work required were unequalled. I found him willing to accept the post.

I sounded Sir John Anderson, then Chancellor of the Exchequer, on whether he would accept the honorary treasurership of the Society, and he asked me to call on him at the Treasury. He was himself a scientist by education, and in his youth had done research on the physical chemistry of uranium. He said that he was an officer of various societies, but could not take on any more commitments since becoming Chancellor of the Exchequer. I found Anderson much less forbidding than his exterior appearance and reputation. Lord Rothschild became Treasurer of the Society. Rapkine found a suitable house for it at 5 Old Burlington Street.

The French scientists whom Rapkine had rescued from France naturally made the Society one of their meeting centres, and Mme Suzanne Auger joined the Society's staff for a time. Rapkine soon became so occupied with matters concerning the reconstruction of French science that he had to give up his post of assistant secretary. I asked him and A. V. Hill for suggestions for his successor, and both suggested Miss Esther Simpson, Secretary of the Society for the Protection of Science and Learning.

The Society had a restaurant in the basement, and a lounge on the first floor. A stream of scientists from all quarters of the world visited the Society, where they could meet and eat with their friends and colleagues, and continue their conversations in the lounge. There were also some bedrooms, where visitors could be accommodated for a few days.

I aimed at making the Society a social centre for scientists visiting Britain in search of refuge, performing governmental duties, furthering international science, strengthening peace, and engaging in the other social activities on which the welfare of science and scientists depends. British scientists, scientific journalists, administrators and politicians concerned with scientific affairs, were to be encouraged to join, and help their colleagues from abroad.

To promote these aims, a series of receptions and discussion meetings was launched. I made a point of getting expert professional stenographers, accustomed to taking down scientific material, to record the proceedings; it was expensive, but endowed the Society with some valuable historical material.

Before the Society for Visiting Scientists was officially opened in October 1944, it held receptions for the press, and for members of the French Scientific Mission. Later in 1944 it gave receptions for Indian scientists, and members of the Parliamentary and Scientific Committee.

During 1945 it entertained Chinese scientists in Britain, Turkish engineers, groups of Belgian scientists, American scientists in Britain, and delegates to the Anglo-French Cosmic Ray Conference. Receptions were also given for the Council of the Royal Society, and its International Relations and Cultural Relations Committees; for the delegates to the United Nations Conference in London for founding UNESCO; and the members of the International Council of Scientific Unions.

Miss Parkinson's second inquiry, on scientific equipment, was in her capacity as Secretary of the Conference of Allied Ministers of Education. In 1940, Britain had found herself with two hundred thousand refugees from invaded Western Europe. Their cultural as well as their other needs required looking after, and the Government requested the British Council to assist with them. The cultural needs of these citizens, mainly of Allied nationalities, included the provision of education. Representatives of the Allied governments in exile met in a committee to deal with these problems. The committee gradually became more and more formalized, and was ultimately established as the Conference of Allied Ministers of Education, under the chairmanship of Mr R. A. Butler, the British Minister of Education. Mr Butler was not only deeply concerned with education, but had a long previous connection with foreign affairs. He was a very able chairman, with great skill in summarizing arguments, and extracting acceptable forms of resolution from them.

As Secretary of the Conference of Allied Ministers of Education in exile, it, and the organization which grew out of it, were nursed into existence largely by Miss Parkinson.

After the Conference had been working for a time, it became evident that relief and rehabilitation of the countries occupied by the enemy was one of its main preoccupations. A considerable amount of help in the discussions on this topic was given by representatives of the United States. The American statesman Mr Fulbright came to England with a draft for a United Nations organization to deal with topics with which the Conference of Allied Ministers was concerned. The emphasis in the early American thought on the problem was on the educational aspect, in particular on teaching. The Americans envisaged a huge illiteracy problem in the devastated countries, where schools had been destroyed.

The representatives in the Conference of Allied Ministers were professors and other personalities in exile, who generally had no special interest in school-teaching but were deeply concerned with cultural matters. They wanted a United Nations Cultural Organization, which would be more effective than the Committees of Intellectual Co-operation of the old League of Nations. They regarded the original American idea of the United Nations Organization for school teachers as quite inadequate, and insisted that the new organization should deal with cultural matters in the broadest sense.

The Europeans proceeded on the political principle that it was essential that the initiative for United Nations organizations should come from the Americans, in order to ensure American participation. At the same time, they did not want merely American-run organizations. In deference to the Europeans, the

Americans expanded their plan for a United Nations Educational Organization into one for a U.N. Educational and Cultural Organization.

Meanwhile, the Conference of Allied Ministers in London had, in discussing the problems of the restoration of educational institutions in the devastated countries, found itself confronted with the question of the equipment of schools and universities, especially in science. Few of the members of the Conference had any special knowledge of scientific matters, so, in 1943 Miss Parkinson consulted me.

The Conference decided to appoint a Science Commission. Its first meeting was presided over by Sir Henry Dale, and I was appointed Secretary. Dr E. F. Armstrong, on the suggestion of Sir Henry and his colleagues, was appointed Chairman. The Science Commission secured the assistance of Brigadier R. A. Bagnold, F.R.S., the distinguished soldier, physicist and geographer, who conceived a plan for inventories of scientific apparatus, suitable for re-establishing scientific work in educational institutions as they were freed from the enemy. His office was set up in the Royal Institution.

The radiological physicist L. G. Grimmett was engaged to assist in this work. Mrs M. Gowing, the author of the official account of the British work on the release of atomic energy, has stated that the first wartime record she had been able to find anywhere of the warning of the dangers to the genetic constitution of man of radioactivity from military weapons was in a paper by Grimmett. His talents were not restricted to science. He was a good musician, and performed excellently on the piano and violin. In his youth he had been very hard up, and when he aspired to become a scientist, he had to secure the necessary education by his own efforts. He helped to keep himself by playing a violin on the kerb in Oxford Street in London.

The Science Commission soon found that there was virtually no information on the amount of scientific equipment which had existed in the occupied countries, or what an adequate equipment should be.

The Conference of Allied Ministers appointed a drafting committee for a future United Nations educational and cultural organization. Dr Armstrong and I pressed for the inclusion of science, which would be essential for implementing the Science Commission's rehabilitation programme after the occupied countries were liberated.

Early in 1945, Dr Joseph Needham while in Washington heard

of the American plan. He saw at once that science should be brought into the new organization on an equal basis, and should be in the title. He wrote to Sir Richard Gregory in England asking for his support for this view. Gregory, as President of the British Association, communicated with Armstrong and myself. We persuaded the Science Commission to pass a resolution in favour of science being included in the title and constitution.

Fortified with the resolution of the Science Commission, we attempted to persuade the Allied Ministers' Drafting Committee to insert 'Scientific' into the title, but this was not accepted, the chief opponent being the American representative Dr Kefauver. It was evident that, in the spring of 1945, the Americans were not in favour of science being included in the new organization. They were still officially thinking only of a U.N. educational organization.

The most Dr Armstrong and I could secure was that the word 'science' should be added to the words 'education and culture' in about half of the places where the latter occurred in the draft.

The national academies of science were also thinking about the creation of a new international scientific organization. They were inclined towards one exclusively for science. The protagonists of this view were opposed to the incorporation of science into the proposed UNESCO.

Needham's long-standing interest in international scientific relations was widened by his special experience in China. He became more and more impressed with the interconnectedness of science, education, and culture. As his ideas and experience grew, he put down his thoughts in a series of memoranda on International Scientific Co-operation. He completed the third version of this document in Chungking early in 1945, and distributed it as widely as his resources permitted. Dr Alexander King in Washington and I in London circulated altogether nearly five hundred copies.

Needham's memorandum was the outstanding personal document on the scope and activities of a post-war official organization for international science. However, in spite of the efforts of a number of far-seeing American scientists the Americans continued to think basically of a purely educational organization, and most of their scientists wanted an exclusive international science organization.

Shortly before the U.N. Conference for the Establishment of an Educational and Cultural Organization met in London in November 1945 this situation was completely changed through the explosion of the atomic bombs over Japan. This event gave science a new order of importance in human affairs.

The inclusion of science in the title of UNESCO now became a political necessity; it provided the best argument of persuading the U.S. Congress to vote the money for U.S. participation in the new body. The leader of the American delegation, Archibald McLeish, now arrived with science in the forefront of the American proposals. He moved that Science should be included in the title. Miss Wilkinson, the leader of the British, and M. Blum of the French delegation supported him.

In 1945–46, S.V.S. provided a unique meeting-place for many-sided discussions on science, and its problems in the new social situation. Among these was one in July 1945, to hear the accounts of the members of the British Scientific Delegation to the U.S.S.R. Academy of Sciences. Speakers included Gordon Childe, E. M. Crowther, Julian Huxley, Spencer Jones, D. M. S. Watson, and W. A. Wooster. The British delegation had departed for the U.S.S.R. under peculiar circumstances. At the last moment, members who might conceivably have anything to do with atomic physics were not allowed to go.

All the members had assembled in the rooms of the Royal Society just before they were supposed to leave for the U.S.S.R., and then learned that a number would not be able to go. One of them had asked me to meet him at the Royal Society, and when I arrived, I found the atmosphere extremely excited. I learned that Sir John Anderson was in another room, delivering the unwelcome news. I saw an eminent physicist striding up and down in a rage, and vowing that he would join the Communist Party tomorrow. This preliminary contretemps made the views of the delegates who were allowed to go all the more interesting on their return.

The account by D. M. S. Watson was particularly interesting. He spoke on recent Soviet paleontological researches. They had discovered the fossils of animals which were intermediate between fossils previously found in Texas and South Africa. They corresponded with what had been forecast on theoretical grounds. One hundred and forty sites had been discovered, of which only six had so far been explored, so one could imagine what further discoveries were to be expected. This work appeared all the more remarkable because it had no economic implication, and was not the kind that Watson had expected to hear about during a desperate war.

In September of the same year, a discussion on 'The Social

Social Needs of Scientists

Implications of the Atomic Bomb' was held, with N. F. Mott in the chair. Speakers included P. Auger, J. D. Bernal, P. M. S. Blackett, M. L. E. Oliphant, Irène and F. Joliot-Curie, and Sir George Thomson.

An early discussion meeting was held in February 1946 on 'The Place of Science in UNESCO', under the chairmanship of Julian Huxley. In opening it, he explained that he had accepted the chairmanship of the discussion before he had been asked to be Secretary of the Preparatory Commission for setting up UNESCO, to which he was appointed four days before. Accordingly, he would limit himself to listening to what others said on the subject, for at the moment he was as much in the dark as anyone else as to what might be done. He remarked that the Society for Visiting Scientists was playing a most valuable role, not only in providing a home from home for visiting scientists, but in providing a platform on which the various subjects concerning the social relations and applications of science could be discussed, with the most interesting and stimulating results; for there was no other organization at present in Britain which was capable of providing these functions. He hoped that the Society would not only continue to exist, but would come to play a new and more important role in relation to UNESCO when this came into existence at the end of the year.

Huxley's involvement in UNESCO was remarkable and fortunate. He was recovering from his departure from the Zoological Society, and he still had no suitable appointment. Then, one day, he was summoned by the Minister of Education, Ellen Wilkinson, and was amazed to be told that the Government wanted him to be Secretary of the Preparatory Commission, which implied that he would be their candidate for the post of the first Director-General of the whole organization.

The effect of this on Julian's health was striking. His load of depression was lifted, his posture straightened, his eyes sparkled, he strode out confidently, and his reactivated mind poured out ideas in its characteristically scintillating manner. Ultimately, Huxley was appointed the first Director General, after considerable political contest. Some of the opponents of his appointment alleged that he had excessively progressive views.

At the S.V.S. discussion, Huxley proposed five aspects of UNESCO for discussion. First, the relation of the new organization to existing international organizations. Second, what kind of central machinery should he set up by UNESCO for the exchange of personnel, ideas and information? Third, the implementation of

Dr Joseph Needham's proposals in his memorandum on the international organization of science, for the spreading of the butter of science evenly over the bread of the earth, and extending what the Needhams had done in China on a world scale. Fourth, there was the relation of UNESCO with the United Nations Organization itself, and its other organizations concerned with science. Finally, there was the problem of the relation of UNESCO to the individual nations which constituted the United Nations.

Huxley then called on me to speak on the historical aspects of the origin of UNESCO. I concluded with a plea for adequate finance for UNESCO, enabling long-term planning and large-scale mass-production programmes for the re-equipment of schools and universities to be carried out.

The next speaker was Professor F. J. M. Stratton; he had become General Secretary of the International Council of Scientific Unions in 1937, but in 1939, because of the war, it had virtually ceased activities. These were revived in the summer of 1945. He mentioned that the U.S. delegation had proposed that discussions should be started with the I.C.S.U. on methods of collaboration. Stratton said it was hoped that UNESCO would give substantial financial support to the I.C.S.U., while the scientific control remained still in the hands of the unions.

They could help UNESCO by advice on such matters as scientific personnel in the international sphere. Then there were questions of international scientific policy. It would have been better if the industrial development of atomic energy had come into the international sphere through such a body as UNESCO, instead of the way in which it did. If UNESCO had existed, the problems of atomic energy could have been referred to the Unions for Physics and Chemistry, and kept on an international basis from the start. If that could have happened, the unhappy development of atomic energy would not have taken place.

Later on in the discussion Professor Solly Zuckerman said that Professor M. L. E. Oliphant, who had been sitting next to him during the discussion, had remarked to him that he had not heard of the International Council of Scientific Unions; he himself had not heard of it until about six months ago.

By 1945, S.V.S. was but one of the activities fostered by the Science Department. *Monthly Science News* was being published in five languages, and reprinted in seven different countries. The world circulation was about 65,000 copies a month. Ten pamphlets on *Science in Britain* had appeared. A monthly compilation of thirty-six pages of science abstracts and reviews of scientific books was produced, with the title of *Science Comment*, for the information of foreign scientists, and those countries in which technical periodicals were in short supply.

The Department gave advice on the selection of scientific books and periodicals for foreign libraries. Advice was given on scientific films, with regard to scenarios, commentaries, and the recommendation of technical experts. The Department advised on the preparation of scientific exhibitions for circulation abroad, on subjects such as Rutherford and Research in Atomic Physics, and Parsons and the Steam Turbine. It assisted in making the arrangements for British scientists to lecture abroad.

Dr Joseph Needham's activities in China grew to such a scale that he acquired the assistance of Dr Dorothy Needham, Professor W. Band, Dr L. E. R. Picken, Dr A. G. Sanders, and numerous supporting staff. The Chinese sought his advice on industrial and military, as well as cultural matters, so the War Cabinet's North American Supplies Scientific Sub-committee recommended that an office in the Ministry of Production should deal with such matters as lay outside the British Council's sphere of activities. This required liaison between the Science Department and the Ministry of Supply, in addition to dealing with the manifold cultural matters arising from Dr Needham's activities.

In 1942 the British Council had been approached by the head of the Book Department of the Academy of Sciences in Moscow, with the request that exchange relations with various British scientific and learned societies should be resumed. This was carried out through the Science Department. A secretariat was provided for the Anglo-Soviet Scientific Collaboration Committee, and hundreds of letters and communications of British scientists were forwarded to Soviet colleagues. A survey of Soviet scientific periodicals in British libraries showed their scarcity, and arrangements were made to centralize Soviet publications arriving through the British Council at the Science Library in South Kensington. Panels of technical translators were formed, and special glossaries compiled. More than two thousand scientific reprints were collected from twenty-one libraries for presentation to the Science Bureau of VOKS in Moscow, which operated under the chairmanship of Joffe.

The growing activities of the Science Department aroused increasing interest, but they also caused concern. Their scope is illustrated by the chart of organization, made in 1945. (see appendix) Sir Henry Dale had an ambivalent attitude, as President of the Royal Society and Chairman of the Science Advisory Committee to the

British Council. The Royal Society had long been entrusted by the Government with the conduct of certain relations with foreign science, and spent funds provided by the Government for these objects. It had taken a leading part in the creation and conduct of the International Council of Scientific Unions. He was naturally concerned that these Royal Society activities should not be trespassed upon by the new and growing Science Department of the British Council, which in 1946 had annual estimates which exceeded those of the Royal Society for that year.

As Chairman of the British Council's Science Committee, Sir Henry appreciated what had been done. He wrote in a foreword to an account of the Science Department's activities, published in 1945, that he wished 'the Director and his distinguished associates, and all their staff of assistants, a continued and developing success, in creating and maintaining a widely based international understanding, which, in science perhaps more than in any other department of human activity, is now so vital to the future of civilization'.

At the same time, Sir Henry told me personally that he was disturbed by the possibility of an encroachment by the Science Department on what he regarded as the domain of the Royal Society, and that he intended to see that this problem was resolved before his term as President of the Royal Society came to an end; this was due in November 1945.

He told me that he intended to leave the Royal Society with its position enhanced and not diminished. He appeared to believe that on all matters pertaining to science the authority of the Royal Society should be paramount. He considered that, in the last analysis, the Royal Society should have the final word on any important scientific matter with which the Science Department was concerned. He seemed to conceive the Science Committee of the British Council with himself as Chairman as an executive rather than an advisory body.

The fundamental difficulty arose from the problem of the relations between a Government department and a private body. There was a widespread but not often publicly expressed belief that the Royal Society should have a *de facto* if not *de jure* control of Government expenditures on science. It belonged to the wide area of British unexpressed but understood gentlemanly agreements as to how things should actually be managed.

Such views were not often published in print, but this did happen with reference to the Science Department, in December 1944. I had reviewed the interesting posthumous book by Sir

Henry Lyons on *The Royal Society* in the *New Statesman*. Lyons had been Treasurer of the Society, and was intimately acquainted with its archives. He described the various crises of policy in its history, and the effects of changes in its character.

When it was founded in the seventeenth century, it was very much a Royal Society, in which the King, leading statesmen and scholars were personally interested. In the middle of the nineteenth century the system of election of fellows was modified. It had the effect of converting the Society into one of middle-class professional scientists. The chief connection with statesmen was virtually severed. By the beginning of the twentieth century the sustaining of standards of quality in assessing scientific achievement had become the Society's chief, and very important, function.

The administration and finance required for such a purpose were small and grew very slowly. Consequently when the large scientific developments of the second quarter of the nineteenth century took place the Royal Society had neither the administrative tradition, staff or finance to cope with them directly. Yet it felt, as the repository of the highest scientific ability, that it ought to have the last word in any important scientific matter.

In my review I asked: 'What relation is the Society to have to those new and immense scientific activities, many of them conducted and financed by the Government? Is it to have a directive function? Is this possible without administrative and financial responsibility? Further, can such directive power be exerted without practice in politics and social affairs? Hasn't the policy of the last hundred years unfitted the Society for the role of statesmanship?'

A fortnight later the *New Statesman* published a long letter from Professor Andrade, saying that my review showed no signs that I had read the book. He contradicted my interpretation of the facts of the history of the Royal Society, but went on to say that 'If Mr Crowther will be instrumental in procuring for the Royal Society an income of, say, £10,000 a year for general purposes, unencumbered by tendentious political conditions, I for one will write a eulogy of which we shall have nothing to complain . . .'

While in the midst of these activities, my contributions to the *Manchester Guardian* did not entirely cease. I reported the conference of the Association of Scientific Workers on the planning of science. It was held under the chairmanship of Sir Robert Watson-Watt. Sir Stafford Cripps outlined the relations between science and the Government, and the improvisatory nature of the existing organization. The organization at the beginning of the war was inadequate, and what had been built up during the war had done as well as any plan could which was not old and mature.

Air Chief-Marshal Sir Philip Joubert explained how the Battle of Britain had been won with the assistance of radio-location, and said that the scientists had indeed an honourable place among the few to whom the many owed so much. Professor Blackett reviewed the possibilities of rehabilitating science after the war. The vast quantities of war equipment could be used for stocking university laboratories at home and in the devastated countries. He pointed out that the planning of science was inevitable, and that it was the business of scientists themselves to see that it was directed to the right ends.

A couple of years later, Blackett helped his young colleague A. C. B. Lovell to do exactly this. Scientific war equipment was collected for cosmic-ray research, and dumped on land in the University of Manchester's botanical gardens at Jodrell Bank. The world-famous radioastronomical observatory grew from this start.

Some days after the conference on the planning of science I wrote on the second annual luncheon of the Parliamentary and Scientific Committee, at which Lord Samuel presided. Lord Cherwell (formerly Professor Lindemann), Sir John Anderson and Sir Stafford Cripps were near him at the high table. Samuel looked down the table and smiled coyly at Lord Cherwell. He said that all our science today was 'Unter den Lindemann'; Cherwell stared icily at the ceiling, while Samuel went on, well pleased with his joke.

When Herbert Morrison became Lord President of the Council, and responsible for the scientific research councils, he was the chief guest at one of these luncheons. In his address he spoke of what he called 'isotopees'; it did not sound too good a qualification for dealing with such things as atomic bombs.

At the end of 1943 Sir Henry Dale in his anniversary address to the Royal Society said that Britain certainly required a new and adequate centre for science, capable of progressive adjustment to changing needs. He hoped that it would be regarded as a necessary feature of the forthcoming plans for national reconstruction.

In 1968 there was still no such centre in London.

I also wrote on the Royal Society's reception for Allied Scientists in August 1944. It was one of the happiest receptions. Hadamard, the great French mathematician, signed the book on being admitted as a Foreign Member; he was the senior member of the French scientific mission led by Rapkine. Among others present was the

Social Needs of Scientists

Norwegian physicist Gunnar Randers and his wife. Randers had spent the early part of the war years in the United States. Shortly after he came to Britain, we met in the Society for Visiting Scientists. I asked him for his comparative impression of the United States and Britain, as a detached Norwegian. He pondered and then said: 'The Americans think they are *it*, and say so; the British *know* they are *it*, and do *not* say so; as for we Norwegians, we know that we are *not it*.'

At the end of 1944 I heard the news of the death of W. P. Crozier, the Editor of the *Manchester Guardian*. He had given me sustained support and understanding. Crozier had never expected to become editor, but was called to the post through the sudden and unexpected death of his predecessor, Edward Scott, in a boating accident. Crozier was a man who had taken a first in 'Greats' at Oxford, but never paraded the fact. He was firm but cautious in his judgment, and in his quiet way, a great editor. He was not a scientist, but he believed science was very important, and he was prepared to listen on the subject. I gained his confidence, and he gained mine; consequently, there was never any difficulty in mutual agreement on difficult points.

His successor A. P. Wadsworth, whom I had first met in Tawney's company, was an entirely different kind of man. He reflected the outlook of the Lancashire textile workers, one of the most conservative sections of the British working class, whose history he had meticulously written.

As the Science Department's activities grew, I received numerous requests to write articles and books, deliver speeches, advise on films, and even distribute prizes on school speech-days. I had to turn most of these invitations down, owing to lack of time.

The most important of these various addresses was the Trueman Wood Lecture to the Royal Society of Arts, in May 1943. It was on Science in Soviet Russia. Sir Stafford Cripps, who had been British Ambassador in the U.S.S.R., and was then Minister of Aircraft Production, took the chair. I explained that the development of science in the U.S.S.R. had been inspired by philosophy, practical value in raising the standard of living in a socialist state, and military value, in that order. I outlined the dialectical materialist philosophy, as expounded by Stalin, and described how the new scientific organization had actually been created under the impulsion of these motives. After this, I gave examples of concrete scientific achievements which had flowed from this development.

The audience at this lecture included numerous diplomats and officials who were not accustomed to listening to such a discourse.

Sir Stafford, in closing the meeting, observed that he did 'not think that we are all dialectical materialists here, but it is quite clear that there is a power and force behind scientific research and development in the Soviet Union which has been demonstrated by the results of the last two years and which also augurs a very great development in the postwar period'.

The lecture was reported in the Soviet press. It received thirty lines, which was quite a lot for a British lecture in the conditions of 1943. Questions about it were raised in the House of Lords. Lord McGowan, the Chairman of Imperial Chemical Industries, asked the Government spokesman, Lord Cherwell, whether he had read it, and what was to be done in the light of the information in it. In his reply, Cherwell showed that he had indeed read it, and added a few additional pieces of information about the progress of scientific research in the U.S.S.R., especially in his own field of low-temperature physics.

I became more and more concerned with the future of the Science Department of the British Council and the various developments that had arisen from it.

I hoped that it might be a permanent focus for thought and action on the social relations of science, but this did not appear to be possible unless the problem of the relation of Government and private scientific organizations was worked out.

After the Labour Government was elected in the summer of 1945, I thought that it might be possible to get the problem resolved in the case of the Science Department of the British Council. A mutual friend gave me an introduction to Hector McNeil, the Parliamentary Secretary at the Foreign Office, under whom the British Council then came. I saw him and explained my views and the reasons for them, and expressed the hope that something would be done. 'Give us six months,' he said. I waited for six months. What I had said was not ignored, but it was not understood. No effective action followed, though an excellent opportunity occurred, as the chairmanship of the British Council itself became vacant.

When it became clear, early in 1946, that no effective action would be taken, I decided to resign from the British Council. My decision was reinforced by incidental actions in the scientific field by Herbert Morrison as Lord President of the Council. It seemed that the new Labour Ministers did not include much socialist principle in their conduct of departments. They acted as if it was sufficient to give orders to the old administrative organization, without changing it.

My connection with the Oxford University Press had continued through the war. In March 1944 I was informed that they were thinking of bringing out a book for the serious general reader on advances in science during the last fifteen years. They had reason to believe such a book would be very useful in post-war Europe. They asked me whether I would be prepared to get a team together to write it.

I soon concluded that a collective work, though attractive, was not practicable under war conditions. Appropriate contributors were too deeply involved in the war effort. Besides this, the core of the book would be the scientific advances made during the war, most of which were secret. The most important problem would be the arrangement of an official procedure for regulating the collection and release for publication of hitherto secret material.

I wrote to Sir Henry Dale as Chairman of the Scientific Advisory Committee to the War Cabinet, explaining the situation, and asking whether the advice of the Committee might be sought. Sir Henry replied that there was a favourable disposition towards the proposal for such a book, and the procedure for preparing it would require further consideration.

This took a long time. Apparently the Prime Minister liked to know about anything to do with history. But the Yalta Conference was in sight, and such a detail had to wait until it was over. Then, after the Conference, it had to wait again, owing to the press of post-conference major business.

Ultimately, the War Cabinet approved the proposal that a book for the general reader on the British scientific war effort should be written. I was to be one of the authors, in collaboration with another. But the administrative arrangements for its preparation and publication were to be carried out by the Department of Scientific and Industrial Research.

C. P. Snow was asked whether he would collaborate, but he understandably did not like the idea of joint authorship. Professor R. Whiddington was presently appointed joint author. He was a senior physicist who was a scientific adviser in the Ministry of Supply. He was familiar with a wide range of war research, especially in electronics. He described his role as 'an opener of doors', and he contributed particularly in the preparation of the material on radar.

The Department of Scientific and Industrial Research under Sir Edward Appleton carried out the administrative arrangements with great consideration. I watched these as a spectator. There was an earnest discussion among the officials concerned whether our office should have a carpet. They consulted higher authorities, and then furnished our room with large desks and a handsome green carpet.

We invited various scientists who had made important contributions in the war to call on us, and explain what they thought were the points that ought to be made in the kind of book we were to write. One of these was E. C. Bullard, who had done brilliant work in combating magnetic mines, and equally brilliant work of another sort in breaking up fossilized routines in certain service research laboratories. He came into our room, paused on the carpet, stared at our desks, and commented: 'Ah! *Admiral's desks, I see.*'

Whiddington and I visited many laboratories and institutions, and called on leading scientists, collecting material and opinions. These were digested and expounded in a suitable style. The book was published with the title *Science at War*. The actual title was suggested by A. V. Hill, at a meeting of a committee which gave advice on the book.

One of the helps I most appreciated was from Government draughtsmen, in preparing the illustrations. They found the task interesting, and entered into it with zest, inventing ingenious ways of solving problems in perspective.

I continued to work on the book after I left the British Council early in 1946. It was completed, and issued with the imprint 1947, though it actually appeared early in 1948.

There was not much opportunity to write for the *Manchester Guardian* during 1945, with the addition of the preparation of the book to the British Council activities, but I continued to contribute occasionally. One note was on the election of Kathleen Lonsdale and Marjory Stephenson to be Fellows of the Royal Society. They were the first women to be elected in the Society's three centuries of existence. It was a historic moment in the development of modern scientific society, for it marked the advance of women as scientific workers. Women form half of the human race. The recognition of their scientific ability emphasized that a vast reservoir of scientific talent in the human population remained almost unused.

An obituary and leader on Sir Ambrose Fleming were published in April. The inventor of the thermionic valve had died at the immense age of ninety-five years. Fleming had in his earlier days been employed by Edison, and then by Marconi, under contracts in which inventions made during his work for them belonged to his employers. He subsequently felt that he had not received adequate financial rewards from his inventions. He acquired an asperity of manner, which was not improved by deafness.

He was for long professor of electrical engineering at University College, London. In the 1920s I had interviewed him there, to ask his advice on whether any of his colleagues might be prepared to write a book for the Oxford University Press. Fleming had rigged up a microphone, to which earphones were attached, to assist his hearing. He sat behind his desk, and I in front of him. The microphone was placed on a bench to my left. I started by explaining the object of my visit, and his first comment was: 'How much do you propose to pay me for this advice?'

I had not thought of this, and hesitated. Fleming then released his grievance. 'Yes,' he said, 'we scientists are always expected to do something for nothing. Now, if I were a lawyer, there would be a proper fee.' He stared at me angrily, and I began to stutter some reply. He roared: 'Don't speak to me! Speak to the microphone!' I turned and spoke to the instrument, so that I was no longer looking at him. I felt more and more like a performing animal. Presently, he said, crisply, 'Good morning.' I jumped up, and ran out of his laboratory and the department.

About twenty years later, I received a charming letter from him. He had just read the Pelican edition of my book on *British Scientists of the Nineteenth Century*, which contained lives of Maxwell and Kelvin. Fleming had been one of the rare students of Clerk Maxwell, and had known Kelvin well. He was very appreciative of what I had said about them, particularly with regard to my critical comments on Kelvin's character which he said he could confirm from his own experience. I do not suppose that he had by that time the slightest recollection of the inexperienced young man who had called on him long before, or in any way associated him with the book.

In June I reported the extraordinary affair of the refusal of exit visas to eight British scientists to attend the 220th anniversary celebrations of the U.S.S.R. Academy of Sciences, to which I have already referred. I remarked that it seemed incredible that the scientists could not have been told at an earlier stage that they could not go. It was particularly unfortunate that such an incident should have occurred on the first post-war occasion on which Soviet scientists had made a major gesture towards the resumption of normal scientific relations, which had been interrupted by the general state of affairs in Europe during the last ten years. The incident clearly raised serious questions concerning the liberty of the subject, and a thorough investigation of it from every angle seemed essential.

About six weeks later, the atomic bombs were dropped on Japan. This explained but did not excuse the cancellation of the visas. The atomic bombing was a stupendous event in the history of mankind, and it also turned out to be of profound importance to me as a scientific writer.

Within forty minutes of the news of President Truman's announcement of the dropping of the bombs, the *Manchester Guardian* rang me in London from Manchester, and asked for articles on the event. I immediately wrote two, with the titles 'How the Atomic Bomb was Born', and 'Neutron's Gigantic Release of Atomic Power'; they were published on 7 and 8 August respectively.

Wadsworth wrote to me on 7 August, concluding his letter: 'Thank you very much for your excellent articles last night and tonight.'

I followed these with a third article on 'The Social Implications of the Atomic Bomb'. I then left London with Whiddington on one of the visits to institutions, collecting material for the book on war science. It was the anti-submarine research establishment on the Clyde. We stayed in Glasgow, and on the morning of 10 August took the local train for Greenock. I bought a copy of the *Glasgow Herald*, and casually opened the pages after we had settled in our compartment. I noticed there was an article on 'Science and Future Prospects for Society', which seemed slightly familiar, and on looking at it again saw that it was the article I had written for the *Guardian*. There was nothing surprising in this, for *Guardian* articles were often published simultaneously in the *Glasgow Herald*.

When I returned to London some days later, I looked through the back issues of the *Guardian*, but could not find it, so I wrote to Wadsworth, inquiring what had happened. He replied that he did not agree with it. I then suggested that he should publish it under my name, but he replied that they did not wish to be associated with the opinions in the article, and would not publish it.

I had said that the importance of science as a social and political factor had been increasing rapidly, and the invention of the atomic bomb had 'suddenly thrust science right into the first place for social and political consideration'. How were we to manage 'this method and instrument that might suddenly terminate the long story of human existence'?

For centuries science had seemed to be 'little more than a

Social Needs of Scientists

fascinating adornment on the fringe of human interests, appealing only to the passion of intellectual curiosity'. This conception of science as primarily an activity of isolated individuals had been gradually modified, as science and the number of scientists had grown. Science had now become a large co-operative activity, depending on many colleagues and large resources.

The co-operation in the invention of the atomic bomb was a stupendous culmination of the tendency. It was accomplished by a corps of physicists of several nations backed by an army of 120,000 men and a budget equal to half of the normal annual budget of the British people for the whole of their government. All this was concentrated on one problem with a result of supreme importance.

Several conclusions were to be drawn. First, the achievement was accomplished only through a tremendous effort of organization and planning. There was no other way of securing the momentous result, 'perhaps the most important that has yet occurred in history'. Any attempt to have done it by individual or private action would have ended in chaos and failure. 'Without planning we should never have had the atomic bomb, and if it had never existed we should never have been free to use it. Thus, under contemporary scientific conditions, planning has become the condition of freedom. Without planning we cannot discover the new knowledge and power that we desire. We must recognize that purely individual action will get us nowhere in the realm of major research.'

A second point was that the bomb was discovered by state and not by private enterprise. Indeed, it was a super-state enterprise, for it depended on the co-operation of several governments. A third point was that the mechanism of the atomic bomb was kept secret. The knowledge was being controlled in the interests of mankind. President Truman was clearly aware that this was a departure from the tradition of comparatively free publication of scientific discovery. He said that the knowledge would not be divulged until methods of protecting the United States and the rest of the world from sudden destruction had been examined.

'Thus we see that the doctrine of completely or anarchistically free scientific research and publication has broken down at its first major test. And this was really to be expected. It would be anti-social to present the secret of the atomic bomb to lunatic individuals who would not scruple to apply it to evil ends.

'The days when we could publish new knowledge on principle and without reflection are past. We were able to do it formerly because science was not yet of absolutely vital importance; now that it is, we cannot afford any risks.'

How was the new scientific knowledge to be kept from becoming the instrument of lunatics and tyrants? There was no simple solution. 'It is evident that in the future freedom will depend on a complicated system of planning, organization, and control. We shall share a limited freedom, but it is one which will be much wider and more potent than anything mankind has known before.'

How was the new system of social control to be created? One feature would be the participation of science in government. In the future, statesmen would have to possess scientific understanding as well as good will.

If we were not to be 'altogether abject', in Bacon's phrase, 'we must go forward and solve our problems of planning and control. There is no hope for us in longing to go back to the old days of individual irresponsibility, with the craven hope that somehow everything will work out all right if we just work and publish as we wish without any organized action.'

The article was the beginning of the end of my service as Scientific Correspondent of the *Manchester Guardian*. However, this did not happen quickly. My articles gradually became less welcome. No doubt Wadsworth's attitude to science in general, his temperament as an editor, and the increasing pressure on space also contributed, but I was most conscious of his disagreement with my views.

In spite of this the number of contributions from me during 1946 was greater than in any of the previous three years. The release of war scientific information provided much first-rate material for copy, especially on radiolocation or radar, with its exciting and varied achievements and possibilities. Having left the British Council, I had more time for writing. But the proportion of copy published to the quantity suitable for publication decreased, and the delays in publishing increased.

After I became involved in the peace movement in 1948 I again had less time for writing. No copy was asked for, and I sent less and less. My last article as *Scientific Correspondent* appeared in 1949. My appointment was never terminated, but just lapsed.

CHAPTER TWENTY-TWO

1944-46
REVIVING SCIENTIFIC RELATIONS

As the Nazis retreated it became possible to revive direct contact between British scientists and those in countries which had recently been occupied. At an early stage, in 1944, Joliot was brought to London by the Americans, and put up in a hotel near Hyde Park. I called to see him, and was shown in by a polite and intelligent American in a captain's uniform.

Joliot was still wearing the clothes he wore during the occupation. I noticed that his shoes were worn out, and disintegrating. I was invited to attend the dramatic meeting between him and his colleagues in exile, and went to the hotel where it was to take place. There were long, enthusiastic and emotional embraces.

The revival of the Anglo-French Society of Sciences had now become possible. I applied in my personal capacity to the British Council for support for it, especially in the form of travelling expenses for participants in its activities, as a contribution to the advancement of Anglo-French cultural collaboration. The proposal was referred to Sir Henry Dale for his advice, who deferred his opinion.

Meanwhile, the first conference of the Society was fixed for 20 January 1945, on the subject of the Solid State, to be held in the Society for Visiting Scientists in London. There were to be morning and afternoon sessions. Auger, who was the Secretary of the French side of the Society, and had returned to Paris, had had difficulties in arranging transport for the French attendants at the conference, who included the Joliot-Curie's, Wyart, Mathieu, Laval and himself. When the American air transport heard that they wanted to come to England, they were not interested, but the British Embassy arranged for a special plane for the party. The arrival of Irène Curie gave particular joy, for there had been little hope that she would be well enough to travel. It was very difficult to get rooms in London then, but the manager of Brown's Hotel, who was French and had connections among French intellectual personalities, said he would be only too pleased to make a special effort for the Curie family, and I secured a suite there for them. I was very much concerned about Irène Curie's health, for she had a disturbing cough.

My anxieties about the arrangements became patent when the Curies and I started the hotel lift, which stuck half-way, and then nearly shot out at the top. They observed this with quiet amusement, but subsequently told me I had looked after them like a parent.

The conference on the Solid State opened in the morning of 20 January, with Mott in the chair. He spoke on the fracture of solids; G. I. Taylor on the measurement of the mechanical properties of metals at very high rates of loading; Andrade on evidence for Griffith cracks; and Orowan on the theoretical yield strengths of crystals. Max Born, Bernal and others took part in the discussion.

Then the participants went downstairs to have lunch in the S.V.S. restaurant. Joliot took the chair at the afternoon session. He referred to the founding of the Society in 1940, and the hope that it would foster Anglo-French scientific relations in the cultural and other fields. He suggested that there should be further conferences, on Cosmic Rays and other topics.

The informal and friendly atmosphere was obvious. Some of the participants, who had been very active in resistance and other non-research activities, were naturally out of practice, and the conferences helped in getting intellectual muscles loose again, and establishing Franco-British scientific friendship on a close personal basis.

On the following day an executive committee meeting was held at S.V.S., at which Dirac, the Joliots, Perrin, Auger, Rapkine, Wyart, Bernal and Blackett were among those present. At this meeting the outline of a constitution for the Society was worked out. It was to become an elective body, with not more than a hundred members, fifty from each country. The first general election was to be held in January 1946. Meanwhile, a number of scientists were to be invited to join. These included Mott, J. B. S. Haldane, Astbury, Max Born, Sir Edward Appleton, Lennard-Jones, Harington, Heilbron, Munro Fox and Sir Henry Dale. All accepted, with the exception of Sir Henry Dale.

The day after this meeting, on 22 January, the Joliots and Sir Archibald Clark Kerr, the British Ambassador at Moscow, dined with the Tots and Quots at Brown's Hotel. Blackett took the chair.

Clark Kerr mentioned that there had been an improvement in Anglo-Soviet relations, particularly with the Kremlin. Churchill and Stalin seemed to get on together. The question of the future of German science then arose. It was suggested that the policy for it should be formulated by Britain, the U.S.A. and the U.S.S.R.

Reviving Scientific Relations

Irène Curie interposed that France could not be left out. The only difference between Britain and France was that made by the Channel, and this could not be a sufficient reason for the exclusion of the French.

Joliot spoke of the future of science in France, and the attitude of French scientists to Germany. He thought that severe control would have to be kept over Germany for at least ten years. He said emphatically that scientists should hold commanding and executive posts in the reconstruction of France. Ministers should have scientific advisers, to control, direct and censor. The question of the German scientific-instrument industries was raised by Bernal, and I spoke on the activities of the Science Commission of the Conference of Allied Ministers.

The group of French scientists were received at the House of Commons by the Anglo-French Parliamentary Committee. The host was Sir Robert Bird, M.P., the head of the food manufacturing firm in Birmingham. Sir Robert spoke in French, and dwelt on the relations between Priestley and Lavoisier, and now they had pleasure in receiving the liberated French scientists.

Joliot replied, and referred to the initiation of the Anglo-French Society of Sciences in 1940, and my part in it, and their intention to develop its work. Joliot told me afterwards that he had spoken to M. Massigli, the French Ambassador, on the importance of the Anglo-French Society of Sciences, and of the need for stimulating Anglo-French relations. A few days later Massigli, in an address at Liverpool University, said that he wanted Anglo-French relations to expand in all branches of culture, particularly in the scientific field.

In his address at the House of Commons Joliot also insisted on the importance of scientists in affairs. They had clearer minds, and used a more rational method. Then he excused himself for pushing his own profession too much; but the scientific method must be introduced into all aspects of a nation's life and affairs, including many in which one would not have expected it according to traditional conceptions. He implied that government had been too much in the hands of lawyers, and was too much addicted to verbalism.

There were about a hundred guests at this gathering. They included Sir Henry Dale, who thanked me several times for being invited.

I paid the bill for the stay of the Joliot-Curies at Brown's Hotel out of my own pocket, and I felt that the finances of the Anglo-French Society of Sciences ought now to be put on a proper basis.

The British Council administration was inclined to favour the provision of financial support for it, but the inquiry they had sent to Sir Henry Dale in November 1944 had still not been answered. I now asked the Council, on behalf of the Society, for a reply.

Sir Henry expressed the opinion that the effort had been a purely private one on the part of myself and certain distinguished friends of mine in this country, and certain distinguished French scientists with whom we happened to be on terms of personal acquaintance and sympathy. After this, the Council turned down our request.

This appeared to me an unconvincing reason for refusing support. All new societies are initiated by a number of like-minded private persons coming together, and deciding that something should be done; the Royal Society itself is the classic example. The question was not who made the proposal, but whether it was a good one, and in line with the Council's aims.

After this I doubted whether the Anglo-French Society of Sciences could have a future. However, we continued, in the hope that some unforeseen source of support might reveal itself.

Rapkine had been outstandingly active in preparing for the return of French scientists in exile to France, and in securing aid for reviving French science, from such bodies as the Rockefeller Foundation. One of his plans was for the invitation of individual British scientists to lecture in Paris, to speak in broad terms of what had been happening in their subjects during the war years. He drew up a list of twenty-six lecturers, with Dirac at the head. The French were to pay them a fee and put them up in Paris, while the British Council was to pay the cost of transport between London and Paris. It was in line with the activity that the Anglo-French Society of Sciences hoped to foster systematically.

Dirac started the series with a lecture in the Palais de la Découverte on 6 December 1945. More than two thousand people tried to get into the small lecture hall. They thought that he was going to reveal the secrets of the atomic bomb. He delivered a brilliant lecture on the difficulties and problems of wave-mechanics, in fluent French. When some of the audience in the lecture-room, which held about two hundred, saw that he was not going to speak on the atomic bomb, they tried to go out; but they could not, because of the large crowd outside, which was listening to loudspeakers. It had been found necessary to call the police to form a cordon.

I was invited to lecture in the Palais de la Découverte on

13 December. I left for Paris on 10 December. It was my first visit to the Continent since 1940, when I had returned with the plan for the Anglo-French Society of Sciences, and I was intensely interested in what I was to see and hear. I crossed the Channel by boat, and on it happened to meet A. O. Rankine, who told me that Sir Henry Dale had spoken to him about Polanyi's lecture at the British Association Conference on 'Scientific Research and Industrial Planning' in terms of unqualified praise.

I was put up in the Hotel Scribe, and had most of my meals at Maxim's, which had been taken over and labelled The British Empire Club. Among the first to lunch with me there were Pierre and Suzanne Auger. Auger was to take the chair at my lecture, and wanted to know the substance of it. I was driven about Paris in the British Council's car, whose chauffeur casually remarked that he was an Englishman who had been interned in France during the war, and had formerly been chauffeur to James Hazen Hyde, the American who had provided the funds for my lectures at Harvard in 1937. I was taken round to call on my French friends, and then to the Palais de la Découverte, where the Augers were waiting for me.

My lecture was to be delivered in English. A French translation had been prepared by A. Béra, the translator of *British Scientists of the Nineteenth Century*, for distribution to anyone who wanted it.

When I entered the hall, there was a little burst of applause, apparently led by Hadamard. He was the senior of the group of French scientists whom Rapkine had brought to England. On another occasion, he presented me with a copy of his book on the psychology of mathematical discovery, with an inscription that my services to Anglo-French friendship were present in all their minds.

Auger introduced me to the audience. He said I was very definitely English: not Scottish or Welsh; I was therefore very 'original'.

My lecture was on 'The Social Relations of Science'. The first three-quarters was given to the exposition of my views on science as a product of the social epoch, with illustrations drawn from classical times, the Middle Ages, the Renaissance, the Age of Exploration, the rise of modern science in the seventeenth century after the decline of feudalism, and the period of the Industrial Revolution. My aim was to show how the science in each epoch reflected fundamental characteristics as a whole. Consequently, the deepest problems in dealing with science could not be understood apart from this general historical background,

I referred to the condition of science in France as I had seen it in 1937, and had written in the following year that 'supporters of democracy are deeply interested in the strength of France. Will the French people be able to withstand attacks by Fascist aggressors?' I had seen the brilliant revival of French scientific research under the Blum Government. But it had appeared to me that the movement was primarily based on cultural idealism; science in France was being pursued too much as an abstract ideal, and not as a social necessity. It had concluded that the Principles of 1789 were an insufficient guide to scientific development in 1937. I supposed that these were an abstract statement arising from the social situation at the end of the eighteenth century.

'Why not compose a new set of Principles of our own, for the year 1945—the first of the atomic era, for our guidance?' They would not be couched in abstract terms, but in socio-historical, developmental terms. They would help us to see that the safe development of civilization depended on a permeation of the scientific outlook through all aspects of life, and especially in agriculture, industry, education and the press.

The pathological social phenomena that we had recently witnessed were the symptoms, not the disease. The horrible exacerbations of racialism and nationalism were not the causes of the present troubles, but arose from failure to solve underlying problems of social organization, and adapt it to new scientific possibilities.

I argued that the current concept of freedom in scientific research was connected with the character of mechanistic science, developed from Galilean and Newtonian concepts drawn from the features of the familiar material world. The new sciences of quantum mechanics and nuclear physics were based on the unfamiliar world of the very small. Limitation was the fundamental idea in quantum mechanics, but, as Niels Bohr had pointed out, this limitation of possibilities was the very condition for the advance of physical knowledge. His introduction of the conception of the atom with limitations on its state had been the key to enormous advances in spectroscopy and other branches of physics. 'Perhaps some parallel kind of stratification of social organization and knowledge, forced on us by the release of atomic energy, would, if we faced it and worked out its implication, lead to a tremendous new flowering of social advance and development.'

The science of 1600–1900 was relatively harmless in its immediate effects, and simple ways of disseminating and handling it were socially sufficient. The new element of catastrophic knowledge might require a more complicated and differential social structure for its control. It might be necessary to organize a special class of persons to promote and manage the new science on behalf of the community, and it might be that the spread of atomic knowledge through an unselected population would be a social parallel to the reversion to the primitive in the biological field.

I concluded my lecture with the observation that our problem was to adapt social forms to present scientific and technical possibilities. Those who succeeded first would lead mankind.

After the lecture, the British Council gave a dinner in the Maison des Alliés. I was placed between Irène and Frederic Joliot-Curie; others present were Leveillé, Bauer, Biquard, the Ephrussis, the Wurmsers and the Béras. It was like a gathering at the Society for Visiting Scientists in London. I had a long and spirited discussion with Irène Curie on my lecture. She argued that limitation of distribution of scientific knowledge was impossible, and if it were, then she would be no longer interested in living in a society in which it pertained.

Langevin invited me to luncheon at *Au Cochon de Lait*. When I arrived there, rather early, I recognized it as the restaurant near the Odéon Theatre at the bottom of the Luxemburg Gardens where, in April 1940, just before the invasion of France, I had lunched with the Joliot-Curies, Francis Perrin, Auger and Thon. Langevin came in almost immediately. Riley arrived a little later, and finally, Langevin's son, André.

Langevin told me how his daughter had survived in Auschwitz. This was due to an S.S. officer who was a biologist, and wished to evade being sent to the Eastern front. He persuaded the German authorities that it might be worth trying to acclimatize Russian rubber plants to Poland. They allowed him to make a laboratory and garden at Auschwitz for this purpose. He picked out a number of biologists from prisoners, whose usual length of life before entering the gas chambers was two weeks, to help him in the work.

One of these was a Jewish woman biologist of some standing. When the list of prisoners was scanned, the name of Langevin was noticed, and she said that Hélène Langevin was a biologist, so she was picked out. Langevin's daughter was in the camp for more than two years but survived, the rubber plant acclimatizers having slightly better conditions.

André Langevin was very critical of my lecture, and my views on a limited class of atomic energy experts.

The luncheon was superb, the best I ever had. It consisted of oysters, fish, chicken, cheese, beautiful apples and pears, and perfect wines. I have a vivid memory of Langevin selecting the pears. He examined them most carefully by eye and hand, and then offered me one which he deemed satisfactory. It had an aroma and flavour unparalleled in my experience.

Langevin remembered the dinner Laugier had given me in 1940. He had a deep feeling for people as well as science; his affection for his son was evident and demonstrated. He told me with reference to his own imprisonment during the occupation that it was necessary to spend some time in prison once in one's life, in order to know oneself. Considering his experiences during the occupation, and his heart-trouble, he still seemed remarkably well. He seemed to me to have effected a better combination of abstract principles with political wisdom than most of the French scientists.

It was the last occasion on which I saw Langevin. His first important research had been done at Cambridge at the end of the nineteenth century, where he had come at the age of twenty-five, at the same time as Rutherford. He had a rich spring of fertile ideas, but his wide interests prevented him from following all of them himself. The method of producing very low temperatures by demagnetization was foreshadowed by him, and de Broglie conceived the wave-theory of matter under the influence of his ideas.

His influence over persons arose from his selflessness; his attention was always directed outside himself. He was a wonderful listener and friend; young men of talent always felt that he was genuinely devoted to them.

A great memorial demonstration for Langevin and Jean Perrin was held in Paris, which I attended. A memorial meeting was arranged in London, over which Sir Henry Tizard presided, and at which Pierre Biquard was the chief speaker.

At the Society for Visiting Scientists the French scientists who had come to the London meeting were received by Mr A. V. Alexander, the Minister of Defence, who said that when he was First Lord of the Admiralty he had had every opportunity for appreciating Langevin's contribution to the development of underwater detection devices, which could be applied not only to anti-submarine detection, but also to the location of fish-shoals, and many other things.

Joliot-Curie had insisted on coming to London, though unwell, to speak in his memory. Langevin had not limited himself to his scientific activities; his universal spirit and precision of judgment enabled him to analyse social problems profoundly. Joliot remarked that the presence of the Minister of Defence showed that the English knew how to honour men of great talent born in other countries. He offered thanks for these marks of regard and affection.

These events led me to suggest that the World Federation of Scientific Workers should organize an international meeting in Paris on the tenth anniversary of Rutherford's death. I will say more about this meeting presently.

After the luncheon with Langevin I went to see Freymann, the Mexican manager of the publisher Hermann, who brought out the French translation of *British Scientists of the Nineteenth Century*. His shop was closed, but I got into it through a side-door from the British Institute, where I was due to speak informally in the evening. Freymann was delighted that I had found the way in. He told me how he had been visited by the Gestapo when he had three million francs from the British Intelligence concealed underneath his desk. His shop was searched for hours, and ultimately he was told he would be taken away and tortured; he was put into a van. He then told the S.S. officer in charge that he was making a very serious mistake. Why, asked the officer. Freymann said that in France there were five Mexicans, two of whom were invalids. That left three. If anything happened to them, then the thousands of Germans in Mexico would be interned, including the German agents who had fled there from the U.S. The officer paused, opened Freymann's shop again, and telephoned headquarters for instructions. Presently he came back, saluted, and told him he was free. 'But you will now work for us.' Freymann knew he had won, and refused, and the man went away.

I ran upstairs to the British Institute, and gave my talk on the social relations of science.

During this first post-war visit to France I met again my translator Marc André Béra. He was staying with his father-in-law, A. Bayet, who came in to see me for a few minutes. Bayet told me he disagreed with my views on secrecy, for he believed that the secrecy of the Pythagoreans and other societies had led to the loss of scientific knowledge, for example, early scientific knowledge of flying. For this reason, he was against my conception on the role of secrecy with regard to atomic energy.

Béra had been offered the directorship of the French Institute in Edinburgh, which he subsequently accepted. He told me of his experiences during the war. The part of the French Army in which he served surrendered in 1940 without fighting, owing, he thought, to treachery. He was interned. In June 1940 he made a speech in the internment camp, in which he forecast that England would win. It was received with general laughter, but as the years passed his prestige rose steadily.

From the camp near Lübeck he saw the bombing of Hamburg and Kiel. In the phosphorus-bomb raid on Hamburg, an enormous number of person were burnt, and the sun at Lübeck was obscured by smoke for three hours. Next day the lettuces in the camp garden were covered with dust. In the attack on Lübeck the bombs were released over the camp. The Germans believed they would not be hit, because there were Rothschilds in it.

He organized a library, in which English literature played a special part. Many learned English just to while away the time, and then became interested in it. Non-British prisoners in this way became influenced by British ideas. There were always radio sets in the camp, and they listened to the British radio.

When the British Army captured Lübeck, Béra became liaison officer. He suggested to the British that they should commandeer ten thousand German blankets and send them for the relief of the Belsen victims; he secured the signature of the British command for the order. Then he acquired the car of an S.S. officer who had been shot, and started off for home.

During my visit, the French Foreign Office gave a luncheon for me at the Maison des Alliés, which used to be the Rothschilds' palace.

When I returned to London at the end of December in 1945, I was already aware that Mr Attlee's Government, which had been greeted so enthusiastically in the previous summer, was showing little sign of altering the character of the administrative organization. In August of that year, scientists had rushed into the Society for Visiting Scientists, almost inarticulate with joy, as the overwhelming victory of the Labour Party was declared. There was great hope of creative action in the scientific and other fields. In fact, the whole nation held its breath for several months, expecting something to happen, but nothing very much did. The requisite ideological preparation had not been done.

While I now saw no prospect of doing creative work for science within the British Government organization, I still thought that useful work might be done in the Society for Visiting Scientists, and I determined to utilize it as much as possible for developing the social and international relations of science.

CHAPTER TWENTY-THREE

1947

INTERNATIONAL SCIENCE, SCIENCE WRITING, AND POLAND

The International Council of Scientific Unions, which consisted of a group of scientific unions in the various scientific disciplines, such as astronomy and chemistry, had been virtually suspended during the Second World War. This demonstrated that a body of such a kind, made up of unions which dealt primarily with technical matters, was limited in its capacity for dealing with international scientific policy. An international scientific organization of a broader character, designed to include the social and political as well as the technical concerns of scientists, was required.

Its germ was to be seen in bodies like the Association of Scientific Workers in Britain, and similar organizations in other countries. The idea arose simultaneously in several of these bodies.

In February 1946, the Association of Scientific Workers organized a conference on Science and the Welfare of Mankind. I reported its proceedings for the *Manchester Guardian*. Sir Robert Robinson, the President of the Royal Society, presided at the first session, and the opening speech was made by Herbert Morrison, Lord President of the Council. The Royal Society's Secretaries, A. V. Hill and A. C. G. Egerton spoke at other sessions, and Julian Huxley made his first speech on Science in UNESCO since he became Executive Secretary of its Preparatory Commission.

Other leading speakers at the conference included the physicists Blackett, Oliphant, Bernal and Watson-Watt; H. L. Richardson, the authority on agricultural chemistry and overseas development; J. M. Burgers, the Dutch applied mathematician and Chairman of the International Council of Scientific Unions' Committee on Science and its Social Relations; Dr T'U Chang-Wang, General Secretary of the Chinese Association of Scientific Workers; Dr James A. Simpson of the Chicago Atomic Scientists' Association; J. P. Mathieu of the French Association of Scientific Workers; P. Bonet Maury of the French Centre of National Research; Miss P. M. Cooke of the South African Association of Scientific Workers; Dr Dorothy Needham and Dr D. Riley, Science

Liaison Officers of the British Council in China and France respectively, dealt with science overseas; C. P. Snow, the Scientific Adviser to the Civil Service Commission; V. Stott, Chairman of the Institution of Professional Civil Servants; Dr Stephen Taylor, M.P.; the nutritionist F. Le Gros Clark; Arthur Horner, President of the South Wales Miners' Federation, and Professor B. Farrington spoke on domestic and cultural aspects of science.

The weight and range of these speakers was impressive. Blackett argued closely in favour of the veto mechanism of the Security Council, as a realistic reflection of political power. J. A. Simpson spoke in the manner which had made his recent evidence before the McMahon Committee of the U.S. Senate on the control of atomic energy so illuminating. We saw what the young American atomic scientists were like. He dealt with methods of inspection and detection of atomic-energy plants, and thought that they would be difficult to conceal. He thought, too, that there would always be a number of scientists brave enough to report illegal atomic activities to UNO. He suggested that an international atomic energy laboratory should be formed. It might attract the best atomic physicists in the world, and surpass the research in national laboratories, so reducing the offensive value of the latter. The achievements of C.E.R.N., the European Centre for Nuclear Research, have provided evidence for Simpson's view.

A paper by Joliot-Curie was read for him. He made a strong plea for the free exchange of information on atomic energy.

Dr T'U Chang-Wang, the Chinese meteorologist, made a particularly interesting speech. He described how he had flown across half the world to attend the conference, and he gave an eloquent account of how the future of China rested on the application of science and engineering to her problems. He spoke with a whole-minded enthusiasm and directness. Subsequently, I was in touch with T'U for a long time, and I learned much from the experience.

Other notable speeches included one by A. V. Hill, who proposed a new ethical code for science, similar to the Hippocratic; Farrington's argument that science is not ethically neutral, the pursuit of science being a factor in the creation of the social conscience; and Arthur Horner on breaking down the barriers between workers and scientists.

Points in speeches which were disturbing included Sir Robert Robinson's personal approval of the Government's decision not to set up a comprehensive Ministry of Science; this was not because science was not sufficiently important, but because it would be better for science to permeate all ministries. After Robinson's opening remarks, Herbert Morrison delivered a confused oration in which he said that the Government were determined now to organize so that no crumb of scientific knowledge which might be of benefit to the community was wasted, and other remarks of a similar kind. Following on the decision not to set up a Ministry of Science, the assertion sounded unconvincing. The organization of science could not be done effectively without one.

The conference, in spite of light and shade in the thoughts uttered, was inspiring. A remarkable range of people with varied social interests had joined in discussing the question of Science and the Welfare of Mankind.

In the course of the conference, a number of the participants decided to move a resolution that a World Federation of Scientific Workers should be formed, and that the Association of Scientific Workers in Britain should be asked to draft a constitution. When this was ready, a meeting of representatives of the various associations should be called to consider its adoption. This was held in London in July 1946. It was attended by representatives and observers from eighteen associations in fourteen countries, and the natural sciences division of UNESCO.

I summarized the aims in the *Manchester Guardian* as including the fullest utilization of science for promoting peace and human welfare; the taking of action to ensure that science is applied to help to solve the urgent problems of the time, such as the peaceful use of atomic energy and the rehabilitation of devastated countries; the application of science to the problems of undeveloped and colonial countries; the promotion of international co-operation in science and technology, and the exchange of scientific workers and knowledge. It was to encourage a closer relation between the natural and social sciences, the improvement of science teaching, the professional status of scientific workers, and it would aim at collaboration with UNESCO and other appropriate bodies.

I wrote that the foundation of the Federation was an historic step towards the time when the scientific workers of the world speak with one voice on the critical problems of the place of scientists in modern society and the proper use of scientific discoveries.

The speech which impressed me most at the foundation meeting was by J. D. Bernal; it was a brief, condensed, deep but transparently clear description of the aims of the new organization. Thenceforth, I always associated the World Federation of Scientific Workers with him.

Joliot-Curie of France had been elected President of the new Federation; N. N. Semenov of the U.S.S.R. and J. D. Bernal of Britain, Vice-Presidents; and Harlow Shapley of the U.S.A., Treasurer. Roy Innes, the General Secretary of the Association of Scientific Workers of Britain, who had been in charge of the preparatory work, became Acting Secretary-General, until the post could be taken over. Harlow Shapley intimated that he could not continue as Treasurer, and W. A. Wooster, the Honorary General Secretary of the A.Sc.W. of Britain was appointed to succeed him.

Wooster asked me whether I would be prepared to accept the post of Secretary-General. It was a most attractive suggestion, though the salary attached to the post was problematical. Joliot was told that I was willing to be a candidate, and he spoke enthusiastically in my favour. So, in 1946 I became Secretary-General designate of the W.F.S.W., and in 1947 Secretary-General.

I was not conscious of having had any part in the creation of the W.F.S.W. I had watched and reported the proceedings as a sympathetic spectator, and I had been very glad that the Society for Visiting Scientists existed and had provided a place where the participants in the developments could meet, rest, formulate ideas, and discuss what should be done next.

I was surprised at the first W.F.S.W. conference in Paris after my appointment to hear Joliot say to the press that I was one of its founders. It appeared that my activities before the war, and in founding the Anglo-French Society of Sciences had influenced the development of his own thoughts on the international organization of scientists.

Nevertheless, I never felt involved in the W.F.S.W. as one of its founders, and consequently was more detached in my relation to it than I was in my attitude to those organizations which I had myself initiated.

The great increase in concern with science, both nationally and internationally, which occurred during the Second World War made it desirable to promote the organization of the exponents of science to the public. The Americans had already recognized the need in the 1930s, by forming their National Association of Science Writers, to which they had elected me an Associate Member.

The Society for Visiting Scientists had provided a place where such writers could meet scientists from home and abroad. It was desirable that both Visiting and British scientists should have the best possible relations and understanding with those who were

International Science, Science Writing, and Poland 275

best able to make their needs, and the needs of science, known and understood through the press and other media of communication to the public.

It was for these reasons that I called a meeting at S.V.S. in 1947, to discuss the formation of an Association of British Science Writers. Among those who attended were Ritchie Calder and A. W. Haslett. I suggested that this Association should be formed with a number of aims, which included the following: 1. The establishment of science writing as a definite profession with appropriate status. 2. Co-operation to the advantage of members, for example in the passing-on of requests for articles and books which cannot be accepted because of other engagements, to those in a position to carry them out. 3. The improvement of the opportunities of members to receive science news from official institutions and laboratories, learned societies, industrial laboratories, etc. 4. The increase in space in the press for science articles. 5. The improvement of rates of remuneration for science writing. 6. The organization of meetings to exchange views on matters of common interest, and attempt to form agreed views on the problems that affect science writers.

It was not to be expected that any or all of these aims could be realized at once, especially in the conditions of the press and publishing then obtaining. The extreme shortage of space in the press made rapid improvement unlikely, but the great extension of interest in science, and appreciation of its importance, to some extent made up for the shortage of space and other difficulties of publication. It was desirable, however, that science writers should be in as strong a position as possible to benefit when the situation improved. Agreed views and corporate action would be of great help in achieving that end.

After discussion, it was decided to establish such an association. Maurice Goldsmith was elected Secretary-Treaurer and I, Chairman.

The post-war situation in 1946–47 involved a good deal of travel. My rapid tour of liberated countries in Europe in 1946 was followed by four visits to France and one each to Poland, Mexico and the U.S.A. in 1947.

I visited Paris in the spring of that year, to join in the discussions on the relations of the W.F.S.W. with UNESCO and other bodies. Almost immediately after I arrived, I was told by a French scientific colleague that the B.B.C. had telephoned to the French Radio Diffusion, saying that they had noticed that Bernal and Haldane were to speak for them; they thought they might like to know that they were Communists. They had no objection to their speaking, of course, but they could suggest other British scientists who were not Communists!

It soon became clear that there was little chance of close collaboration between UNESCO and W.F.S.W. The Americans especially were against W.F.S.W. Heavy cuts in the UNESCO budget had entailed a severe reduction of staff. The appointments of no less than twenty per cent of the senior staff were being terminated.

Freymann gave me a vivid account of the general atmosphere in Paris, and expressed some interesting speculations on the nature of French government. He said there were two governments in France: the legal, which was powerless, and the illegal that had the real power. He told me that a rich Mexican, a friend of the President of Mexico, had been in Paris recently. He wanted to buy a motor-car to tour France. The best the Mexican Embassy could do was to get the promise of one in three months' time. The Embassy chauffeur overheard the discussion. He offered to get one by the same afternoon. It was a new Packard, and the Mexican paid 1,200,000 francs for it. There were three other new Packards in the garage where it came from; these had all come into France through the customs without being examined.

I heard similar stories in descriptions of the setting-up and equipping of the new scientific institutions being founded in the liberated country. Here again, chauffeurs often seemed to be more influential in securing scarce scientific equipment than the conventional sources of supply. It applied in every direction. On one occasion, when a leading scientist had to keep an important engagement abroad at short notice, after the normal means of transport had failed to provide a passage in time his chauffeur said he would secure the loan of a minister's private aeroplane. 'How on earth can you do that?' he was asked. 'You just leave it to me,' he was told. The plane was produced, the journey made, and no more questions were asked.

In October 1947 I visited scientific institutions in Poland. In that early post-war period the air service between London and Warsaw was run by the R.A.F. Travellers stayed overnight in Berlin. The aerodrome at Gatow beside the Havel and the Berlin lakes was subject to mists. The mist prevented us from leaving for Warsaw for three days. As our impatience grew one of the R.A.F. pilots told us that his motto was: 'If you have time to spare, travel by air.'

After two days of morning mists and incarceration, the R.A.F.

took pity on us and ran us round Berlin on a sight-seeing expedition. This was very interesting to me, for I had not seen Berlin since the early 1930s. We ran down Heerstrasse to Reichskanzlerplatz. The great modern building there had become Military Government House. Running down the Kaiserdamm, about three-quarters of the houses seemed uninhabitable. The enormous factory at Spandau was gutted, and I saw no smoke from the chimneys of Siemensstadt. The destruction of the Charlottenburg Hochschule seemed very great. The Tiergarten looked like a desolate heath with a few odd trees and bushes, a blown-up flak tower, and all around, gutted houses.

We were taken to the Reichskanzlerei in the Russian zone. While we were there, a party of hundreds of sightseers visited it. We were told that the marble was being taken down and used in the reconstruction of Stalingrad. The court with Hitler's dugout was open. An old man was sitting there, and boys offered to show us around for a consideration. We could not get into the dugout because it was full of water. The Reichskanzlerei looked like an immense tomb, and the burnt-out Kaiserhof and the streets around made the most forbidding appearance I had ever seen — until I reached Warsaw and Wroclaw. Most of the Adlon was destroyed, but a restaurant and some rooms were in use. I asked to see the district in which I had lived in 1930.

My first general impression of post-war Berlin was of destruction and desolation so great that it looked more like rocks contorted by nature than a human construction. The population looked yellow and pinched, and there was no sign of reaction against the foreigner. There seemed to be no relation between the occupying powers and the German people who looked as if they were resigned to waiting for their ultimate withdrawal. I thought there would be great pressure to get the huge, empty, gutted factories going again. Who would do it, and could Berlin revive without a single government?

I was glad to have had this unexpected view of familiar places in Berlin. It turned out to be useful as well as interesting, for it provided a measure of what I was to see in Poland.

When I at last arrived in Warsaw, three days late, I was met by Joseph Barbag, Director of the Minister's office in the Ministry of Education, and the physiologist J. Konorski. Barbag was a geographer, who had been on a British Council course at Cardiff. We drove into Warsaw through miles of ruins. They looked different from those in Berlin. This was due to their having been made less by bombing, and more from being set on fire or blown up by destruction squads, who went about their job systematically.

The zoologist Jaczewski, formerly the director of the Zoological Museum, told me what had happened to some scientists during the war. He had become a leader of the underground university, and the military underground. His company was among the last three to surrender in the Warsaw Rising. He had been interned at Lübeck, and like Béra, released by the British, became a liaison officer, commandeered a German car, and made his way home in it.

In listening to everyone I met I tried to discover the foundation of Polish enthusiasm. I noticed a deep disillusion with the French. The Poles had not foreseen the defeat of France, and it had come as a much greater shock to them than to the British. I was told that American Business was the No. 1 enemy of the human race. The Americans were very primitive; they had no interests and conversation, in such contrast with the simplest Russians.

Among the interesting new activities was the planned development of the Vistula. All universities were collaborating in a scientific survey of the river.

On the following day I left by train for Lodz. I noticed the flourishing and picturesque peasantry. Many country railway stations were decorated with beds of autumn flowers, and creepers turning a brilliant copper colour. Lodz is a flat, rectangular city, rather like those in the American Middle West. The buildings were almost undamaged, but the population had been decimated. It contained 250,000 Jews before the war; now there were only 30,000. Consequently, the city had been left with many empty buildings and for this reason numerous Polish scientific institutions had been set up there in the first post-war period.

At Lodz I was the guest of Konorski and his wife, the physiologist L. Lubinska, a former pupil of Laugier in Paris. Their recent research had been on the mechanical excitability of regenerating nerve-fibres, using a method of Young and Medawar. Konorski was using two rooms in his flat as a laboratory; his research was continually interrupted by a stream of student callers.

I was taken to meet the eminent philosopher T. Kotarbinski, then Rector of the University. This highly gifted, charming man was very thin in figure, with splendid frank eyes, and a courtly manner. Before the war he had long been at Warsaw, where he was one of the few professors who fought the rampant anti-Semitism. He told me that he had been at work on a large book on Francis Bacon. He thought that Bacon had made the greatest contribution to the development of logic since Aristotle to his own time, anticipating to some extent von Wright and other modern logicians. He had expounded his evidence for this in a large manuscript, four copies of which were made in 1939 and deposited in different parts of Warsaw; all of them were destroyed. He referred especially to Bacon's shorter papers which contained original contributions to methods of thinking.

Kotarbinski's conversation impressed me profoundly, for I was still under the misapprehension that Bacon was an expositor and propagandist, and had not made discoveries in the realm of intellectual technique. His remarks strengthened my intention to write on Bacon, which I was not able to do until ten years later. Kotarbinski expounded the history and origin of the Polish school of logic, one strand of which derived from Catholic medieval scholasticism.

Kaminska, who was his second wife, had been one of his pupils. She was a mathematical logician, with a quiet, gentle manner, large head and impressive black eyes. She had been imprisoned in Auschwitz, and branded on the arm. Kotarbinski took the chair at one of my lectures, and Konorski interpreted. The subject was International Science.

Konorski told me that he became interested in conditioned reflexes when he was a young student. He sent a note of his work to Pavlov, who replied. Then he sent a longer paper, to which Pavlov replied at greater length. He was enormously proud of this, and was promoted in the university. He went to Pavlov's laboratory in Leningrad, where he worked for two years.

One of the most interesting calls at Lodz was at the Polytechnic. It was housed in a 20-acre textile factory, which had been looted by the Germans. The Vice-Rector was the chemist Achmatowitz, who had been a pupil of Perkin and Robinson at Oxford in 1929–31. He was a Polish Tartar, and a Mohammedan.

After Lodz I visited Wroclaw, Mrs Konorski acting as my guide. The devastation in Wroclaw, the former German city of Breslau, was fantastic. The Germans and Russians had held respective halves of the city, and then fought it out for three months. We were met by Szymanowski and Baranowski. The former had recently returned to Poland after thirteen years at Pittsburg. He worked in short-wave radio, and was a friend of E. U. Condon. Baranowski was the leading biochemist in Poland; he was a tall, pale, bespectacled, tireless man. He had the best equipped biochemical laboratory in the country, and felt responsible for the development of Polish biochemistry.

Then we went to the laboratory of the eminent bacteriologist

L. Hirschfeld. It was one of the medical institutes which had previously formed the Breslau Medical School. Most of the heavy German brick buildings had survived.

I saw Baranowski's laboratory. He had done well-known work on the chemistry of proteins. With him was the X-ray crystallographer Chrobak. I was told that the Germans had offered him an institute if he would work on the manufacture of synthetic quartz used in the directing mechanism of V.2s. He refused, and disappeared for the next three years, disguised as a woodworker.

My next visit was to the famous pure mathematician H. Steinhaus. He showed me the apparatus he had invented for helping surgeons to locate foreign objects in the human body. He pointed out to me a young woman physicist who had worked in an iron foundry during the war. She and three other Poles had escaped being shot by the Germans by hiding for two years in a hole in the foundry floor near the furnace. They were not found, even though the place was searched with dogs. Nobody thought of looking right in the middle of the foundry, where iron was being smelted.

Steinhaus showed me photographs of the house in a little village, where he and his wife had hid for four years during the war. They had a false name and papers; he bought the birth certificate of a dead peasant. He taught boys in an underground secondary school, and his wife sold chocolates. Her sister and brother-in-law died in Auschwitz. Their infant girl, who was with relatives in the Pas de Calais, was sought by the Gestapo, but hidden by two old French ladies who were devoted to her, and were bringing her up. She had since become a medical student at Lyons.

Baranowski told me that the University Library was intact when the Germans officially surrendered, but set on fire by them three days afterwards. The garrison in Wroclaw disobeyed orders. When Baranowski himself entered the town, there were still some seven hundred fires burning.

After my visit to Wroclaw Baranowski drove me for six hours without a break through Katowice to Krakow. We went past lead mines, coal mines, zinc refineries, oil refineries, and other factories. The region was covered in smoke, and reminded me of the West Riding of Yorkshire. It was evident that the Nazis had not destroyed the industrial regions. We passed a Soviet airfield with scores of fighter planes, and companies of Soviet troops marching and singing military songs. Their officers were smartly dressed, in new uniforms.

The contrast of Krakow with the Katowice industrial area was very striking; it was rather like Oxford after Rotherham. The city was undamaged, and already had twenty thousand students in its various higher institutions. I was taken to the Mining Academy, which had been the headquarters of the gauleiter Franck. I was told that Krakow had survived with so little damage because it was not in one of the zones that changed hands, so that its population was not disturbed as a whole. When the Russians liberated the city, they carried this out by a flanking movement, which caused the Germans to retreat before they had time to destroy it. The Director of the Mining Academy, Goetel, told me that he and his students rushed into the burning building as the Germans left, and succeeded in putting the fires out. They saved part of the Gestapo archives, and found a plan for a system of crematoriums, with an estimated output of six million corpses a year.

The physicists Weyssenhoff and Niewodniczanski took me to the Physical Laboratory, where they had recently held an exciting conference on cosmic rays. I had met them at a Bristol conference in 1937. Weyssenhoff reminded me of the chapter in *The Progress of Science*, in which I had written on Bohr's institute at Copenhagen, and he told me that foreigners often read this chapter when they first went there. Niewodniczanski had worked in the Cavendish Laboratory in 1934-35, and was friendly with Cockcroft and Shoenberg. He and his wife were living in laboratory rooms, which had been adapted to the purpose.

I saw the black granite slab in the old physics laboratory, in the heart of the university quarter, marking the place where oxygen and nitrogen were liquefied for the first time, by Wroblewski and Olczewski in 1883.

The Polish physicists had been greatly stimulated by the visit of the cosmic-ray physicists, including Blackett and Powell, and Joliot-Curie. The Governor of Krakow Province, Dr Pasenkiewicz, was a mathematical physicist, and had entertained them. Joliot-Curie told him of my forthcoming visit, so the Governor gave a dinner in my honour as the representative of the World Federation of Scientific Workers. During it I heard details about Auschwitz from Davidovski, and discussed W.F.S.W., UNESCO and the World Federation of Trade Unions with Dr Bienkowski, the Vice-Minister of Education. It was concluded that one big union of scientific workers would be the most suitable for Poland.

While I was in Krakow I heard more of the work of Choyonowski, the founder of the *Life of Science* journal, and the promoter of the notion of a 'science of science'.

The atmosphere of Krakow was rather bourgeois, compared

with Lodz and Wroclaw. The city had been ruled by the Austrians from 1794 until 1918, and there was still a residuum from their tradition.

I found the scientists in Krakow coldly critical of Heisenberg. They told me that he had accepted an invitation from Franck to lecture in Krakow at the end of 1943. The Poles regarded it as part of a programme for establishing the domination of German culture in the new Nazi province.

On 6 October, four days before I had left London for Poland, the *Manchester Guardian* had published my obituary notice of Max Planck, who had died at Göttingen in his eighty-ninth year. As the head of German science during the Nazi régime, his deportment had been a matter of controversy, and the subject had not left my mind. I had called on him in the early 1930s to seek his permission for the English translation of a contribution of his to a book I was editing, *Science Today*.

When I was shown into his study, he awaited me, seated behind a long wooden bench, carefully and neatly prepared for work. He was a quiet man, dark-haired and pale, with a slightly severe, penetrating look; he appeared to be a German scholar of the nineteenth-century type. He worked to the highest standards, and attacked the most difficult problems. The German respect for authority was conveyed in his demeanour.

When Nazi Germany arose he was torn between love of science and love of the Fatherland. He tried to be loyal to the Nazi State and to science, an impossible line, which involved him in Faust-like conflict between duty and truth, and he suffered Faust-like consequences. He became the instrument of harm to some scientists, but in his person lost one of his sons, a Foreign Office official who was executed in the sequel to the Gordela–von Stauffenberg conspiracy to assassinate Hitler.

Planck visited England just after the war, in 1946, when he was a very old man. He attended a reception at the Society for Visiting Scientists. The famous French mathematician E. Borel, who was not much younger, and had suffered during the Nazi occupation of France, was also a guest, and he did not conceal his irritation at finding himself in Planck's presence on a social occasion.

I left Krakow for Warsaw, where my first visit was to the ruins of the Ghetto. They covered about two square kilometres, with a fairly uniform layer of pulverized rubble, about fifteen feet deep. They looked like a desert of individual bricks and dust, and I did not notice a single piece of wood in it. A few streets had been cleared, to leave the track for street-cars. The pre-war population of the Ghetto had been about 250,000. The Germans herded 600,000 into the area, which started quite near the centre of Warsaw. I was told that Hirschfeld had spent two years in it. Some had escaped through the sewers. There were said to be 200,000 Jews buried under the ruins. The Germans sealed the Ghetto and used it as an artillery and bombing practice target for three months —this was why it had such an extraordinary pulverized appearance.

I looked forward to visiting the institute of the eminent physicist Pienkowski with particular interest, for about twelve years before Cockcroft told me about Pienkowski's laboratory, organized with military precision and authority. He described how the professor clapped his hands, and instantly assistants came running from all quarters. This did not happen in the Cavendish Laboratory.

Pienkowski was wonderfully fresh and well preserved. He had a round, fat Polish face with a continual smile and bright grey eyes. I was then quite delighted when he made a shattering clap with his hands, and instantly an assistant shot forward.

Pienkowski's laboratory had been one of the best organized in the world. The Germans stripped it completely, and Pienkowski became a gardener in the Municipal Gardens. Later he became Rector in the Underground University organized for the whole of Poland during the occupation. He had great influence, though his political views were conservative. His laboratory had been excellently restored, and a new wing was being added for a Cockcroft-Walton atomic disintegrator.

I returned to London shortly afterwards. Very strict precautions were taken about entering the British plane at the airport, and there was some difference of opinion between the pilot and the airport authorities about protocol. It appeared that Micholaijczik, the Vice-President of Poland, had left the country in obscure circumstances, and special care was being taken about the departure of British planes.

While I was in Poland I heard that I was to write a report on the conservation of natural resources for UNESCO. In connection with this, I was to proceed to the UNESCO Conference in Mexico City. My visit to Mexico City also enabled me to represent the World Federation of Scientific Workers at the UNESCO Conference.

Before leaving for Mexico I attended the Rutherford Memorial ceremony, organized in Paris on the tenth anniversary of his death.

CHAPTER TWENTY-FOUR

1947
HOMAGE TO RUTHERFORD AND CONFERENCE IN MEXICO

THE CEREMONY IN honour of Rutherford, inspired by the tenth anniversary of his death, began in Paris on 7 November 1947, under the patronage of the President of the Republic, M. Vincent Auriol, and the Academy of Sciences. It was organized by the World Federation of Scientific Workers, the French Association of Scientific Workers, and the French Union of Workers in Higher Education and Research. The main administrative work was carried out by Biquard.

The Chief guests were Lady Rutherford and her grandchildren, Miss Elizabeth Fowler and P. H. Fowler, who has since become an authority on cosmic-ray physics, and a Royal Society professor. Fifty-nine scientists and personalities from sixteen countries accepted invitations.

The functions consisted of a Reception by the President of the Republic, a Memorial Meeting in the Great Hall of the Sorbonne, and two scientific meetings, on Rutherford's Scientific Work, and on the Development of Modern Sciences and its Social Consequences. After these, a reception was given at the Maison de la Pensée Française, and J. D. Bernal delivered a lecture in the Palais de la Découverte.

Considerable financial support was needed for this international ceremony. UNESCO was among the donors, and the Cultural Relations Department of the French Foreign Office, the French National Centre for Scientific Research, and the French High Commission for Atomic Energy also helped.

M. Bidault, the Minister for Foreign Affairs, presided at the public meeting at the Sorbonne, which was attended by more than three thousand people. Among the speakers were Teissier; Møller, who brought a message from Niels Bohr; Maurice de Broglie; Urey; and Joliot-Curie, who referred to a host of messages, from Einstein to S. I. Vavilov.

Louis de Broglie presided at the session on Rutherford's scientific work, which was held at the Collège de France. The speakers were M. N. Saha, J. S. Foster, Irène Joliot-Curie, and Sir George Thomson.

Saha had come from India to participate in the ceremony. We happened to arrive at the Elysée rather early for the President's reception, so we waited in the anteroom. I started a conversation on the influence of autodidacticism on scientific thinking, as illustrated by Raman's ideas. I waited for his observations, and after a while he said: 'I also am an autodidact.' I learned that he was a student of law when he made his famous contribution to the theory of the ionization of gases by heat.

As a fellow-member of the committee of organization, I met Maurice de Broglie, and was struck by the perfection of his English, and his insight into the English physicists of his own generation. He would have been very much at home in the intellectual world of the Rayleighs. When he discussed Rutherford, he spoke of him in the way an aristocratic English scientist of the first intellectual rank would have done; he saw him in English rather than French terms.

Blackett opened the discussion on the Social Consequences of Modern Physics. He spoke of Rutherford's methods, and how their results had destroyed the old organization of science, in which the pure scientist in his laboratory could ignore the world around him. The scientific heirs of the Rutherfords and the Curies lived in a sterner world, where political consciousness and activity, which to Rutherford's epoch would have been merely a diversion from the pursuit of new discovery, had become the conditions for the survival of human society. Scientists now had the double task of not falling too short of their predecessors as discoverers, and at the same time attempting to be more conscious and active citizens.

In March 1947 the U.N. Economic and Social Council undertook to organize a Scientific Conference on the Conservation and Utilization of Natural Resources. It invited the specialist U.N. organizations, such as UNESCO, to join in the Conference.

UNESCO commissioned me to write a report, with suggestions of ways in which it might usefully participate. I was to attend the General Conference of UNESCO in Mexico City in the following November, call at the U.N. headquarters, and visit conservation authorities in New York and Washington, to collect relevant information. I also attended the UNESCO Conference in the separate capacity of Observer for the World Federation of Scientific Workers.

I left for Mexico City a few days after the Rutherford commemoration in Paris. In the meantime, I saw Max Nicholson, then Herbert Morrison's chief aide in the Lord President's office, and ascertained his views and advice on resources, and whom I should see in the U.S. He emphasized the importance of human resources in making natural resources fertile, and that UNESCO should attend particularly to that aspect of the problem.

I travelled by air, and as I had several hours' wait in New York for the plane to Mexico, I taxied to a hotel I knew in the city, to make some telephone calls and have a wash and shave. The hairdresser in the barber's shop perceived that I had just arrived from Europe, and asked me to sign his visitors' book. I looked through the names, and the first that caught my eye was 'Lise Meitner'. She had written after her signature: 'Atomic Bomb.' Another was that of Mrs Eleanor Roosevelt. The hairdresser said that he had deduced I was a writer from the length of my hair.

The flight from New York to Mexico was over hills and mountains most of the way, but the Mexican mountains were an entirely new experience. Around the aerodrome at Monterrey they jutted huge and jagged into the sky. Their edges looked as sharp as newly cast steel from a furnace. I had an impression of the newness of the New World compared with the Old, and its primeval inhumanity. Later on, when I looked at the Aztec architecture and sculpture, it seemed to me that the landscape had contributed to its inspiration.

As the plane approached Mexico City, I saw the reflections of mountains in lakes, and terraced fields around the cores of circular hills. Near the city I saw the first green fields, apparently market gardens; clouds looking like black smoke hung over the environs.

I had had a lively conversation in the plane with the lady who sat next to me. After breakfast in the aerodrome, we exchanged a few further words, when a pleasant gentleman approached us, and said he was going to take her away from me. I exclaimed 'splendid', meaning that I did not wish to prevent him. He at once said: 'I see you are not a diplomat.' 'No,' said the lady earnestly, 'he is a scientist.' This was my first meeting with Arthur (Tex) Goldschmidt, of whom more later.

Mexico is a rectangular city, in a valley about twenty miles wide, encircled by mountains nearly 20,000 feet high, from which volcanic smoke ascends. The large lake, on an island on which Montezuma had had his palace, had been greatly reduced by a tunnel driven through the mountains to drain flood water. This had dried large areas of lake bottom, which was covered with fine dust; this was whirled into dust clouds by the wind. In the suburbs were metal refineries covered with perpetual fumes, so that the sun was obscured.

The city then had a population exceeding two million. There had been a building boom, financed out of the huge sums spent by the United States on the All-American Highway and other semi-military works, built to ensure the defence of the Americas after Pearl Harbor. There were numerous idyllic villas put up by speculators, and painted in various pastel shades. Many of them were empty.

There was a mixture of old buildings and new offices, schools, etc. The UNESCO Conference was held in the new Training College for Teachers, built in reinforced concrete, black glass, light metals, and polished slippery tiles. It had fluorescent lighting, but no lifts. Mexico is 7,100 feet above sea-level, and those acclimatized to sea-level find themselves short of breath after any exertion. Walking up and down the stairs of the conference building was a strain. In meetings, unacclimatized persons suffered from sudden loss of memory. One of the first I attended, the working party for science, was presided over by H. J. Bhabha, normally most alert and quick-minded. I was astonished to see him stand up to say something, try to remember what it was, and sit down without saying anything. It was most unlike Bhabha, and made me wonder how complicated diplomatic negotiations might be faring.

My hotel was a large construction, started three years before and not yet complete. It was sited beside the railway goods yard. There was a magnificent chapel with a cross lit by concealed lighting; glass stars were let into the floor of an enclosure in the formation of the Great Bear, and lit from beneath, and there was a fresco on the war of 1814. L. G. Grimmett, then of the Natural Sciences Division of UNESCO, was staying in the hotel. In the evenings, according to his habit, he played for an hour or two on the lounge piano. When the manager heard him, he bought a stock of music, and people began to come in to hear the free piano recitals.

At the first meeting of the Natural Sciences Working Party that I attended, Adrian occupied the British place, and the Chairman and Rapporteur were supported by Needham and Grimmett. The Hylean Amazon scheme for a scientific institute for studying the Amazon region, and the Social Implications of Science, were two topics on the agenda. I had suggested that fellowships for research in the latter topic should be created. Ultimately, this was agreed.

On that day, I accompanied Auger and Bhabha to lunch. Our car was hit by another, but fortunately only a glancing blow. The roadsides of Mexico were littered with the wrecks from fatal car crashes. Bhabha was indeed to suffer violent death later, when his plane crashed into Mont Blanc.

During the meal we discussed the latest Nobel Prize award for physics; it had been given to Appleton. Bhabha said it ought to have been given to Blackett, who, however, was to receive the prize in the next year, 1948. When I returned to New York I heard American physicists saying Appleton was lucky. They thought the award had been made to get away from nuclear physics; but, said one of them, nuclear physics was more important than all the rest of physics put together.

When I arrived in Mexico City, Joseph Needham had already acquired a considerable knowledge of Mexican archaeology. He invited me to join Grimmett, Zhukova and himself on a day's strenuous archaeological tour. We went to Copilco and saw the burials of 3000 B.C., which had since been covered by a thirty-foot layer of lava. We visited the canals at Xochimilco, all that was left of the great lake. We saw one of the pyramids, built in A.D. 1143 and added to every fifty-two years, in order to prevent, according to Aztec ideas, the destruction of the world. Then we drove for an hour to see the great courtyard and pyramid at Teotihuacan, a temple dedicated to the Plumed Serpent or Rattlesnake God, Quetzalco.

The extent of Mexican architecture was enormous. The country contained sixteen thousand pyramidal constructions. The stark violence of Mexican art appeared in remarkable contrast with the external characteristics of Mexican people, who were unusually gentle, smiling, and often beautiful and charming. The Mexican population was overwhelmingly of native descent. Its culture was fundamentally Mexican, with a surface of Spanish culture. Now Americanism was being slapped on top, in the form of factories, cars and baseball.

The crucial event for science at the conference was the passing of the grants-in-aid budget of $240,000 for the Natural Sciences Division. After the rejection of a disturbing proposal by the Americans, it was moved in an able speech by Adrian. It provided the funds for the Division to carry out a satisfactory programme of work.

During the conference I met one of the Mexican experts on conservation, the protozoologist Professor E. Beltran. He entertained me at his home, a beautiful modern villa four miles from the centre of the city, and at Sanborn's, the remarkable restaurant where intellectual Mexico met between 7 a.m. and 9 a.m. for breakfast. It was in the hall of what looked like a Spanish mansion. I was told that the milk and drinking water used in it were flown in one thousand miles, every morning, from Texas.

During breakfast people moved about from table to table, transacting business, making appointments, and meeting friends. I was introduced to Professor Mesa, the agricultural adviser to the Bank of Mexico. He invited me to speak on Soviet Science at the Mexican-Soviet Society. I was told of the political meeting which was to be held in the circus stadium, Arena Mexico, at which Lombardo Toledano was to found his new Popular Party.

Toledano was an Observer at the UNESCO Conference on behalf of the World Federation of Trade Unions. I called on him in his fine office in a modern building. He was an intellectual, a philosopher and a jurist, an extreme contrast with Anglo-Saxon trade union leaders.

On the day of the Arena Mexico meeting I set off by myself, to get there in good time, an hour before it was to start. When I arrived, the Arena, accommodating about seven thousand, was already nearly full, and I could not find a seat; I stood for the first three hours. The audience was good tempered and well behaved. There were many Mexican Indians. Husbands and wives sat side by side, husbands with a long tail of hair down the back. They were quite undemonstrative, apart from laughing in the right places. I had the impression that they were politically acute.

Toledano began to speak at about 11 p.m. and went on for an hour and a quarter. He outlined the programme of the new Popular Party. It was to be 'integral', not Communist, not Socialist, free in religion, patriotic and modern. It was to do much for science, to which Toledano repeatedly referred.

The meeting was divided by an interval, in which a very big man with a very small voice addressed the meeting. This was Diego Rivera; his task was to collect funds for the New Party. He asked those present to defend their country by presenting 1,000-peso notes to the Party fund. He paused and waited, but no 1,000-peso note came. 'Well,' said Diego, 'it is evident that we are all workers here—no rich people.' 4,565 pesos were collected in smaller amounts.

Mesa talked to me about Mexican agriculture. The country was potentially rich, but there were many difficulties, principally from climate and mountains. The peasants lacked education and machinery. There was little rotation of crops, which was hindered by the climate. In general, rotation could be practised only on irrigated land. In Mexico, agriculture and husbandry had not been properly combined as in Europe and the United States. Farmers did not construct yards for preparing manure. In general, fertilizers were not used, except for sugar cane.

It seems that the backwardness in animal husbandry had a very long history. The Aztecs had no domesticated animals, except a few dogs. They represented these in art as fighting for food. One explanation of the Aztec's eating of human flesh was agronomical. There was a shortage of protein owing to the lack of domesticated animals. It was most easily obtained from human bodies.

I met several of the Spanish Republican scientists who had settled in Mexico, including the entomologist C. Bolivar and F. Giral, the son of the last Republican Prime Minister, José Giral, and Spanish translator of my *Science in Soviet Russia*. F. Giral was engaged in the manufacture of synthetic sex hormones. He worked in a firm founded by Hungarians and Germans. There were abundant sources of material in Mexican plants, and they were able to market products at much lower prices than the Americans.

I was taken round the unique Institute of Cardiology. I was told that there were 300,000 cases of heart trouble in Mexico City owing to the altitude, which explained why so great an institute had been created. The main programme of research was on rheumatic fever.

The Institute was decorated with remarkable murals by Diego Rivera. They contained vivid portraits of all the famous medical scientists, from Galen and Harvey up to Röntgen and Sir Arthur Keith. Servetus was portrayed being burned at the stake, in a lurid yellow light.

I was taken to the department of physiology under Rosenblueth. There I found Norbert Wiener, who told me how he divided his time between M.I.T. in Boston and Rosenblueth's laboratory, trying to apply cybernetics to the explanation of nervous conduction, and the mechanism of the heart. He expounded his view that biological physics had been dominated by ideas of energy, whereas the really important feature of a biological system was the history of its entropy. Wiener was accompanied by the mathematical logician Walter Pitts from Chicago, and other research colleagues. The Institute of Cardiology was indeed one of the most interesting I had ever seen, for its size, design, qualities of research, and artistic merit.

I soon met again the charming Texan whom I had seen at the airport and who had decided that I was not a diplomatist. He proved to be Arthur Goldschmidt, universally known as 'Tex', and the American resources expert. His face was thin, pale, and narrow-boned, and he had a beautiful Southern drawl. He had been one of the leading men in the New Deal, and had had much to do with the development of the theatre in the U.S. after the crash of 1932. This development discovered Orson Welles and other talented artists, in the course of finding jobs for thousands of actors. Later on, he was concerned in the production of the Krug report, which was the basis of the Marshall Plan.

I was impressed in Mexico by the great fame of the Huxley family, as I had observed years before in the U.S.S.R. It seemed that the battle Darwin and Huxley had won in Britain in the previous century was still undecided in Latin America. The Huxleys were regarded by many Latin American biologists and rationalists as the heroes of the struggle of science and reason against the Church and obscurantism.

On one of the days when the Science Commission had no business in the afternoon I found myself outside the headquarters building before lunch. Stratton suggested that we visit a club in the suburbs for lunch. UNESCO delegates had been invited, and a bus sent to collect any who cared to accept.

The club was in the mountains above Mexico City, and was not yet quite finished. The proprietor told us that he was of Danish origin, and had been the *New York Herald* Correspondent in Berlin in 1918. Then he worked with Count Kalergi for the idea of the United Nations of Europe. He had thought of inviting all the Nobel Laureates to stay in his club when it was finished, to place the problems of the world before them, so that they could be solved by their combined genius.

The club had not yet any members, but Stratton and I were given lunch in the big, complete but as yet unpatronized restaurant. There was a stage on which the *corps de ballet* from the city could perform, and there were many fine decorations. Some of these were strange, especially in the suites for private parties; it was interesting to see what kind of environment was thought suitable for Nobel laureates, engaged in the solution of the fundamental problems of mankind.

On my last evening in Mexico City I gave my lecture on Soviet Science. The meeting was crowded. It was the only lecture I have delivered with members of the audience packed round me on the rostrum, besides standing in the body of the hall, which was full to the walls. Beltran interpreted my lecture most expertly.

Afterwards I was taken for drinks in an enormous long, empty bar. For the first time in Mexico, I took a chance, and had a fruit cocktail. Alas! I felt the effects during the night and morning. The plane was due to leave early, but was delayed for seven hours because of a defective magneto, and there was nowhere to sit down at the aerodrome. It got hotter and hotter as the sun rose; the American Airlines were in the end reduced to borrowing a magneto from the Mexicans, and we left for Dallas in the afternoon.

In New York I called on I. I. Rabi at Columbia. He was optimistic about American physics. They had restarted academic research very quickly after the war. The U.S. Navy had been very liberal and broadminded in subsidizing research. Without their aid, there would have been much hardship. Senior scientists were worse off than before the war through inflation and taxation, but the workers were better off. Rabi told me about his brilliant pupil Schwinger, who was later to receive a Nobel Prize. However, he did not think that the U.S. produced geniuses. He believed that science was flourishing in Britain because it was kept clear of politics, whereas in France it was hindered by cliques, in spite of the French having the finest minds. Rabi agreed with those who considered that Chadwick would have been the most appropriate successor to Rutherford in the Cavendish Laboratory. As to many others, X-ray analysis did not appear to him concerned with as fundamental problems as the atomic nucleus.

William L. Laurence, the science writer of the *New York Times*, and his wife Florence invited me to stay with them in New York. He had been appreciative of my work as a scientific journalist, though his political views were very different from mine, and I had admired his passionate enthusiasm for science and the communication of this through his articles to a large public. Laurence had the confidence of the redoubtable General L. R. Groves who had entrusted him with reporting the dropping of the atomic bomb on Nagasaki.

Laurence expressed views on the U.S.S.R. with which I disagreed. He had left Tsarist Russia as a young opponent of the régime, and told me that he observed that I was still not disillusioned, and wished that he could say the same of himself.

With this degree of mutual understanding, I was able to appreciate the Laurences' generosity very particularly. They had an apartment near the East River, and entertained me splendidly: The climax was a splendid party they gave for me. Pierre and Suzanne Auger, who had arrived from Mexico at three o'clock that morning, were among the first of the guests to arrive. Francis Perrin came, and Robert K. Merton, whom I had first met at Harvard. The biochemist Evans, who was going to London as head of the U.S. Scientific Office there, the geneticist L. C. Dunn, the biochemist Michaelis in whose field Louis Rapkine worked,

Homage to Rutherford, and Conference in Mexico

and the cartoonist Mauldin, were also among the Laurences' guests.

In the middle of the party the *New York Herald* rang up, said that the fourth Lord Rayleigh had just died, and would I write an obituary notice of him? I said that unfortunately I was otherwise engaged.

I was particularly interested to meet the Laurences' friend, Count Ilya Tolstoy, the grandson of Tolstoy. In 1922–23 my old schoolfriend Ralph Fox was in 'a little band of Anglo-Saxon oddities', consisting of 'Pacifists, Socialists, faithful Christians, rootless Intellectuals', who were engaged in giving relief to the starving peasants in a remote part of south-eastern Russia. A young Russian, expert with horses, was sent to assist them in their difficult work. This was the nineteen-year-old grandson of Tolstoy, Ilya Andreyich. Ralph gave a fine description of him in the chapter entitled *A Russian Patriot*, in his first book, *People of the Steppes*, which was published in 1925.

Ralph had talked to me about Ilya Andreyich, and their life and struggles among the starving in the steppes, and now, in 1947, ten years after Ralph had been killed in Spain, I met Ilya Andreyich in New York City. I saw the big strong man whom Ralph had described to me twenty-four years before.

He had then been enormously popular with the ladies of the steppes, and now I heard he was equally popular with the ladies of New York City. He told me how, after he and Ralph had been eating hashish, Ralph had wandered out into the desert in a dream, and was lost. Ilya went out to search for him, circling round the region with horses; he found him sleeping in a hollow in the moonlight, with the happy expression of the hashish eater.

After the party I accompanied the Laurences and Ilya Tolstoy to a restaurant. They made some anti-Soviet remarks to which I took exception. Tolstoy, who had worked for Stilwell and the State Department in the Office of Strategic Services, and thence in Tibet, took the line that the U.S.S.R. had been a disappointment. Laurence condemned it on the ground that in it there was interference with personal liberty. I held that no stable civilization could be built on egotism.

After we left the restaurant, Ilya Tolstoy walked with us for some way, and then went into a hotel. Laurence looked at it and him, and said sadly: 'Ilya, I'm surprised at you.'

CHAPTER TWENTY-FIVE

1947
CONSERVATION OF RESOURCES: NEW YORK AND WASHINGTON

THE COLLECTION OF material for the report for UNESCO on natural resources involved twenty-five interviews with authorities on various aspects of the problem. They included Carter Goodrich on population; Fairfield Osborn on fauna and flora; Hermann Mark on wood; Boyd Orr on food and agriculture, Thomas Blaisdell on trade; John E. Doerr on national parks; W. Vogt on biological conservation; Charles E. Kellog on soil; S. Raushenbush on economics, and Arthur Goldschmidt on power.

Goldschmidt was Vice-Chairman of the U.S.'s Inter-Departmental Working Committee, making preparations for the International Conference on conservation. He was Director of the Power Division in the U.S. Department of the Interior, that is, the department dealing with the country's natural resources.

The idea of the Conference had been inspired by Gifford Pinchot, a close friend of both Theodore and Franklin Delano Roosevelt. He convinced F. D. Roosevelt before the Second World War of the importance of the resources problem, and the dependence on it of future efforts to raise the standard of living. Roosevelt requested Ickes, the Secretary of the Interior, to investigate it. The President was influenced by the conception of science as an international resource which could be used to improve the general condition of man; for this reason he desired that mankind should develop and master science to the utmost. He was also influenced by Hitler's agitation for access to resources, which emphasized their bearing on war and peace.

It is probable that Pinchot pointed out that the depletion of resources in the Second World War was increasing the urgency of the problem. The President took a memorandum on conservation to the Yalta Conference; President Truman became interested in this memorandum. Raushenbush was asked to explore the possibility of organizing a post-war meeting on resources, and Truman suggested to the United Nations that they should call such a conference.

Meanwhile, the economic motive for conservation had gained

Conservation of Resources: New York and Washington 295

on welfare and peace, which had had so large a part in the original impulse.

I incorporated the results of my inquiries in a 30,000-word report. It contained fourteen chapters, reviewing the situation with regard to the main renewable and non-renewable resources. I concluded with four pages of suggestions of points on which UNESCO might consider taking action. I summarized the various topics in forty-five points, which might be considered as items for agenda for the forthcoming Conference. For example, one point was the cultivation of 'the surface of the globe as a World-Park'.

Among the resources experts whom I consulted, Arthur Goldschmidt was particularly interesting. His grandfather had been a German immigrant who left Germany for religious reasons, as he was a free-thinker. He went to Texas, where he started a German language paper. Goldschmidt himself had grown up in San Antonio, where he had delivered newspapers at the age of seven. On his way to the UNESCO Conference in Mexico City, he had called in San Antonio to see a barber to whom he had delivered newspapers as a boy. He had not seen him for twenty years, and now he found that his son was at Harvard, after studying in Texas and at Columbia; he was doing research on Greek drama. Goldschmidt thought that this was a good illustration of what was so wonderful about America: the opportunity for all.

Goldschmidt went to Washington from Columbia in 1932, and became one of the leading administrators in the New Deal. He was one of those subsequently accused of 'Un-American activities' by the committee started by Dies of Texas; each case was examined by a committee of investigation. The committee which investigated Goldschmidt had a member who always asked 'do you believe in God?' Goldschmidt's friends gave him all kinds of advice on how he should answer. However, on the day when he was examined, this man was taken ill. 'So God came to the help of the atheist!'

The U.S. had already spent $100,000 on investigating Goldschmidt, and it was still going on. When he was examined, he was asked to outline his career. He described how he had worked his way up, and through college. When he came to that, each Congressman on the committee then made a speech on how he also had worked his way up. Ultimately, they were all buddies together. Someone then suggested that they had obviously got the wrong man, and a report was produced saying that Goldschmidt was a loyal American and an excellent citizen. Since then, he had been in a free and privileged position. He was a civil servant who had been given a kind of public guarantee; he was in the position of the man who had been released from an asylum, and ran for Congress on the ticket that he was the only candidate in the state who was certified sane.

The point on which he was most severely pressed was his subscription to the Spanish Republican cause. He was asked how much, and he said it was large compared with his income. He was asked if he could remember any particular sum, and he said that he could remember one: it was equal to the value of his dog. How so? Well, his dog had puppies which were very valuable. What was to be done with this family, which was too large for him? He could not distribute them among his friends, because there were not enough to go round, and that would be unfair. As one does not take money for the sale of one's dog, it could not be found another master in that way. Then it struck him that if the family were sold, and the proceeds given away, that objection would not arise. So all were sold and the proceeds given to the Spanish Republican cause. This tale of a boy and his dog was regarded as truly American, and his home state, Texas, considered that that was the right attitude to a dog, and from that point the case against him broke up completely.

Goldschmidt's office in the department of the Interior was bigger and more comfortable than any minister's I had seen in Whitehall.

Writing the report increased my consciousness that the fundamental problem was political; the technical facts were reasonably clear, the difficulty was to get appropriate action on them.

When I arrived in New York after the strenuous days in Mexico, I went to my hotel and fell into a deep sleep. On the following morning, while breakfasting in bed, my telephone rang. I was told that a car, belonging to the French Delegation to the U.N., was waiting to take me to Lake Success. I was puzzled, for I had intended to make my own arrangements to go to the U.N. offices later in the day. I thought that Laugier might have arranged it, though this seemed unlikely, as I had not yet telephoned him. I finished breakfast, dressed hurriedly, and found a most determined driver, who ignored what I said, and was purely interested in getting me to Lake Success. After he had delivered me there, he insisted that he would take me back to New York, no matter what I wanted.

I stopped arguing, went into the building, and found the officials concerned with the working party on the conservation of resources. When I came out many hours later, he was still there, contentedly waiting for me. He drove me back to my hotel, and brought out a paper to sign. It turned out that the car had been ordered for a period by a French diplomat who had occupied my room. He had left unexpectedly for Europe, and had forgotten to cancel the car.

I called on Laugier who, as Secretary of the Economic and Social Council, was responsible for the proposed conference on resources. He told me he would never forget what I had done at the time of the invasion of France. He invited me to luncheon in his home at Great Neck. He had had a bad accident, and his leg was still in plaster after eleven months. We discussed the peculiarities of American and French life. He thought the United States might collapse within five years. De Gaulle and Pleven were responsible for the financial situation of France, and the dominance of the black market. France was neither Gaullist nor Communist, but left-centre, and there was no one to express that point of view. On another occasion, Laugier remarked that Joliot was a good fellow, but he ought not to have displaced himself from the direction of the National Centre of Scientific Research.

In Washington I stayed at the Gralyn Hotel, much used by British scientists. It was a little old-fashioned, and like some other things in Washington, reminiscent of provincial Yorkshire before the First World War. My room had no bath and washbasin. Main meals were taken at a restaurant across the road, in the reconstructed stables of a general who had chased Red Indians.

After my first busy day I went to bed quite late, and fell asleep. Presently the phone rang, and E. U. Condon told me he was coming to see me immediately, as he was leaving Washington on business. So he called and told me of his views on resources, and other matters. Since I had first met him in Pittsburgh in 1938, Henry Wallace had appointed him Director of the U.S. Bureau of Standards. The smear campaign against him was under way, but he was bearing it courageously.

Three days later, as I walked from my room downstairs, I saw the back of a familiar figure; it was Cockcroft's. He had arrived from England at that very moment, and had not slept. He had breakfast with me in my room at seven-thirty on the following morning. He told me that Lady Rutherford had appreciated the Rutherford celebration.

The headquarters of *Science Service* was a few doors from my hotel. I called on Watson Davies, and heard about his organization. He then took me to his home and switched on a powerful radio set, which picked up the B.B.C. very well. We listened in silence to a fatuous broadcast on how Nazis were being trained at Wilton

Park to be good democrats; Watson Davies's face was blandly expressionless.

I called at the headquarters of the Federation of American Scientists. It was in a house with many tiny rooms. The tall Californian, R. L. Meier, was in charge. He and his wife and a few assistants seemed to be working on very limited resources. He explained that the F.A.S. would gladly collaborate with the W.F.S.W., but would not affiliate, as that would cause it to be accused of having Communist contacts. He was very much interested in the impact of scientific advances on sociology; he and his wife had investigated the social composition of the scientists in the U.S. Their results appeared to throw light on why so many of them were reactionary, which made it impossible for F.A.S. to affiliate to W.F.S.W.

I visited W. Vogt, chief of the Conservation Section of the Pan-American Union. He was a distinguished authority on the biology of birds, and had given important advice on the conservation of the millions of birds on the west coast of South America, which created the enormous guano deposits. It appeared to me that he tended to apply knowledge of bird and other animal life too directly to human society.

His exposition of the dangers arising from wastage and misuse of resources was acute and vivid, but his proposals for human society were a simple form of Malthusianism. He thought wholesale reduction of human population, including that in the United States and Britain as well as elsewhere, the main answer to the conservation problem.

On my way from New York to Washington, I called on P.A.M. Dirac and his wife, who were then at the Institute of Advanced Studies at Princeton. With them I met Mrs Dirac's brother, E. P. Wigner, today also a Nobel Laureate, and J. R. Oppenheimer. It was my first meeting with Oppenheimer. He was a thin, rather tall man, with a sharp ingratiating manner; he looked as if he had some kind of internal conflict.

He said that the trouble with American society was that not enough Americans were anti-American. He thought that the Anti-Red witch hunt would get much worse before it got better, probably up to the next Presidential election. When I described how the British UNESCO Delegation behaved at Mexico City, systematically carrying out a brief, while the American lurched first one way and then another, he remarked that the delegations at the U.N. Atomic Energy Commission behaved in exactly the opposite way. The Americans had a rigid position, whereas the

British had none at all. Oppenheimer was himself the chief architect of the Lilienthal Plan.

Harry Grundfest and his colleagues in the American Association of Scientific Workers, which was affiliated to the World Federation of Scientific Workers, entertained me in New York. I met Melba Phillips, the distinguished theoretical physicist, who was victimized in the period of McCarthyism. After being appointed Assistant Professor in Brooklyn College, she was dismissed on account of her opinions. Subsequently, she was elected a full Professor in the University of Chicago, and in 1967 became President of the American Association of Teachers of Physics. Her colleagues particularly esteemed her as a person of good judgment, besides scientific distinction.

She came from a farming family in Indiana, and was a pupil of R. H. Fowler when he lectured in California, and a colleague and collaborator of Oppenheimer. She contributed notable work on the quantum theory of alkali metals, and became an author of well-known textbooks.

Melba Phillips had been one of the chief creators of the original atomic scientists' organization. Our discussion of the organization of scientists in the United States gave me a vivid impression of intensity and fragmentation of thought and action. Grundfest told me about the American Association of Scientific Workers' agitation against bacteriological warfare, which was supported by a substantial body of biologists and bacteriologists. They published a report on the subject which became classical. In 1968 the subject again became a very live one.

Melba Phillips invited me to a party at her apartment in Greenwich Village. Among her guests were Theodore Rosebury, author of the well-known book on biological warfare, and L. A. Berne of the United Office and Professional Workers of America.

Melba Phillips is now one of my oldest and most valued friends, and the American personality who inspires in me the greatest hope for the future of her country.

CHAPTER TWENTY-SIX

1948
SCIENCE IN BRITISH OCCUPIED GERMANY

EARLY IN 1948, after my return from America, the Polish Ambassador took the chair at a meeting in the Society for Visiting Scientists, at which I spoke on my visit to Poland in the previous autumn. Shortly after it the Polish Cultural Attaché called on me, and said that Polish intellectuals were joining in the organization of an international meeting in the cause of peace, which would be held in Wroclaw during the coming summer. He asked me whether I could suggest British scientists who might be interested, and consider accepting invitations to attend. This meeting turned out to be the famous World Congress of Intellectuals for Peace, held in Wroclaw at the end of August 1948. I suggested about a dozen scientists, and invitations were sent to them. Most of them accepted. Before the Congress met, however, I visited scientific institutions in the British Zone of occupied Germany.

In April 1948 the Association of Scientific Workers sent a commission to report on the condition of science in the British Zone of occupied Germany. Its members were R. A. Watson-Watt, R. C. Murray and myself. We travelled via Harwich to Göttingen, where the British office for the control of science in the zone was established. Our journey went through the Ruhr, which was littered with ruins almost as terrific as those in Poland. In the warm April weather, numerous fruit trees in the region were covered with blossom, which made the destruction look like the remains of a dead civilization.

The atmosphere of occupation struck me at once as very unstable. At Göttingen we stayed in an old commercial hotel taken over by the T-Force, for interrogation on science. Our first call was at the Research Branch headquarters; this was in the group of about twenty buildings that had been built for aerodynamical research around Prandtl's original laboratory dating from the first years of the present century. The buildings had been taken over by the new Max Planck Gesellschaft, which had replaced the old Kaiser Wilhelm Gesellschaft. There were the remains of an enormous wind tunnel, with reinforced concrete walls one metre thick, and containing two thousand tons of steel.

When this famous group of aerodynamic laboratories first came into British control, we were told that the young British scientist Goodey, who was then about twenty-three years old, was put in command. The senior German scientists then there were astonished to find themselves in the charge of such a youngster, but apparently he handled them excellently.

We had a long discussion with Colonel R. K. Blount, the British scientist in charge of the research branch of the Control Commission, who entertained us to lunch in what had been the Kameradschaft House. He explained the change of policy from that of the invasion forces, whose aim was to defeat the enemy, to the programme for the rehabilitation of German scientists and science.

We called on Otto Hahn, the first president of the new Max Planck Gesellschaft. As such, he was in effect the head of German science. Hahn was an open, attractive man, a modern scientist in the old German academic tradition, dedicated to research and free from non-scientific ambitions. He did not like to be bothered with politics and did not know much about them, though he had a sensitive conscience. He did not make his great discovery until he was sixty-three, after a life-time of first-class research. It was a just reward for his rejection of vulgar ambitions. He had the strength, and also the weakness of the old tradition. Though he avoided political disaster he was unable to contribute much effective opposition to Nazism.

The new Max Planck Gesellschaft seemed to have started with an authoritarian outlook. Its Secretary, Dr E. Telschow, had been secretary of the old K. W. G. Hahn himself, von Laue and Heisenberg were the new authorities. These eminent German scientists seemed to slip back naturally into, not the new Nazi authoritarianism, but the old German academic authoritarianism. We heard what the leaders of German science thought, but no one told us what the young scientists of, say, the age of twenty-eight were thinking. When we put the question directly, there was no reply; apparently it was not known, nor indeed had it been imagined that they might have had views. It seemed that German scientists of that age had no way of expressing their opinions.

We discussed the problem, and the opinion was expressed that a German scientist of twenty-eight would have been working under war conditions for eight years, which would leave him with the mind of a man of twenty. One of our party remarked that, whether or not this was the true explanation, they did have the minds of men of twenty.

We were told that whereas in America eighty out of every thousand attended a university, in England the figure was forty, and in Germany at this time, fifteen. Of these, only twenty per cent were studying natural science. The immediate post-war students were distinctly not inclined to science.

Conversation with the senior German scientists seemed to indicate that they were, on principle, against the planning of scientific research. This was in remarkable contrast with the possibilities of the situation, for the military control had made detailed records of the state of science in the occupied zone, thus providing the data upon which the planning of scientific research could have been carried out better than in almost any other country.

Our visit to Göttingen coincided with the celebration of the ninetieth anniversary of the birth of Max Planck. It had been expected that he would be present, but he died shortly before, and was buried in the Göttingen cemetery. There were about one hundred and seventy representatives at the celebrations, including twenty from abroad. C. G. Darwin and E. N. da C. Andrade represented the Royal Society and the Physical Society. The representatives visited the grave, on which there was a slab bearing just the words: MAX PLANCK. Hahn was the centre of the group.

A memorial meeting was held in the Hall of the University. On the platform were rows of professors, and representatives sat in the stalls. Darwin spoke shortly on Planck, describing him as a modest and correctly behaved man, and perhaps the greatest Foreign Member of the Royal Society. Von Laue's paper was read for him, as he was hoarse. The most striking message came from Einstein, which was read for him by Hahn, who particularly emphasized his plea for freedom in science.

Heisenberg gave a general talk on the Quantum Theory and its consequences. I thought he looked less impressive than he had done fourteen years before. As a middle-aged man of forty-seven, he did not look a great personality; when he was young, his youth emphasized his genius. His prestige now was obviously enormous.

The most remarkable part of the proceedings was the playing of magnetophone records made by Planck on his eightieth birthday in 1938, and on the occasion of the award of the first Planck Medal to de Broglie in the same year. He spoke with vigour and made a series of jokes about the celebrations of his seventieth, seventy-fifth and eightieth birthdays, and then said with great force that he would tell them something on his ninetieth.

Science in British Occupied Germany 303

On the occasion of the award to de Broglie, he had spoken on how Germany and France must be reconciled before it was too late. It was strange to hear jokes from the dead man, whose life was being celebrated. At lunch afterwards a German at our table said how courageous Planck had been in 1938, to have spoken as he did about France. It struck me that Planck's remarks might be interpreted as a hint to the French to collaborate.

There was a bierabend, which gave an opportunity to meet more of the representatives. In particular, there was Rompe from the Soviet zone, who said that Berlin was now the most interesting city in the world. All events were reported on differently by the presses in the four zones of the city, and comparison was very instructive.

I joined a group with him which included Bothe, who had superintended Joliot's laboratory during the occupation of France, and Jensen, who was one of the bravest of the anti-Nazi German scientists. Jensen told me that he saw Bohr two weeks before he left Denmark, and gave him an extensive account of the state of Nazi research. He was certain that Heisenberg's actions during the war were well intentioned, and that he probably knew nothing about concentration camps; he thought him rather naïve in politics.

Rompe told me that about a hundred Germans had returned from England to Berlin. It was very noticeable how much England had influenced them. They had generally become better and simpler speakers, and more competent at managing meetings.

During our stay in Göttingen we had some interesting strolls through the old university town. We saw the tower in which Bismarck had lived when a student, and we walked on the ramparts which had been frequented by the famous Göttingen mathematicians, where they combined philosophical argument with gentle exercise.

On the day following the Planck ceremonies we were taken to the Mining Academy at Clausthal, and the Gmelin Institute. The latter had been moved from Berlin. It produced the great Handbook of inorganic chemistry, planned in one hundred and thirty volumes, fifty of which had appeared. It was now housed in a disused explosive works in the Hartz mountains near the Brocken. The Director was Dr Pietsch, whom I had last met at the Mendeleev Congress in Kharkov in 1932.

We found that Pietsch had got out his photos of Dnieprostroy at the time of the Mendeleev Congress. They included snapshots of Liesegang, A. P. M. Fleming, Neill Greenwood and myself.

He said that the other German delegates at that congress, Hess and Klages, were no longer in science; Hess was in the hills in Bavaria, and Klages was retired.

Pietsch was an unusual man. Besides being a distinguished chemical editor, he was very religious. He had organized the evacuated institute like a monastery. Many of the staff lived in rooms resembling cells. There was a cellar refectory, similar to that in the Society for Visiting Scientists. The larder was controlled with detailed efficiency, very much like the index of the great Handbook. The Institute was then receiving a grant of $20,000 a year from UNESCO.

Pietsch believed that man was to be regenerated through religion, and organized discussion in his institute on science and religion, and on science and politics. He had invited Zwillig from Berlin to put the Communist case, 'just to show how unsuitable it was to the Western Zone'.

The mountain views from the institute were magnificent. They enhanced its isolation, and the unusual combination of Christian ideology with German editorial efficiency.

Murray and I visited Elberfeld, Düsseldorf, Cologne and Dortmund, while Watson Watt went to Berlin. We were accompanied by Dr Marcus Francis, then Regional Research Officer, H.Q. He was a Stoke man, who had fought in the First World War, then worked in industry for a number of years. He had been in Mme Curie's laboratory for three years, when Joliot and Irène Curie were making their greatest discoveries, and had seen Joliot rise from obscurity to fame.

Francis's thorough knowledge of German as well as science was very helpful, especially as so many of the British officials were former civil servants from India. Their traditions, combined with ignorance of German, were conspicuously unsuitable qualifications for ruling Germany. It enhanced the atmosphere of social instability that was so pervasive in the zone.

We visited the research laboratory of the huge chemical works at Leverkusen, which then had 10,500 workers. We met the Director, Professor Bayer, a big man with sword-cuts, in what I suppose was the Heidelberg or similar style. Bayer told us that chemical education had reached only thirty per cent of the old standard. He thought that forty per cent of chemists' time was being wasted through lack of housing, poor clothing, lack of boot repairs, searching for food, etc.

We were entertained by Mr Fowles, a chartered accountant, who was in control of the Leverkusen works. He was familiar with

German industry, and had lived in Berlin from 1923–39. He had been controller of Krupps before he came to Leverkusen, and was very discontented with the British policy on the control of German industry. He lived on the ground floor of Professor Bayer's handsome house, the professor himself occupying the upstairs rooms.

This visit reinforced the impression of the superficiality and temporariness of the British control.

After Leverkusen we went to Elberfeld. The tradition of research on pharmaceutical drugs there had culminated in Domagk's discovery of the sulphonamide drugs. We visited his laboratory, which was quite a modest one. It was alongside the tramway overhanging the river. Besides this tramway, there was also a railway and a road. In the laboratory one noticed a continuous rattling of traffic. It was well equipped, but the great discovery had apparently not been made under ideal research conditions.

At Düsseldorf we visited the former Kaiser Wilhelm Institute for Iron and Steel. It had been a magnificent institute, but only half had been repaired. Then we visited the German trade union headquarters for the British zone. While we were there, an extremely excited member came in with the news that a new appointment to the central board for iron and steel in the Ruhr might precipitate a working-class explosion.

Our next visit was to Cologne. We noted the destruction, and recollected the famous forecast by Blackett, that the thousand-bomber raid would not kill more than six hundred Germans. When the city was captured, it was learned that four hundred and fifty had been killed in the centre, and one hundred and fifty in the outskirts.

I went to see the Rector of the University, Professor Kroll, a German scholar in the romantic tradition, who delivered an eloquent sermon on the inability of the English to understand the German education system. Kroll had been Rector in 1931, and told me how the Nazis had infiltrated the University. First, roughs from the streets in good clothes came in, but this failed. Then, after 1933, pressure was exerted from outside. He attributed the facts of German history to the position of Germany in the centre of Europe, with pressure from all sides.

The English had never suffered this, and had been able to evolve free institutions through the centuries. He said the old relation between professor and student was returning; by this he appeared to mean paternalism.

A colloid chemist, who was also present, whispered to me when he believed that nobody else could hear, that he thought the Germans could not govern themselves, and needed the English to govern them.

One of our most interesting visits was to the famous Institut für Arbeitsphysiologie at Dortmund. The Director, Lehrmann, told us they could deduce how much food people were getting from the black market from their physiological performance. He also said that the German actors were the hardest-worked and worst-fed class. Their union had consulted him, and now he received a stream of free theatre tickets.

An interview which made a deep impression on me was with the director of a steel works. He had a familiar, impudent attitude, and implied that he took his real orders from the Americans, not the British. I had observed during the war, especially in connection with the plans for the re-equipment of the scientific institutions of devastated countries, the penetration of American organizations by American business. Americans who were employees of instrument firms in civil life appeared in military uniforms, acting as technical experts. Their advice was often more like that of a manufacturer's salesman than a technical expert.

I returned to England from the British zone with the conviction that the British control of science was inadequate to any fundamental reshaping of the German scientific and industrial tradition. It was true that a number of scientists with compromising Nazi pasts were being excluded from direct government employment, but then, I noticed that some of these were being given jobs in former Kaiser Wilhelm Institutes for industrial research. Already, in 1948, Germans were laughing behind the Englishmen's backs.

CHAPTER TWENTY-SEVEN

1948
A REDISTRIBUTION OF EFFORT

THERE HAD BEEN disapproval in certain quarters of my management of the Society for Visiting Scientists, which became acute after the war. Some said it was unbusinesslike; others referred to the Society as 'that Communist cell round the corner'. Finally, in May 1948, Sir Harold Spencer Jones resigned from the presidency on the ground that he did not agree with the way the Society was run. Without his support it was impossible to continue, so I immediately told him that I would resign from the secretaryship at the earliest moment convenient to the Society. He then withdrew his resignation, and presently proposed that there should be two Secretaries, the other being Professor F. J. M. Stratton. This was what various people had wanted for a considerable time, as Stratton was the General Secretary of the International Council of Scientific Unions, which was sponsored by the Royal Society. Shortly afterwards, I resigned.

I was sorry to depart from office in the Society, but it had been clear to me since 1946, when I resigned from the British Council, that the conditions for adequate development were uncertain. Sooner or later, the scope of its activities might be curtailed and it might gradually become moribund. Presently, the Society had to close down its house, and most of its activities.

In June 1948 I attended a conference of the Hungarian Trade Union of Engineers and Scientists in Budapest, on behalf of the British Association of Scientific Workers. Harry Knight, the General Secretary of the Association of Supervisory Staff, attended on behalf of his union. We shared a room in our hotel. He had previously been a psychiatrist in a mental hospital, and every night before going to sleep, he read a little from one of G. F. Stout's works on psychology.

The twenty-two thousand Hungarian engineers and scientists who belonged to the union had been mobilized in the reconstruction of their country after the war. Their spirit and enthusiasm were reflected in their speeches; among the most enthusiastic were engineers who had been politically very reactionary. Their co-operation had been obtained by giving them large and interesting tasks in design and construction.

In the middle of my speech, the chairman asked me whether I would be kind enough to allow it to be adjourned and completed later, as Erno Gero had to deliver his speech, and an aeroplane was waiting to take him to a very important meeting. I agreed immediately, and Gero started on his speech at once, making the first public announcement of the Hungarian First Five Year Plan.

I resumed and finished my speech. One of my aims was to secure the affiliation of the Hungarian scientists and technologists to the World Federation of Scientific Workers. The conference decided to found a federation of appropriate Hungarian scientific bodies, which might seek such an affiliation.

This was my first visit to Budapest. I was impressed by the beauty of the city and the cultural sensitiveness of its inhabitants. The magnificent river, the pursuit of sport in an artistic spirit, and the cuisine were very attractive.

After the conference I travelled to Prague to discuss arrangements for the First General Assembly of the World Federation of Scientific Workers, to be held there in the following September. I was struck by the difference in climate and atmosphere between Hungary and Czechoslovakia. In Hungary it was warm, in Czechoslovakia cold and dull. The Czech towns were definitely Western European, neat and tidy. Budapest reminded me of Marseilles in comparison, with the light grey rocks, warm still air, lights and music. The sharp division between Western and Eastern Europe was marked by the hills, from which the Elbe flowed from their northern side to the North Sea, while on the south side the Hungarian rivers flowed into the Danube, and to the Black Sea. On the western side the countries were Protestant, and on the eastern Catholic by tradition.

The World Congress of Intellectuals for Peace, which met at Wroclaw in August 1948, was attended by some four hundred participants from thirty-seven countries. The British group contained forty-three members, at least ten of whom were scientists, a much higher proportion than in any other national group, owing to some extent, I suppose, to my suggestion of scientists for invitation.

In the final session of the Congress, a permanent international committee of intellectuals for peace was proposed. It was to include two representatives for each of the main delegations; in general, one for arts and one for science. The British had in mind

A Redistribution of Effort

Louis Golding and J. B. S. Haldane, but at the very last moment, Haldane refused to accept nomination. The organizers of the British delegation were at a loss, and there was a hurried consultation. What was to be done? Presently, I was proposed, and I accepted. The news was rushed off to the platform.

I had been enthusiastically in favour of the Congress, but it had never occurred to me that I might become a member of the international committee of the movement. I was received civilly by my international colleagues, but obviously not without surprise. When the new international committee was first called to meet in Paris, the secretariat sent the British invitations to Louis Golding and J. B. S. Haldane; they still had Haldane in their minds. Haldane sent the invitation on to me.

The World Federation of Scientific Workers became effectively established by the holding of its First Assembly in Czechoslovakia in September 1948, at the invitation of the Czechoslovak Association of Scientific Workers. It was held in Dobříš, a country palace forty kilometres from Prague, owned by the Czechoslovak Union of Writers.

Thirty-three delegates and observers participated, from thirteen countries, and UNESCO. The membership was then twenty-four thousand, quite a small aggregate. The largest body was then the British Association of Scientific Workers. The U.S.S.R. Union of Cultural Workers had not yet affiliated.

The proceedings were opened by Joliot-Curie, who said that the scientists grouped in the Federation were convinced that the gigantic problems that faced mankind could be solved by the application of the general scientific method, by the application of scientific rationalism. The false idea that science was responsible for the present difficulties of mankind was to be combated.

We were among the new kind of men, whose notion of self led to a co-ordination of individual efforts. As we did not wish to be a detached élite, we should occupy ourselves with the critical situation in which the scientists of several countries found themselves, and with the problems of peace, and the full utilization of science for the benefit of mankind. We must work to put our knowledge at the disposal of the great community of workers.

The biologist Professor J. Bêlehrádec, who had been chairman of the Czechoslovak organizing committee, was elected Chairman of the Assembly. He guided the proceedings with great skill, in four languages. Business was conducted in six sessions. In the first, the Federation's Constitution was discussed and adopted. In the second, I gave my first report as Secretary-General.

I referred to the foundation of the Federation in 1946, when there was the possibility of bringing together the new technical and administrative experience of the scientists who had been engaged in large-scale military scientific work, with the new political experience gained by scientists who had taken an active part in the resistance movements.

The need for the combination had been demonstrated during the previous three years. The release of atomic energy had made many scientists think about the social significance of science for the first time. But not having had the advantage of having taken part in resistance movements, they had suffered from lack of political experience. They had underrated the political and overrated the technical aspects of the control of atomic energy.

The war against Fascism had produced a degree of unity among groups and peoples with very different social ideas. By 1948, however, much of it had been dissipated. An attack on socialism was being promoted under the guise of a struggle against totalitarianism. Under the slogan of the Freedom of Science, the organization and planning of science was being attacked, and irresponsible individualism in research advocated.

The huge scientific activities of the modern world could not be carried out without planning; in fact, planning was now the condition of freedom. It was essential to give the scientist the maximum possible freedom and opportunity to make his contribution to the stock of human knowledge.

The attack on planning was closely associated with that on every from of rational social organization. The political complexion of this movement was to be particularly clearly apprehended, for its protagonists were among the loudest of those who asserted that there was no connection between science and politics; a sentiment that had such a strong though mistaken attraction for many scientists. The scientist's drive to get back to the laboratory and the study was very understandable. He wished to concentrate on his particular research problems, and his absorption in the quest for knowledge was his deepest joy and pleasure. How strongly was he tempted to believe that science had nothing to do with the rest of life, and that he should have the right to do his work without any relation to other people. The prevalence of such ideas, and the retrograde propaganda in their favour, were conditions and forces with which the World Federation would have to contend.

I reported on the Langevin and Rutherford ceremonies, and my visits to Mexico, U.S.A., Poland, Hungary and Germany; on the influence of the existence of the W.F.S.W. in inspiring the formation of new associations in various countries, and on activities on behalf of the W.F.S.W. in connection with UNESCO and the World Federation of Trade Unions. The Federation had issued a questionnaire to affiliated bodies on secrecy, and had acted on behalf of scientists subjected to persecution in various countries. Owing to our intervention, the Conference on Human Rights, held in London in June 1947, passed a resolution demanding full opportunities for native scientists in colonial countries. I mentioned the need for the Federation to produce its own journal on the social relations of science.

The most interesting financial point was that the first two years' activities of the W.F.S.W. had been carried out on a total income of about £500. There had been no paid officers.

Among other business, the Federation considered the general topics of the Organization and Social Responsibility of Scientists; Atomic Energy, Secrecy and Peace; and Reconstruction, Colonial Countries and National Resources. They debated and approved the Charter for Scientific Workers, drafted by J. D. Bernal.

The First Assembly was an encouraging occasion. The Czechoslovak Association of Scientific Workers made efficient and comfortable arrangements. Dobříš had the housing resources of a modernized baronial castle. During the German occupation Nazi leaders used to stay there. They had had remarkable plumbing installed. Joliot-Curie was slightly taken aback when he learned that his splendid suite had at one time been used by Heydrich.

Nejedly, the Czechoslovak Minister of Education, Science and Culture, was an interesting personality. He was typically Czech; small in stature, with dark hair turning grey, moustaches, and spectacles through which peered bright, enthusiastic, wily eyes. He was an authority on music, and especially on the role of music as a medium of national integration in Czech history during the centuries when the country had been ruled by foreign powers. He knew how to talk to the people, and he gave weekly broadcasts on affairs which were accomplished presentations of Government policy, in a form that the general population could appreciate.

Soon after the foundation of the Peoples' Republic of China, the Chinese scientists invited the W.F.S.W. to hold its next General Assembly there. They offered a generous contribution towards the travelling expenses, besides looking after delegates in China. I was disappointed that a majority of my colleagues were not in favour of accepting it. The emergence of the Peoples' Republic of China seemed, from the beginning, to be welcomed with a qualified enthusiasm. I thought this an unpromising sign.

The Second General Assembly of the W.F.S.W. was held in Paris. In its proceedings, attention was focused still more on shaping policy to meet requests from associations of scientific workers in the capitalist countries.

CHAPTER TWENTY-EIGHT

1949
LONDON, PARIS, MILAN AND MOSCOW

THE DEATH OF Louis Rapkine from lung cancer, on 13 December 1948, at the age of forty-four, marked the premature extinction of one of the most creative personalities of the age. A Memorial Meeting was held in March 1949, at the Society for Visiting Scientists. The chief guests were Mme Sarah Rapkine and the French Ambassador, M. Massigli. Sir Robert Robinson, then President of the Royal Society, took the chair. Among the speakers from France were A. Lwoff, F. Perrin, J. Hadamard, Mandelbrojt, and Magat. Joseph Needham and A. V. Hill were among the English speakers.

Rapkine's contributions to science, and his unique rescue of French scientists were vividly described. Hill also mentioned that Rapkine had inspired the foundation of the international Society for the Protection of Science and Learning, under the presidency of Niels Bohr.

J. D. Bernal read a particularly moving appreciation of Rapkine from P. M. S. Blackett, who wrote that 'vision and intelligence alone would not have led him to the success he attained; what made this success possible were his gifts of character—such a combination of intelligence, persistence and charm with first-class scientific ability, are rarely found in one person'.

Blackett said that he had received a congratulatory letter from Rapkine on his receipt of the Nobel Prize, a few weeks before his death. It was one of the most vivid letters he had ever received, and contained absolutely no hint of his mortal illness. His qualities in science, in organization, and his inspiration to all in different countries to work together, were taken too much for granted. He was part of the atmosphere of the century, 'one of the men of the century on whom the model of the scientific world will be based'.

I was asked to speak on Rapkine's part in founding the Society for Visiting Scientists. I concluded my remarks with the observation: 'Whenever we wonder what S.V.S. *ought to be*, we have only to remember the aims, spirit and character of Louis Rapkine.'

In June–July 1949 I attended the Second Congress of the World Federation of Trade Unions as the representative of the World

Federation of Scientific Workers. It was held in Milan, and attended by two hundred and fifty delegates. In my address on behalf of the W.F.S.W. I said that there had never been a time when unity between workers by hand and by brain had been of such profound importance for social progress. The discoveries of scientists had created tremendous possibilities for human progress and also for disaster. The implications of the release of atomic energy were impressive enough, but we could expect in the future biological discoveries of still greater implications. These might be 'conceived not with the invention of mere machines, but with the possibility of directly changing the nature of living things, including man himself. When we remember what Nazi scientists have already done in their extermination camps, we would be utterly horrified by what would happen if these future discoveries fell into the wrong hands'.

While the need for the scientists and the people to come closer together increased, the growing complication of science tended to drive them apart. Scientists were under pressure to separate themselves from other workers, and in capitalist society, to attach themselves to the ruling classes. The effect of this was to place the control of the new science in the hands of the capitalist monopolies, and make scientists their intellectual slaves.

The W.F.S.W. had the task of opposing these tendencies, especially by developing the co-operation between organized workers and organized scientists.

V. V. Kuznetsov of the U.S.S.R. and Lombardo Toledano of Latin America were two of the leading figures. I was much impressed by representatives from Cuba, such as Lazaro Pena, and from Algeria. They were remarkable men, rich in personality and political subtlety. After seeing them, it was evident that the workers in those countries had formidable leaders; this made later events in Cuba and Algeria easier to understand.

An Indian delegate made a long, complaining, excessively boring speech. After a while, one of the Cuban delegates let forth a Gargantuan yawn which reverberated through the hall. I have never heard an interjection expressing the feelings of an audience so perfectly, and at the same time transforming the atmosphere from tense irritation to amused relief.

V. V. Kuznetsov, President of the U.S.S.R. Council of Trade Unions, with a membership of 28,000,000, and leader of the Soviet delegation, was then forty-eight years old. He was by training an iron and steel metallurgist. He had visited Britain in 1945, in connection with the foundation of the World Federation of Trade

Unions, and had discussed the organization of scientists with Blackett, who was then President of the British A.Sc.W.

The Soviet scientific workers, who constituted a section of the U.S.S.R. Council of Trade Unions, had not yet affiliated with the W.F.S.W. We were very desirous of securing their affiliation, which I hoped to facilitate in discussions with Kuznetsov. This was greatly helped by his being a scientist himself. In addition, he had in his delegation Mme Kuznetsova (no relation of his), the general secretary of the union containing the Soviet scientists. The contrast of discussing problems of the organization of scientists with them and with Western European trade union leaders was startling. They were already perfectly familiar with the scientific as well as the trade union aspects of the problem.

Kuznetsov invited me to lunch in the pleasant small garden of a restaurant. His party included, besides Kuznetsova, the General Secretary of the U.S.S.R. Council of Trade Unions, Soloviev. In conference, Kuznetsov spoke little, and only when he had something very clear and definite to say; he had himself very much under control. At lunch, his conversation was extensive and lively. He had a logical, cool, witty mind, and could be fierce and cutting; very far indeed from the bonhommie of the traditional trade union leader from the capitalist countries. In committee he had a quiet manner, great mental force and intellectual power of negotiation.

Some weeks later I saw him again, in Moscow, and had further talks about the affiliation of the Soviet Union of Scientific and Cultural Workers with the W.F.S.W., which eventually took place. In the course of conversation, he commented on the effect of the Second World War on the U.S.S.R. He said that in 1940 the country was advancing very rapidly. Major problems had been solved, and the way had become clear; the construction of the new socialist and technological society could proceed with ever-increasing speed.

Then their positive constructive effort suddenly received a terrific blow. A vast part of their new creation was smashed to pieces. I thought Kuznetsov spoke like an engineer whose great project had been wantonly destroyed in the midst of construction. This destruction of construction had outraged both his engineering sense and his patriotism. No doubt the events of the war caused the iron to enter the soul of many Soviet technologists.

Kuznetsov spoke English well, though he used an interpreter on official occasions. In 1931–33 he had worked abroad on the production of special steels. S. P. Tambovtzev, who had conferred with my committee of technical educationists in 1929–30, told me he had given Kuznetsov lessons in English.

Later on, Kuznetsov was transferred to the Soviet Foreign Office, where he became a Deputy Foreign Minister, and Soviet Ambassador to China. In 1968 he had a prominent role in Soviet relations with Czechoslovakia, as political representative of the U.S.S.R. in Prague.

Our Milan friends had a beautiful House of Culture, in which appropriate meetings could be held. Antonio Banfi took the chair for me when I spoke there on The Social Relations of Science. He was specially interested in Galileo, and we exchanged opinions on the contributions of Galileo and Francis Bacon in this field. A special performance of scenes from Offenbach's 'Orpheus in the Underworld' was put on for the delegates at the Scala.

I was particularly interested in seeing the walls of the Sforza castle in Milan, for Leonardo da Vinci applied to Ludovico Sforza for appointment as a military engineer. I took the opportunity to see Leonardo's *Last Supper* in a convent in the outer part of the city. It was being heavily repaired; I gathered that it had suffered during the war. Two features that were particularly impressive in the original were the rhythm of the thirteen figures, and the colouring. The reproductions I had seen were very inadequate in these respects. I found Milan a much more sympathetic city than Rome. There was an attractive air of competence, down to details.

After the W.F.T.U. Congress in Milan, I travelled in the following month to the conference of Soviet Partisans of Peace. It was more than fourteen years since I had been in the U.S.S.R., and the changes were great. War damage in Moscow had been quickly repaired, and the minds of the people had already become deeply engaged in the future.

Several of us visited T. D. Lysenko. We held discussions with him for three hours in his office. He answered our questions promptly, showing that he was familiar with them, and had given thought to the topics before. He distinguished sharply between the actual work of Mendel, and the theory of genetics developed from it by Morgan and his successors. He said that after he abandoned Morgan-Mendelism his experimental researches in the improvement of crop plants and animals had become much more fertile. He showed us his experimental farm and his greenhouses near Moscow, and many specimens of his tomatoes, potatoes, plant 'chimeras', and vernalized wheats.

When Bernal told him that he was himself the son of a farmer, their discussion became most hearty and cordial. But in general, he received questions rather suspiciously. He appeared to be highly strung, and a chain-smoker. In 1950 I met him again, when I saw another side of his temperament.

I met again S. P. Tambovtzev, whom I had not seen since 1935. He had become a professor of machine-tool design in the Baumann Institute, and was engaged in teaching and research on machine tools. During the war he had been concerned with tanks, and after the defeat of Germany, had been responsible for the dispersal of the German tanks in the Eastern Zone. He told me that the Germans were in general terrified of the Russians. The consequence was that when they met one like himself (he had a charming shrewdness and humour), they were so relieved that they became easy to handle. They kept telling him how much they liked him, and would do anything for him. He had received the honours appropriate to such services.

He was now devoted to academic work in the technological field, and was working on a thesis on machine-tools for a doctorate in science.

Tambovtzev most unfortunately died a few years later. His widow wrote of his good will towards me, and added that he had been happy to have completed his doctorate.

CHAPTER TWENTY-NINE

1950-54
FORESHADOWINGS

IN 1950 I WAS invited to attend a Soviet congress of Partisans for Peace in Moscow, at which the delegates to the Sheffield–Warsaw Congress were to be elected. After my speech at this congress, Lysenko came to me and offered congratulations. He had appreciated my remarks, and I saw that he could be charming when he thought that what had been said was correct. I gained a better understanding of his personality, and his political earnestness; I saw the weight he accorded to the political aspects of scientific questions.

However, it seemed to me that the Peace Movement was developing into a middle-class pacifist organization. I saw no future in such a movement, so I decided to resign from the presidency of the British Peace Committee at the end of 1950.

After my resignation, I had more time for scientific writing. I resumed work on *British Scientists of the Twentieth Century*, which was published in 1952. After this, I was commissioned to revise *Discoveries and Inventions of the Twentieth Century*. It was then in its third edition, and out of date. I carried this out, the fourth edition appearing in 1955.

Meanwhile, the W.F.S.W. continued to expand. But the fundamental difficulty of combining unity with content soon became acute. Majority opinion was in favour of avoiding problems on which there were deep differences, and concentrating on those on which there was wide agreement. This precipitated in the W.F.S.W. the same kind of difficulties as those which had afflicted the peace movement.

In 1951 I went to the first New Zealand Congress for Peace, and addressed other meetings in New Zealand and Australia. On the return journey I was to address peace meetings in Colombo and Bombay, but I was not permitted to enter Ceylon or India.

After Auckland, I spoke in Christchurch, Timaru and Wellington. I arrived at Christchurch in South Island while the Seventh New Zealand Science Congress was in progress, and conferred with officials of the New Zealand Association of Scientific Workers. While I was there, I was asked to record a broadcast on *Science and Peace*. I made it in relation to Rutherford, who had been a student at Canterbury College, Christchurch, where he had done his first research. An appeal for a Rutherford Memorial was in progress, and Lady Rutherford had spoken in support of it on the previous day.

My host in Christchurch was Mr P. J. Alley, a lecturer in engineering at Canterbury College, and a brother of Rewi Alley. He brought me to see Dr H. R. Hulme, then Rector of Canterbury College, whom I had first met in Bohr's institute at Copenhagen in 1932. Hulme took me downstairs to see the place where Rutherford had done his first research on radio waves; it was in a cellar half below ground. The place had been marked by a tablet, unveiled by Sir Henry Dale after delivering a Rutherford memorial lecture. It was being used as a cloak-room for student's hats and coats.

In Brisbane I spoke on *British Scientists of the Twentieth Century* and *The Responsibilities of Scientists*, and gave two addresses on *Science and Peace*, one of which was in the City Hall. I was asked whether I was prepared to speak in the 'suburbs' of Brisbane, and on agreeing discovered that this was at Townsville, seven hundred miles up the coast. The flight there along the coast was very interesting; there was a succession of silted rivers, and the immense possibilities for irrigation and hydraulic development were evident. I was struck by the magnificence of the Australian beach, extending for two thousand miles.

During my visit to New Zealand and Australia I had been impressed by the conscious effort of their scientists to study their own scientific history, and create their own scientific traditions.

In 1952 the Czechoslovak Government founded the Czechoslovak Academy of Sciences. Its task was to plan and organize theoretical and applied science, to serve by way of creative work the welfare of the people and the building of socialism in the country. It was based on the reconstruction of the Royal Czech Society of Science founded in 1784, and the Czech Academy of Science and Art.

The founding was celebrated in Prague in May 1952. Joliot-Curie and I were invited to represent the W.F.S.W. Unfortunately, Joliot's plane became fog-bound in Brussels, and I had also to stand in for him. Besides ceremonies in the fine restored old Carolineum building, and the National Theatre, the delegates were invited to the country mansion at Liblice, which the Czechoslovak scientists had acquired as a rest house. At dinner there, the Chinese delegate, the Chinese Deputy Minister of Education,

Wei Hseuh, was asked to preside, and I was placed next to him. He spoke excellent English, and we exchanged courteous sociable remarks. After a while, he suddenly turned, looked at me intently, and asked most politely who I was. I explained, and he said: 'I am very happy to meet you. I regard you as an old friend. I have read your books on Soviet science. They were early, too.' He expressed surprise that I had not been to China.

Though at least two of my books have been translated into Chinese, that was before the foundation of the Chinese People's Republic. So far as I know, none of my books has been translated into any of the languages of the other Socialist countries.

The W.F.S.W. held its Third General Assembly in September 1953 in Budapest. Among those who participated, besides the officers and delegates of the W.F.S.W., were General Sokhey of India and Sir Robert Watson-Watt of Britain.

The policy of avoiding questions on which there were deep differences of opinion, which appeared in the peace movement in 1950, had now become dominant in the W.F.S.W. I followed the administrative rules which they had inspired, but I disagreed with them. When it became clear in 1953 that I would not find a useful channel of work in the W.F.S.W., I decided to resign in the following year.

I had to attend once more mainly to scientific writing, for economic as well as ideological reasons. When I told my friends that I would have to depend in the future mainly on books, I could not help noticing their scepticism.

However, my old and valued friend, my literary agent, David Higham, immediately fixed up two contracts. One was for a book on *The Sciences of Energy*, and the other on *Six Great Inventors*. These were the first of about a dozen books that I wrote during the next few years mainly for financial reasons. Among them was *Radioastronomy and Radar*. J. A. Ratcliffe and Martin Ryle read the typescript, and I had the benefit of their comments. The sales, as with two of the other books, ran into five figures.

These books were written specifically for the broad public, and had a considerable sale in schools. I enjoyed and learnt more from writing them than I had expected.

CHAPTER THIRTY

1955-57
THE POST-WAR EPOCH ENDED

THE UNITED NATIONS' International Conference on the Peaceful Uses of Atomic Energy held at Geneva in August 1955 marked the end of the post-war epoch in science. The conditions which had been generated in science during the previous twenty years by war began to change, when it had become evident that the major powers had command of atomic weapons that would be mutually destructive in any future atomic conflict. The situation implied that these powers must have much the same nuclear information, whatever efforts they might have made to keep their own discoveries secret from each other.

There was therefore a strong case for publishing as much of this information as possible, and applying it to peaceful purposes, in the creation of unlimited supplies of energy, and in research in science, medicine, agriculture, and industrial production. The abolition of secrecy, now of much less use in the atomic field, and the diversion of attention from the aims of war to those of peace, would both increase international confidence, and make future atomic war less likely.

President Eisenhower indicated in 1953 that U.S. policy was veering in this direction. The United Nations decided in 1954 that an international conference on the peaceful uses of atomic energy should be called, and it instructed its Secretary-General, Dag Hammarskjöld, with the assistance of an advisory committee of representatives of seven nations, to arrange it. The seven advisers were J. C. Ribeiro of Brazil, W. B. Lewis of Canada, Bertrand Goldschmidt of France, Homi J. Bhabha of India, D. V. Skobeltzyn of the U.S.S.R., J. D. Cockcroft of the U.K., and I. I. Rabi of the U.S.

The Conference was fixed for August 1955 and topics and procedure agreed. Hammarskjöld appointed Bhabha to be its President, and a large expert scientific and administrative secretariat was organized.

The Conference proved to be the biggest organized by the United Nations, and the biggest and most important scientific conference that had ever been held at any place at any time.

It was attended by one thousand four hundred delegates from seventy-three nations, more than three thousand observers from academic institutions and industry, and covered by more than nine hundred journalists.

It was as remarkable for its atmosphere as for its size. The scientific attendance, headed by Niels Bohr, was very distinguished and comprehensive. The papers and discussions were conducted in excellent style. The scientists were so happy to meet again, in a number of cases for the first time for twenty years, that controversy had no place in their thoughts.

I covered the Conference as one of the journalists, collecting material for my book *Nuclear Energy in Industry*. I found myself in the Beau Rivage Hotel, the headquarters of the British Delegation, and others, including W. L. Laurence of the *New York Times*. This was very convenient, but I was glad after a few days to move into a pension, which was less expensive. This was run by a professor of Calvinist theology, where post-graduate students stayed. Its educated Calvinist Genevan background provided an interesting contrast with the spirit of the new age manifested in the Conference.

The scientific meetings were held in the Palace of the Nations, which was sufficiently big and well-appointed to accommodate comfortably the thousands involved. While walking in the main conference hall, I suddenly saw before me Weisskopf, whom I had not met for nearly twenty years. He stopped, gazed upwards, closed his eyes and murmured: 'At last all our dreams are coming true,' and we then went our various ways.

Kurdumov, now an academician, and one of the scientific leaders of Soviet metallurgy, joined me at dinner in the restaurant at the top of the Palace building. We beamed at each other, while he repeated several times that it was just twenty years since we we had dined together in Dniepropetrovsk. We exchanged reminiscences.

There were hundreds of such incidents, which gave the Conference its unique atmosphere. The Polish radio asked me to record a broadcast for their country. I said that one of the first feelings of any participant was joy at meeting old friends, in a renewed atmosphere of confidence and optimism. I referred to meeting Infeld, Niewodniczanski and Soltan again. It was natural to compare the present conference of atomic physicists with those of the past. More than twenty years had gone by since the atomic physicists of the world met together in anything like the same happy collaboration. While the restoration of trust and co-

The Post-War Epoch Ended

operation had begun magnificently, there were very great differences of the past. Those had been purely academic, concerned only with the conception of ideas, the solution of mathematical problems and the making of experiments to find out the facts of nature. In those days very few atomic physicists had had any concern at all about the implications of their discoveries, and the bearing they would have on the destinies of mankind. Indeed, some of the most brilliant atomic physicists were in the habit of saying that their work had no bearing whatever on general affairs.

One of the most striking features of the present conference was the complete disappearance of this attitude. Atomic scientists had been forced to learn a good deal about social and political affairs. In such a conference as this, when two decades of scientific progress were being looked at comprehensively for the first time, one was apt to be fascinated entirely by the present and forget the past. It was a kind of explosion of technical scientific information, hitherto generally secret, though, as it proved, known in considerable detail to a few leading scientific powers.

Bhabha had emerged as a leading nuclear physicist in the 1930s. He belonged to the Parsees, who have such a large role in finance in India, comparable with that of the Rothschilds in Europe. His family were associated with the Tata's. Bhabha was a handsome man, always well dressed, in perfect but expensive taste; his fellow-scientists, when he stayed with them, were impressed by the quiet magnificence of his wardrobe. A considerable painter in the modern style, he was the only nuclear physicist who ever had one-man shows of his pictures in Paris and London. As he had first graduated as an engineer, he also had practical sense. Besides being a valued citizen of the world, he was a particularly valued citizen of India. He had just the combination of scientific and engineering qualities to lead atomic development in that country. He became the chairman of India's atomic energy commission, and he decorated its headquarters with his own paintings.

Bhabha's self-possession, abilities, and independence of Europe and America enabled him to steer the Conference admirably. In his opening address he reviewed the prospects of the supplies of nuclear energy for mankind. He remarked that it was a matter of regret that several areas of the world were not represented at the Conference; consequently, the plans of about a quarter of the world's population for fuel requirements and power needs were not before them. Bhabha was referring in particular to the absence of the People's Republic of China from the Conference.

From his review of the resources of the other three-quarters of the world's population, Bhabha concluded that the known resources of coal and oil were insufficient to enable the whole of that population to reach the standard of living of its most advanced sections. The development of atomic energy was therefore an absolute necessity, if the light of civilization was not to be extinguished by the exhaustion of fuel reserves.

The basic ideas of atomic energy were simple. They had all been discovered before the Second World War, but its technology was sophisticated and difficult. Unlike coal or oil furnaces, which are all fundamentally the same, nuclear furnaces might be of many different kinds. For this reason, research on reactor design was of particular interest.

Then there were the ancillary, and only partly solved problems, such as the dangers of the radiation from radioactive materials in atomic plants. On this matter, the first duty of scientists was to establish the truth, and their responsibility to humanity transcended their allegiance to any state.

Bhabha referred to Einstein, who had died four months before the Conference opened, as having perhaps 'done more than anyone else to lay the scientific foundations of the modern age'. He had added his authority to the warning of the dangers of misusing atomic energy, and there was reason to hope that man's intelligence would overcome his fear and weaknesses.

Near the conclusion of his address, Bhabha said: 'I venture to predict that a method will be found for liberating fusion energy in a controlled manner within the next two decades. When that happens, the energy problems of the world will truly have been solved for ever,' for the heavy hydrogen in the oceans would provide almost inexhaustible fuel. Bhabha's forecast excited world-wide interest, for it implied that he believed that fusion-energy might become available as a source of power by 1975.

E. O. Lawrence and V. I. Veksler appeared together to speak on particle accelerators. Lawrence was twenty years older than when I had first met him. I had never seen Veksler before. He was very short, dark and mobile, with exceedingly wide trousers. Beside the tall, substantial fair figure of Lawrence, he darted like a falcon, his trousers flapping like wings. His discovery of a method of focusing beams of particles had made possible the accleration of particles to very high speeds, providing a new order of experimental penetration into the nature of matter.

The Americans Weinberg and Zinn I also had not seen before. Weinberg spoke on the variety of nuclear reactors, of which there are several hundred different types; perhaps a dozen of these might prove to be economically practicable. The choice of types to develop is a most important exercise in scientific and technological decision. I was deeply impressed by Weinberg's sense of judgment in this matter, and his emergence as one of the best writers on the problem of policy decisions in science has not come as a surprise.

Zinn gave a fine illustration of the special American ability in the field of scientific engineering; Americans seem to combine science and engineering more naturally than Englishmen. It is this quality which enhances more than ever the importance to Britain of such a man as Cockcroft, who combined science and engineering to such a high degree.

In his discourse, Cockcroft reviewed the scientific and technical information that had been delivered. He said that for many of the participants the Conference had been 'a celebration of a very great achievement of the scientific world—the harnessing of atomic nuclei to serve the future needs of man'.

Cockcroft entitled his discourse: The Future of Atomic Energy. Among the points he stressed was the importance of capital expenditure. The main requirement in under-developed countries was the provision of capital and the development of technology and agriculture. Nuclear energy would ensure that there was enough energy for these countries to be brought up to the level of the leading countries. He mentioned Oliphant's suggestion of widespread air-conditioning, to make areas such as the inner regions of Australia more habitable, and the prospect of desalination of sea water to provide unlimited supplies of fresh water. Nuclear energy would in the long run lead to large-scale shifts of metallurgical plants. It would also contribute to transport, which at present absorbed about eight per cent of the world's output of energy. He thought that the development of nuclear propulsion in ships would not be rapid, owing to the potential hazards of leaks of radioactivity in confined quarters. He glanced at various other scientific, medical and industrial possibilities, and mentioned in particular that much more was known about the effects of radiations on animals than on human beings. Research on the latter topic should be pressed on, and in the meantime a cautious policy should be adopted.

Cockcroft concluded: 'Having looked into the cloudy crystal ball with my imperfect eyes I feel quite sure that the real picture twenty-five years hence will be very different. For scientific and technological progress is today so rapid that our predictions must be subject to great uncertainties. Rutherford in 1937 could not predict fission and did not believe that atomic power was likely. Now power from fission reactors is assured. I would like tonight to have been able to predict when the exciting prospect of power from fusion reactions would be achieved. But although we are working seriously on this problem in Britain, my vision is not good enough for that. I am not as bold as our President. The experimental physicist must inevitably have a greater appreciation of the problems and difficulties than the theoretical physicist. However, my faith in the creative ability of the scientist is so great that I am sure that this will be achieved long before it is essential for man's needs.'

He ended by hoping that the statesmen who controlled the destinies of mankind, and whose vision had made the Conference possible, would continue to liberate the creative ability of the scientific world, and so enable it to produce those benefits he felt he had so imperfectly attempted to predict.

A large international exhibition of the peaceful uses of atomic energy was organized in the City of Geneva. Some one hundred and fifty exhibitors from nine countries illustrated their contributions on a generous scale. After it was declared open, Bhabha, and Hammarskjöld, who was accompanied by Gunnar Randers, his personal scientific adviser on atomic matters, toured the exhibits. I was walking through one of the aisles when Bhabha saw me. We exchanged the broad smiles characteristic of the conference, and he left his party to exchange greetings for the first time for a number of years.

The numerous reception rooms and passages of the Palace of Nations provided splendid opportunities for informal meetings and conversations. Half the scientific world seemed to be streaming through; scientists, engineers, industrialists, journalists, publishers and others. I was particularly struck by the scientists and engineers. They kept together in different groups and did not mix. The scientists were of varied aspect and dress, while the engineers were far more formal and stereotyped in appearance, mostly carrying brief-cases. They were the designers for the multi-million-pound plants, different men from the scientists of the laboratory and the study.

I returned to England loaded with material, advice and help for my book, which was published in 1956. I then received a contract to write *Six Great Doctors*. This was another interesting task. I had not previously made any detailed study of medical scientists. My subjects were Harvey, Pasteur, Lister, Pavlov, Ross and Fleming. As usual, I found them all more interesting than I had expected. The influence of Padua, Galilean mechanics and the theory of pumps on Harvey; Pasteur's ignorance of medicine, and interest in the asymmetry of organic molecules, and his attempt to prove experimentally that it arose from the asymmetry of the universe; Lister's Quaker and social background; Pavlov's penetrating experimental powers; Ross's poetry and imagination; and Fleming's extensive scientific competence were aspects I had not sufficiently appreciated before.

When this was done, I wrote two more books, both of which were thoroughly instructive to me, and a pleasure to compose: *The Story of Agriculture*, and *Radioastronomy and Radar*, which I have already mentioned. I learned, in the course of writing the former, of the magnitude of the Agricultural Revolution, of the immense extension of the area of cultivated land through the draining of the Fens, and the increase of production through the introduction of new cultivable plants, and improved animal stock. The effect of the invention of agricultural machinery, especially the tractor made versatile by combination with the hydraulic lift, were particularly impressive. I was interested in the saying of Arthur Young, that Sir Richard Weston, the Royalist refugee who brought the turnip to England from Holland after the Restoration, and started a revolution in British farming, was 'a greater benefactor than Newton'. I learned about Robert Bakewell, who doubled the nation's supply of mutton through his animal-breeding, and provided the protein supply that was essential to the workers in the industrial towns, who could not have carried through the Industrial Revolution on a low protein diet.

Then I got a contract for a book on *Six Great Engineers.* The subjects were De Lesseps, Brunel, Westinghouse, Parsons, Diesel and Hinton. Much was already known to English readers about the first four, but not so much about the last two. Diesel was a brilliant and significant figure. He invented his great engine under scientific inspiration. His university studies in thermodynamics had taught him that there should be an engine more efficient than the conventional four-stroke internal combustion engine; he set out directly to invent it, and succeeded. He was excessively conscious of his genius, and felt he was not sufficiently recognized. He rightly considered that he ought to have been awarded a Nobel Prize. Diesel disappeared from a Harwich cross-channel boat, presumably having committed suicide by jumping overboard. One of Nietzsche's books was found in his cabin, so perhaps that pessimistic philosopher had overturned his already uncertainly balanced mind.

Hinton proved to be a particularly original personality. I had conversations with him, and he gave me material on his earlier years.

The most important scientific event of this period was the Soviet achievement in 1957 of the first Sputnik, or artificial satellite. I was struck by the surprise with which it was received in British scientific circles, showing a lack of insight into a decisive aspect of scientific and technological development. Besides opening up the direct controlled exploration of space, it modified the balance of world power. A misjudgement on such a matter was of cardinal significance.

The most important personal event for me as a scientific writer at this time was the contract from the Cresset Press to write a biography of Francis Bacon. This and the succeeding books commissioned by the Cresset Press gave me the opportunity for broadly working out my view and interpretation of the history and recent developments of British science.

In 1958 I was invited to visit Czechoslovakia and deliver two lectures. The subjects I chose were The Role of Personality in Science, and The Dissemination of Scientific Knowledge.

The theme of my first lecture was that scientists were becoming a very important and growing group in society. With the increase in their contribution to the creation and operation of modern social life, the problem of how best to foster them and benefit from their talents, grew in importance, and became one of the chief concerns of our time.

For science to advance most rapidly, and make the biggest possible contribution to social progress, it was necessary to encourage and employ scientists in the best possible way. To do this they must have good organization and resources, but equally important, careful consideration must be given to the needs of different kinds of scientific talent, and the immense variation in the personal psychology of different scientists.

I illustrated different scientific types from prominent scientists in history, and also from scientists whom I had met. I drew the conclusion that 'the range of types of scientist is evidently very wide. It should be made the subject of systematic studies, to discover how to make the best use of each kind of personality and talent for the benefit of society. We discuss the planning and organization of science at length, but do not make sufficiently thorough investigations of what scientists are really like'.

I subsequently expanded the theme of this lecture in my book on *Scientific Types*, published in 1969.

CHAPTER THIRTY-ONE
1958-59
PLANNED SCIENTIFIC WRITING

THE SCIENTIFIC WRITER'S view of science is not the same as the scientist's or the scholar's. He is primarily concerned with the bearing of science on human welfare, in the realm of values as well as things. His function is expository and critical, from the scientific and the social perspectives.

In order to understand the contemporary problems of science adequately, in themselves and their relation to society, it is necessary to understand their origin, as products of a historical situation, as well as the end of a logical chain of scientific ideas. For this reason, the scientific writer has to look at *all* science, ancient as well as modern, and not restrict himself to the spectacular advances of the day, such as the release of atomic energy, space exploration, radioastronomy, or the resolution of the biological mechanism of heredity. He must learn from the past in interpreting the present, and from the present in interpreting the past.

Within one life-time, aspects which were less conspicuous fifty years ago, such as state organization of science, may become as important as any other. Good administration is now a condition for the progress of research, and still more recently, the determination of the direction of scientific effort has become an overtly political problem. The enormous growth of science has stimulated in it the vices as well as the merits of large organizations. Fifty years ago the scientific world was so small that scientists were known rather intimately to each other. Consequently, everyone was more easily valued and accorded his proper place.

Science has become more penetrated with the techniques of advertisement and public relations, and their capitalistic values. This has been observable in the change of attitude to Nobel Prizes. In the early years of this century distinguished scientists hesitated about accepting them. By the second half of the century, especially after the emergence of American science in the first rank, a Nobel Prize became a mark of professional success: the 'brand image' of a world scientist.

In the 1950s I asked an able young scientist whether he would write a book. He replied that he had noticed that in his subject few

Nobel laureates had written a book before they had been awarded their prize. He proposed to regulate his writing activities accordingly, and for the present there would be no question of writing a book.

When under Hessen's inspiration I thought of writing a series of books on British scientists, my first aim was to relate their achievements to the social conditions of their times. This involved the consideration of their views and actions in politics and general affairs. Increasing acquaintance with political actions of contemporary scientists made the politics of earlier scientists more illuminating and, also, the scientific views of earlier politicians.

By far the most important politician concerned with science whom I had known was N. I. Bukharin. As I have already mentioned, he was a rival of Stalin for the leadership of the Soviet State. It seemed to me that, in spite of Bukharin's personally attractive qualities, Stalin was far better endowed with the qualities required for the consolidation of the new Soviet socialist society. The greatest action that could be made for science at that time was the consolidation of a socialist society, in which the development of science was an integral and organic part.

When I wrote the first two volumes in my series, *British Scientists of the Nineteenth Century* and *British Scientists of the Twentieth Century*, I did not have this observation in mind, for the subjects were all technical scientists rather than statesmen of science. Its relevance struck me in 1957, when I started work on scientists in the sixteenth century. This led me to study Bacon. I saw a parallel between the problems of principle and behaviour in which he became involved, and those which involved Bukharin and Stalin. I found that Bacon's conduct and career threw light on theirs and conversely.

Bacon believed that science was the chief instrument by which mankind could overcome its limitations. It was therefore necessary to secure political power, in order to make the correct decisions on the development and utilization of science. Bacon gave the reason for this when he said that the proper aim of knowledge was 'the benefit and relief of the state and society of man; for otherwise all manner of knowledge becometh malign'. Bacon went beyond the bounds of discretion in trying to secure power and place to carry out his conception.

I was profoundly impressed by the depth and detail of Bacon's plan for science, and astonished by his insight into modern ideas on propaganda and public relations, such as his use of the word 'image' in the modern sense. He wrote of 'minds washed clean', and he conceived the notions of communications and information theory. He was aware of the weakness in now fashionable methods of ascertaining public opinion, when he remarked that it is hard when 'voices shall be numbered and not weighed'.

I found that his views on the history of science were far in advance of his time, and that he had conceived the discovery of 'the act itself of discovery', or what is now called 'the science of science'. I was also surprised by the depth of his views on the implication of the velocity of light, of the existence of 'a real time and an apparent time', and the significance of the small range of specific gravities of substances on the surface of the earth. He held that matter must in some sense consist of 'quanta', for otherwise its stability could not be explained. I had learnt from Kotarbinski of Bacon's technical contributions to inductive logic.

When I had completed my reading, I concluded that Bacon, like Stalin, had accomplished ends so important for the benefit of the human race, that his departures from morality, however regrettable, were more than counterbalanced by his positive contribution. In Bacon's conception of the planned use of science he had made a contribution greater than any technical discovery. As John Playfair put it, 'more substitutes might be found for Galileo than for Bacon. More than one could be mentioned, who, in the place of the former, would probably have done what he did; but the history of knowledge points out nobody of whom it can be said, that, placed in the situation of Bacon, he would have done what Bacon did; no man whose prophetic genius would have enabled him to delineate a system of science which had not yet begun to exist!'

I therefore cast my account of Bacon into two parts: *For Mankind*, and *For Himself*. I aimed at showing that in his effort for mankind, he had accomplished, as he said himself, as much as Christopher Columbus and Alexander the Great, by envisaging modern science and a scientific social order. In the second part, Bacon's departure from morality, and its consequences, were interpreted as the price paid for his unique positive contribution. *For Mankind* was the positive contribution, and *For Himself* the negative.

Bacon's contribution was so great that a whole volume on him was desirable. I decided to make him the exemplar of sixteenth century science. As the quatercenary of his birth was approaching a book on him was welcome, so I wrote *Francis Bacon* with the sub-title 'The First Statesman of Science', which, I thought, drew attention to that aspect of his achievement which was most relevant to today's situation.

In writing the book I began to understand better why the founders of the Royal Society felt so indebted to him. I also learned that he was the inspirer of the French Encyclopaedists, and that Scottish scientists of the eighteenth century, who were distinguished by their width of culture as well as the depth of their discoveries, were profoundly influenced by him.

As Mr Peter Morrah observed, I attempted to interpret Bacon's 'public acts in terms of his abstract theories'. When the actions of the succeeding scientists and politicians are studied in this perspective, they become more intelligible and instructive.

After sketching Bacon's political-scientific contribution, I pursued the impact of his influence on his most important successors. This became very great through the circumstances of the seventeenth century. The social changes in Britain manifested in the subordination of the Monarchy to Parliament, and reflecting the emergence of commercial dominance, in agriculture as well as trade, provided the opportunity for adopting a new scientific outlook and organization. Bacon provided the plan. He had succeeded in formulating what was immanent in the historical situation, in spite of the great external difference between his England and the England of 1660.

CHAPTER THIRTY-TWO

1960-68
COMPLETING THE ACCOUNT

I THOUGHT THAT the book on Bacon was a sufficient introduction, from my point of view, to the British scientific outlook in the sixteenth century. After finishing it, I prepared to write on the scientists of the seventeenth century, entitled *Founders of British Science*. I dealt especially with leading early members of the Royal Society, who gave the first institutional coherence to British science. It was published at the time of the Royal Society's tercentenary celebrations. The subjects were John Wilkins, Robert Boyle, John Ray, Christopher Wren, Robert Hooke, and Isaac Newton. They had a major role in creating the new order in the scientific field. They all made their indebtedness to Bacon plain by their references to his ideas and suggestions.

The contribution of John Wilkins was of peculiar importance. He was the brother-in-law of Cromwell, and was made a bishop by Charles II. His central political position, besides his intellectual and scientific accomplishments, enabled him to take the lead in founding the Royal Society. He converted Robert Boyle from an alchemist into a scientist, and suggested to Robert Hooke that he should write his classical *Micrographia*. He initiated the moves that led to Hooke becoming a life-long professor at Gresham College, and to Newton becoming Lucasian professor at Cambridge. In the life of science he was to prove Bacon's lieutenant; not a genius, but an effective statesman.

Robert Boyle was widely saluted as the scientific heir of Bacon, 'shewing philosophy in action'. He was the youngest son of the richest British earl who had made his wealth by colonial development in Ireland. He communicated his own social status to the new science.

While Boyle helped to bring the ruling classes into the experimental laboratory, his almost exact contemporary John Ray, the son of a village blacksmith and woman herbalist, was bringing the indefatigable industry of a countryman into the classification of bilogical knowledge. Boyle and Ray were of about equal weight, and had similar Puritan views.

Christopher Wren was five years younger. He had been a

brilliant pupil at Oxford under Wilkins. At the age of twenty-five he was appointed professor of astronomy at Gresham College; this was in 1657, before the Restoration. His inaugural lecture was very remarkable. He outlined the future of science and astronomy specially in terms of Bacon's programme, and the interests of the City of London, one of whose professors he now was.

After Wren's lectures at Gresham some members of his audience used to meet for further discussion. At one of these meetings, in which Wilkins took the chair, the Royal Society was conceived. Wren was one of those who sketched a preamble for the charter of the proposed society. His draft was not accepted, but it contained highly significant passages. He said that the aim of good government was to ensure that 'the whole body may be supplied by a mutual commerce of each other's peculiar faculties . . .' and 'wealth and industry diffused in just proportion to everyone's industry, that is, to every one's deserts'. This may be compared with Marx's 'from each according to his abilities, to each according to his needs', and Stalin's 'from each according to his abilities, to each according to his work'.

Robert Hooke was a scientist of the first magnitude in inventive experimental physics. He developed and applied microscopy to great effect, discovering the biological cell, and suggesting in the most direct way the invention of synthetic fibres, by analogy with the method by which the silk-worm makes its own silk. He invented optical sights for telescopes, and the watch-spring. He discovered the law of elasticity, essential for the development of scientific civil engineering. He improved the air-pump, converting it into one of the most important instruments of research, and a step towards the invention of the steam-engine. Hundreds of pregnant ideas and experiments are recorded in his works. He derived from Bacon and his own research procedures a notion of a general scientific method.

The influence of Bacon is least explicit on Newton, but is still there. It is seen particularly in his first published paper, on his great discovery of the spectrum of light. He uses Bacon's scheme for the logical presentation of discovery, and especially the notion of the 'crucial experiment'. There is no sign in Newton's published writings that he was influenced by Bacon's views on the planning of science, but he left an unpublished paper on the planning of science, on Baconian lines.

Newton was younger than the other five scientists. Whereas they received their higher education before the Restoration, he received it afterwards; he belonged to the next generation. He was a typical English country grammar school boy. He never left England in his life, and rarely travelled even in England, except between his Lincolnshire home, Cambridge and London. He did not visit Oxford until he was seventy-eight.

After completing *Founders of British Science*, I started on the major scientists of the next historical and social period: the *Scientists of the Industrial Revolution*. The main figures in this period were Joseph Black, James Watt, Joseph Priestley, and Henry Cavendish. Though these men and their associates led an extremely vigorous, creative, almost tumultuous life, they formed a remarkably calm, stable culture. It was more professional, industrial, and middle class.

These five books completed my sketch of the five centuries of British science in its social setting. When they are compared, the form of the evolution of British science and society becomes plainer.

I then devoted more attention to features that had recently become more prominent. In the second half of the twentieth century science had become so large, expensive, and immediately potent, that its government became of prime importance. How was science to be used, what science was to be advanced, and how much resources were to be devoted to it? The statesmanship of science became as weighty a matter as the making of discoveries. I therefore wrote *Statesmen of Science*, dealing with the work and social background of nine British figures who have contributed most to the statesmanship of science during the last century and a half, from Brougham to Cherwell.

After this, my publisher brought out a new edition of the *Social Relations of Science*. Then I wrote *Science in Modern Society*, dealing with the developments in the subject, especially with regard to Britain, since the publication of the earlier book in 1941.

Among the conclusions that arose from the writing of this book was that since the Second World War a relative decline had occurred in British science and technology as a whole, in spite of brilliant advances in certain sciences, such as molecular biology and radioastronomy. These have raised great expectations, but they have not as yet led to the discovery of fundamental new laws of nature comparable with, say, relativity and quantum theories, or the theory of evolution.

In writing *Science in Modern Society* I became aware of some of the dangers in the current developments of sociology. It is at present predominantly an American science, penetrated with the ideology of American capitalism, and one of its most successful instruments in the intellectual colonization of Europe. It appears to be a way of obtaining deep knowledge by easy means. It is particularly seductive to the large numbers of new types of student, drawn into higher education by the extension of the education system. Many of these students have not had much intellectual training before entering the institutions of higher education, and are led to believe that there are short cuts to knowledge. The emphasis of American sociology on quantities rather than values was particularly dangerous with regard to science.

In view of the headlong pursuit of statistics about science and scientists, it seemed desirable to devote more attention to scientific values, especially to quality in scientists. I therefore wrote *Scientific Types*, to draw more attention to the different kinds of scientists, and the problems of organization raised by their particular talents and personalities.

Returning to *Science in Modern Society*, the question is not only how much science there should be, but how good it should be. For the immediate future, the latter may prove to be much the more important. A study of British government scientific policy since the Second World War showed that able advice had been given for two decades, with inadequate effect. The situation was not due to lack of scientific ideas or talent, and its cause lay outside science, in the social system. It was not clear that a mixed economy could provide a stable basis for rapid scientific and social advance. Indeed, it seemed doubtful whether any stable social system could be based on a mixed economy. Until that was superseded by a more advanced socialist system, it did not seem likely that the potentialities of science and technology could be fully realized; their existing condition was confused. It appeared that the pursuit of profit was incapable of providing an integrating social objective. A predominantly competitive social system was not suited to the development and utilization of science and technology with full efficiency.

It appeared that a fundamental social change was necessary if the best use was to be made of science and technology. The most suggestive sign was outside Britain, Europe, and America, and manifested in the Great Cultural Revolution in China, which signified, among other things, a revolt against the values of contemporary capitalism, which exploits science and technology primarily for profit.

BRITISH COUNCIL IN 1945

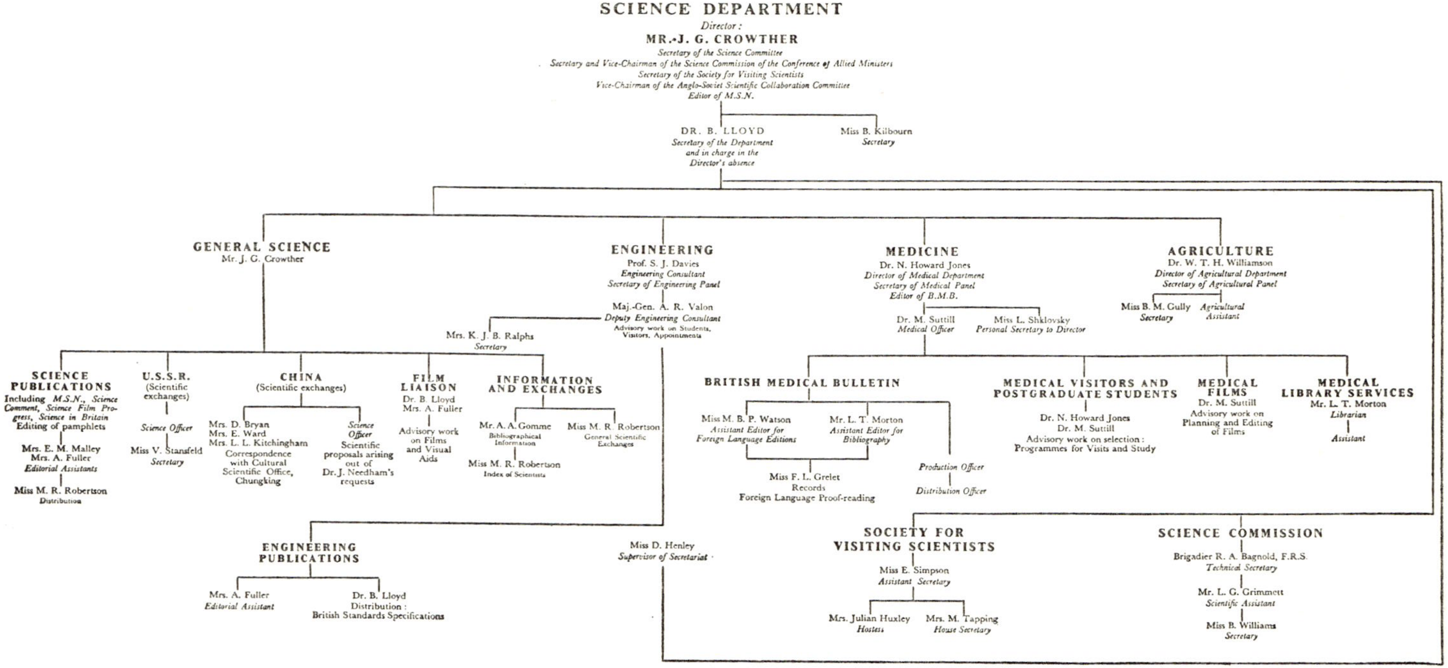

SCIENCE DEPARTMENT

Director:
MR. J. G. CROWTHER
Secretary and Vice-Chairman of the Science Commission of the Conference of Allied Ministers
Secretary of the Society for Visiting Scientists
Vice-Chairman of the Anglo-Soviet Scientific Collaboration Committee
Editor of *M.S.N.*

DR. B. LLOYD
Secretary of the Department
and in charge in the
Director's absence

Miss B. Kilbourn
Secretary

GENERAL SCIENCE
Mr. J. G. Crowther

SCIENCE PUBLICATIONS
Including *M.S.N., Science Comment, Science in Britain* pamphlets, *Science Film Programmes,* Editing of pamphlets

Mrs. E. M. Malley
Mrs. A. Fuller
Editorial Assistants

Miss M. R. Robertson
Distribution

U.S.S.R.
(Scientific exchanges)

Dr. L.
Science Officer

Miss V. Stansfield
Secretary

CHINA
(Scientific exchanges)

Mrs. D. Bryan
Mrs. E. Ward
Mrs. L. L. Kitchingham
Science Officer
Correspondence with Cultural Scientific Office, Chungking
Scientific proposals arising out of Dr. J. Needham's requests

FILM LIAISON
Dr. B. Lloyd
Mrs. A. Fuller
Advisory work on Films and Visual Aids

ENGINEERING PUBLICATIONS
Mrs. A. Fuller
Editorial Assistant
Dr. B. Lloyd
Distribution: British Standards Specifications

ENGINEERING
Prof. S. J. Davies
Secretary of Engineering Panel
Maj.-Gen. A. R. Valon
Deputy Engineering Consultant
Advisory work on Students, Visitors, Apparatuses

Mrs. K. J. B. Ralpha
Secretary

INFORMATION AND EXCHANGES
Mr. A. A. Gomme
Bibliographical Information
Miss M. R. Robertson
Index of Scientists
Miss M. R. Robertson
General Scientific Exchanges

MEDICINE
Dr. N. Howard Jones
Director of Medical Department
Secretary of Medical Panel
Editor of *B.M.B.*
Dr. M. Sutliff
Medical Officer
Miss L. Shklovsky
Personal Secretary to Director

BRITISH MEDICAL BULLETIN
Miss M. B. P. Watson
Assistant Editor for Foreign Language Editions
Mr. L. T. Morton
Assistant Editor for Bibliography
Miss F. L. Grelet
Records
Foreign Language Proof-reading
Production Officer
Distribution Officer

MEDICAL VISITORS AND POSTGRADUATE STUDENTS
Dr. N. Howard Jones
Advisory work on selection:
Programmes for Visits and Study

AGRICULTURE
Dr. W. T. H. Williamson
Director of Agricultural Department
Secretary of Agricultural Panel
Miss B. M. Gully
Agricultural Assistant
Secretary

MEDICAL FILMS
Dr. M. Sutliff
Advisory work on Planning and Editing of Films

MEDICAL LIBRARY SERVICES
Mr. L. T. Morton
Librarian
Assistant

SOCIETY FOR VISITING SCIENTISTS
Miss D. Henley
Supervisor of Secretariat
Miss E. Simpson
Assistant Secretary
Mrs. Julian Huxley
Hostess
Mrs. M. Topping
House Secretary

SCIENCE COMMISSION
Brigadier R. A. Bagnold, F.R.S.
Technical Secretary
Mr. L. G. Grimmett
Scientific Assistant
Miss B. Williams
Secretary

INDEX

Achmatowitz, 279
Adam, N. K., 42
Adams, W. S., 179
Adlon, Hotel, 25f.
Agriculture, The Story of, 327
Alexander, A. V., 268
Alley, P. J., 319
Allibone, T. E., 93, 120
Allied Ministers of Education, Conference of, 240-5
American Association of Scientific Workers, 299
American science, emergence of, 199f., 233
Anderson, C. D., 105, 153
Anderson, Sir John, 241, 246, 252
Andrade, E. N. da C., 251, 302
Andrews, C. F., 25
Anglo-French Society of Sciences, 214-18, 261, 263ff.
Anrep, G. V., 42
'Anti-matter', 67, 106
'Anti-universe', 106
Appleton, Sir E. V., 210, 255, 262, 288
Arbeitsphysiologie, Institut für, 306
Armstrong, E. F., 244f.
Armstrong, H. E., 148, 150
Aron, R., 220
Artificial radioactivity, 130
Astbury, W. T., 109f., 262
Athenaeum, 230, 237
Atomic bomb, 246, 258ff.
  Energy, UN International Conference on the Peaceful Uses of, 321-6
Auger, P., 188, 213, 215f., 262, 265, 267, 287, 292
Auriol, Vincent, 284
Auschwitz, 267

Bacon, Francis, 47, 82, 178, 278, 316, 328, 330ff.
Francis Bacon: The First Statesman of Science, 330ff.
Bagnold R. A., 244
Baker, Sir John, 31f.
Bakewell, R., 327
Baldwin, Stanley, 116f.
Ball games and physics, 62, 176
Band, W., 249
Banfi, A., 316
Baranowski, 279f.
Barbag, J., 277
Barker, Sir E., 206
Barkla, C. G., 127
Bauer, E. H., 267
Bauhaus, 62, 125, 195f.
Baxter, 177
Bayer, 304
Bayet, A., 269
Beaverbrook, Lord, 21, 140
Bêlehrádec, J., 309
Beltran, E., 288
Benes, E., 232
Bennett, Arnold, 140
Béra, M. A., 265, 269f., 278
Berlin, 60-9, 276f.
Bernal, J. D., 84, 94, 110, 188, 201f., 204, 211, 218, 220f., 231, 262, 271, 273ff., 284, 311, 316
Berne, L. A., 299
Bethe, H. A., 126, 206
Beryllium, metallurgy of, 60

Index

Bhabha, H. J., 287f., 321, 323-6
Bidault, G., 284
Biquard, P., 60, 116, 134, 184, 188, 267f., 284
Blaisdell, T., 294
Blount, R. K., 301
Biology, molecular, 109
Biophysics, 43
Blackett, P. M. S. (Lord), 62, 88, 95, 105ff., 120f., 133, 188, 200, 217f., 226f., 231, 252, 271, 281, 285, 288, 305, 313, 315
Bloch, F., 99
Blum, L., 183, 246
Bodenstein, 81
Bohr, Mrs., 107
Bohr, Niels, 81, 98ff., 102, 136, 188, 266, 303, 322
  Copenhagen conferences, 98ff., 166
  on racialism, 158f.
Bolton, L. C., 15
Bombing, effects of, 220f.
Bone, James, 35, 78, 96, 111, 195
Borel, E., 282
Born, M., 126, 128f., 262
Bothe, W., 97, 188, 303
Bowden, F. P., 111, 155
Boyd-Orr, Lord, 201, 294
Boyle, Robert, 333
Bragg, Sir Lawrence, 110, 188, 200
Bragg, Sir William, 83, 156, 221, 226, 228-31
Brayshaw, G. L., 210
Breuer, M., 125
Bridgman, P. W., 169
British Association, 32, 34, 37, 81f., 167, 200ff.
British Council, 226-35, 240f., 254, 264, 267
British Scientists of the Nineteenth Century, 79, 81, 154, 168, 172, 214, 269, 330
British Scientists of the Twentieth Century, 318, 330
Broglie, Louis de, 215, 284

Broglie, Maurice de, 284f.
Brougham, Lord, 335
Bukharin, N. I., 76, 80, 86, 143, 330
Bullard, E. C., 256
Burch, C. R., 120
Burgers, J. M., 271
Burnham Committee, 27
Bush, Vannevar, 180
Butler, Dr. Montagu, 9
Butler, Lord, 243
Butler, N. M., 176

Callender, H., 64
Cannon, Miss, 177
Carrel, A., 171
Cassin, R., 239
Cavendish Laboratory, 153, 159, 283
C.E.R.N., 272
Chadwick, Sir James, 91-8, 101, 292
Chamberlain, Neville, 187, 204
Chapman, S., 13, 232
Chemical industry, views on, 63f.
Cherwell, Lord, 94, 122, 132, 192, 211, 252, 254, 335
Childe, V. Gordon, 20, 200f., 246
China, Peoples' Republic of, 311, 323
  Great Cultural Revolution of, 336
Choyonowski, 281
Churchill College, 104
Clark, F. Le Gros, 272
Clark, W. Le Gros, 84f.
Cleator, P. E., 170
Cloud-chamber, 105, 162
Cockcroft, J. D., 84, 87f., 93, 101-5, 119f., 136, 142, 171, 180, 188, 218, 226, 231, 283, 297, 321, 325f.
Cohen, Rose, 19, 21, 57, 73, 142
Cohn, A. E., 171f.
Cole, G. D. H., 20
Colloids, 42
Compton, K. T., 180

Index 341

Conant, J. B., 168f., 171, 173, 175, 211, 224
Conditioned Reflexes, 42
Condon, E. U., 102, 192, 297
Conservation of natural resources, 283, 285f.
Cook, J. W., 108, 200
Cooke, P. M., 271
Corona spectrum, 42
Cosmic rays, 97, 105
Crew, F. A. E., 34, 112
Cripps, Sir Stafford, 251ff.
Crowther, E. M., 246
Crowther, J. A., 44f., 175
Crowther, James, 28, 74
Crozier, W. P., 99, 118, 225, 253
Curie, Marie, 93, 97, 130
Curie tradition, 133

Dale, Sir Henry, 235, 237f., 244, 249, 252, 255, 260–5, 319
Darlington, C. D., 218
Darrow, K. K., 198f.
Darwin, Sir Charles, 99, 127, 200, 216, 302
Davies, S. J., 58
Davis, Watson, 297
Davisson, C. J., 199
Davy, H., 81
Dean, W. R., 9
Debye, P., 131, 164
Découverte, Palais de la, 186f., 264
Dee, P. I., 101, 120
Delbrück, H., 100, 160
Desch, C. H., 232, 236
Destouches, 215
Dewar, J., 163
Dialectical materialists, 254
Dickinson, H. D., 84
Diesel, Rudolph, 327
Dietrich, Marlene, 60
Dietz, David, 191
Dirac, P. A. M., 39, 67f., 99, 106ff., 218, 262, 264, 298
Discoveries and Inventions of the Twentieth Century, 318
Dnieper Dam, 54

Dniepropetrovsk Physico-Technical Institute, 145f.
Dobb, M. H., 20
Doerr, J. E., 294
Domagk, G., 305
Donnan, F. G., 65, 241
Dunn, L. C., 171, 292
Dunning, J. R., 205
Dunstan, A. E., 119

Ebbutt, N., 66
Eddington, A. S., 37f., 46, 105, 108f.
Ede, R., 19
Eden, Anthony (Lord Avon), 233
Edison, T. A., 189
Egerton, Sir Alfred, 33, 271
Egiazaroff, 54
Ehrenfest, P., 166
Ehrenfest, Mrs., 166f.
Ehrenfest, Paul, 166f.
Einstein, A., 67, 81, 194f., 324
Eisenhower, D. W., 321
Electron, positive, 106
Eliot, T. S., 198
Ellis, Sir Charles, 92, 103
Energy, the Sciences of, 320
Engelhardt, V. A., 56
Engineers, 31f.
English, Basic, 45
Ephrussi, 188, 267
'Espinasse, P. G., 84f.
Evans, Lord, 226f.
Ewald, P. P., 127
Ewing, Sir Alfred, 157
Excellent, H.M.S., 9ff., 15f.

Famous American Men of Science, 162, 169, 173, 189
Faraday, M., 80f., 155
Farrington, B., 272
Feather, N., 97
Federation of American Scientists, 298
Feldenkreis, M., 186
Fermi, E., 135, 205
Finkelstein, B. N., 80, 145

342   *Index*

Fission, 203, 205, 326
Fleming, A. P. M., 54, 210, 303
Fleming, Sir Ambrose, 256f.
Fletcher, Sir Walter Morley, 112
Flexner, A., 193f.
Fock, V. A., 136
Fondation Universitaire, 240
Forster, E. M., 107
Foster, J. S., 284
Founders of British Science, 333
Fowler, Elizabeth, 284
Fowler, P. H., 284
Fowler, R. H., 9, 11, 96, 99, 299
Fowles, 304
Fox, H. Munro, 262
Fox, Ralph, 19, 21, 52, 57, 144, 154, 168, 293
Francis, Marcus, 304
French science, 183–8
Frenkel, J., 138f.
Freud, S., 152
Freundlich, H., 123
Freymann, 214, 269, 276
Frisch, O. R., 121, 203
Frumkin, A. N., 56
Fulbright, Senator, 243
Fusion, 326

Gaitskell, H., 94
Galileo, 316, 331
Gamow, G., 102
Gaposhkin, S. I., 179
Garnett, David, 154
Gasser, H. S., 171f.
Gero, Erno, 308
Giauque, W. F., 164
Gibbs, Willard, 176f., 189
Gillie, Darsie, 66
Giral, F., 290
Goetel, 281
Golding, Louis, 309
Goldschmidt, A., 286, 290, 294f.
Goldschmidt, R., 65
Goldsmith, M., 275
Goodey, 301
Goodrich, C., 294
Graaff, van de, 121, 180

Gray, Sir James, 57, 112, 290
Greene, 175
Greenwood, Neill, 303
Gregory, Sir Richard, 46, 122, 201f., 207, 210, 232, 236, 239, 245
Grimmett, L. G., 244, 287f.
Gropius, W., 62, 125f., 183
Groves, General L. R., 192, 292
Grundfest, H., 299
Guest, David, 78f.
Gurney, R. W., 102

Haag, J., 16
Haas, de, 164
Haber, F., 63ff.
Hadamard, J., 252, 265, 313
Hague, B., 31
Hahn, Otto, 102, 203, 301
Halban, H., 185, 216
Haldane, J. B. S., 45, 47, 83f., 171, 180, 188, 218, 236, 262, 275, 309
Hammarskjöld, D., 321, 326
Hardy, G. H., 11, 90, 226
Hardy, Sir William, 41, 47, 112, 164
Harington, C. R., 262
Harnack House, 65, 240
Harrisson, T., 222
Harrod, Sir Roy, 200, 212
Hart, Henry, 182
Harteck, P., 135
Hartree, Colin, 15
Hartree, D. R., 9, 14, 180
Hartree, W., 14
Harvard, 171–182, 189
Harvey, W., 327
Harwell, 104
Haslett, A. W., 275
Hastings, Somerville, 84
Hayek, F. A., 236
Haworth, W. N., 35f., 108
Heilbron, Sir Ian, 262
Heisenberg, W., 81, 99f., 107, 128, 155, 287 301ff.
Heitler, W., 161

Index

Henderson, L. J., 174ff.
Hessen, B., 77, 79f., 86, 142, 330
Hevesy G. C. de, 123
Heylandt, C. W. P., 115
High-altitude research, 120
Higham, David, 51, 180, 320
Hill, A. V., 9, 12, 14f., 18, 42, 83, 112, 139, 172, 176, 208, 218, 232, 241f., 256, 271f., 313
Hinshelwood, Sir Cyril, 132
Hinton, Lord, 328
Hirschfeld, L., 280, 283
Hulme, H. R., 319
Hutton, Sir Leonard, 202
Hutchins, R. M., 196f.
Huxley, Sir Julian, 34, 84, 87, 112, 212, 238f., 246f., 271
Huxley, Juliette, Lady, 87
Huxley family, fame of, 291
Huxley, T. H., 46f., 87
Hogben, Lancelot T., 112, 168
'Holes', theory of, 67
Holst, G., 165
Holveck, F., 188
Hooke, R., 334
Hopkins, Sir F. G., 68, 147, 156, 219
Horner, Arthur, 272
Horrabin, J. F., 20
Houtermans, F. G., 62
Howarth, O. J. R., 234

Industry and Education in Soviet Russia, 73
Infeld, L., 194, 322
Innes, R., 36, 274
Institute of Advanced Studies, 193, 298
Irvine, Sir James, 36

J-phenomenon, 127
Jaczewski, 278
Jansky, K., 118f.
Jeans, J., 38, 46, 50, 70, 108f., 168
Jeffreys, Sir H., 168
Jensen, J. H. D., 303
Joliot-Curie, F., 68, 97, 130, 133ff., 172, 184ff., 205, 211, 215ff,. 261ff., 267ff., 274, 281, 284, 297, 304, 309
Joliot-Curie, Irene, 68, 89, 97, 130, 133ff., 183f., 211, 261, 263, 267, 284, 304
Jones, B. Mouat, 58, 70–73
Jones, H. Mumford, 179, 215
Jones, Sir Harold Spencer, 246, 307
Jones, Howard N., 227
Joffe, A. F., 55, 77, 120, 131ff., 155, 231f.
Jordan, E. P., 128
Joubert, Sir Philip, 252
Journalism, scientific, 35, 41–51, 204
Jouvenal, H. de, 187

Kaempffert, W., 193, 201
Kaiser Wilhelm Gesellschaft, 65
Kant, E., 124
Kapitza, Mme., 140
Kapitza, P., 39, 44f., 53, 60, 68, 87, 91, 103, 115–18, 139–43, 147–51, 184, 232
Club, 94f.
Kefauver, 245
Keiller, A., 111f.
Kellog, C. E., 294
Kelvin, Lord, 154
Kennaway, Sir Ernest L., 108
Kennelly, A. E., 174
Kepes, G., 62, 126
Kerr, Sir Archibald Clark, 211, 262
Kerr, W., 31, 58
Kharkov Physico-Technical Institute, 113, 136–9
King, A., 245
Kirov, assassination of, 141
Knight, H., 307
Knipper, Olga, 87
Kohts, 87
Kolmann, E., 77
Konorski, J., 277f.
Koo, Wellington, 232
Kotarbinski, T., 278f., 331

Koteliansky, 196
Kowarski, L., 185, 216
Kramers, H. A., 99
Kriloff, A. N., 53
Kroll, 305
Kronig, R., 99
Kronig, Mme., 100
Krzhizhanovsky, G. M., 58
Kuhn, R., 82, 188
Kulik, 53
Kurdumov, G. V., 145, 322
Kurti, N., 122, 164
Kuznetsov, V. V., 314ff.

Labarthe, A., 220
Labour Government, collapse of Second, 81
Lacassagne, A., 44
Landau, L. D., 87, 139, 151
Lange, 56
Langevin, A., 267
Langevin, Hélène, 267
Langevin, P., 34, 164, 213, 216, 267f.
Lankester, Sir Ray, 46
Laski, H., 169, 173, 197
Laue, W. von, 301f.
Laugier, H., 216ff., 297
Laurence, W. L., 191f., 292f.
Lawrence, E. O., 131, 324
Lehrmann, 306
Leiden, Kamerlingh Onnes Laboratory, 163
Lemaître, Abbé, 111, 188
Lennard-Jones, J. E., 262
Lenin, V. I., 73
Leonardo da Vinci, 316
Leveillé, 267
Leverkusen, 304f.
Lewis, W. B., 321
Liesegang, 303
Lindemann, F. A., (see Cherwell, Lord)
Linstead, R. P., 209
Lippman, Walter, 181
Little, D. A., 175f.
Longchambon, 217f.

Lonsdale, Dame Kathleen, 256
Lovell, Sir Bernard, 205, 252
Low temperature physics, 113, 122, 163
Lowell, A. L., 175
Lucas, C. E., 236
Lwoff, A., 313
Lyons, Sir Henry, 251
Lysenko, T. D., 316ff.

Macdonald, J. Ramsey, 83f., 122, 156, 196
McGowan, Lord, 254
McLeish, Archibald, 246
McLennan, J. C., 96
Macmillan, Harold, 210
McMunn, N., 23ff.
McNeil, Hector, 254
Magat, M., 313
Magnetic poles, unit, 68
Maisky, I. M., 232
Manchester Guardian, 35, 49ff., 54, 56, 195, 256
Mangold, O., 65
Mandelbrojt, S., 219
Mapother, E., 84
Mark, H., 62f., 294
Martin, Henry, 236
Massigli, R., 263, 313
'materialization', 133
Mathematical instruction, new methods of, 23f.
Mathieu, J. P., 271
Maton, Commander, 17
Maury, P. Bonet, 271
Max Planck Gesellschaft, 300f.
Maxwell, James Clerk, 154, 189 centenary of, 80
Medawar, Sir Peter, 236
Meier, R. L., 298
Meitner, Lise, 286
Mendelssohn, K., 122, 164
Menon, Krishna, 68
Merton, R. K., 178, 292
Mesa, 289
Métadier, 214f.
Meteorite, Siberian, 53

Index

Millikan, Glen, A., 84
Millikan, R. A., 82, 155, 208
Milne, E. A., 9, 12f., 42, 108f., 161, 168
Mitchinson, Naomi, 236
Mitford, Jessica, 183
Milford, Sir Humphrey, 32, 51
Mitkevich, W. H., 77
Moholy-Nagy, L., 20, 62, 124f., 195f.,
Mond Laboratory, 103, 115–8
Morgan, T. H., 153
Morize, 215
Morrison, Herbert, 211, 252, 254, 271, 273
  shelter, 31
Moseley, H. G. J., 208
Mott, N. F., 153
Muggeridge, M., 143
Muller, H. J., 188, 144f., 153, 168
Murdock, K. B., 179
Murray, R. C., 300
Mysterious Universe, 70f.

Nagaoka, H., 81
Nash, A. W., 119
Needham, Dorothy, 28, 271
Needham, Joseph, 112f., 132, 188, 223, 234f., 244f., 249, 287f., 313
Negrin, Juan, 233
Nejedly, 311
Neumann, J. von, 194
Neutron, 189–200
New Statesman, 35
New Zealand Association of Scientific Workers, 318
Newton, Isaac, 327, 334f.
Newton's Principia, The Social and Economic Roots of, 79
Nicholson, Max, 285
Niewodniczanski, 281, 322
Nimitz, Admiral, 211
Northrop, J. H., 188
Norton, W. W., 171
Nuclear Energy in Industry, 322
Nunn, T. P., 22f., 26

Obituary notices, 130
Occhialini, G. P. S., 105
Ogden, C. K., 43, 45f., 50
Oliphant, M. L. E., 120, 135, 200, 248, 325
Onnes, Kamerlingh, 163, 165
Oparin, A. I., 114
Operational research, 9
Oppenheimer, J. R., 298f.
Orowan, E., 262
Osiris and the Atom, 43
Oxford University Press, 25f., 28, 33f., 42, 210, 255

Paneth, F., 123
Parkinson, Nancy, 240ff.
Pasenkiewicz, 281
Pasteur, L., 327
Paul, W., 20
Pavlov, I. P., 42, 139, 142, 151f., 279, 327
Payne, Cecilia, 179
Peacock, Ralph, 26
Peano, G., 34
Pearson, K., 23
Pegram, G. B., 199
Peierls, Sir Rudolf, 126, 150
Petrovsky, D. A., 57ff., 71, 73, 142f.
Perrin, Francis, 188, 217, 262, 292, 313
Perrin, Jean, 184f., 187, 216, 268
Petroleum, 119
Petroleum, Science of, 119
Philips Laboratory, Eindhoven, 164f.
Phillips, Melba, 299
Physical Problems, Institute for, Moscow, 150f.
Physics, International Series of Monographs on, 39
Piccard, A., 130
Picken, L. E. R., 249
Pienkowski, 283
Pietsch, E., 303f.
Pilley, J., 84
Pinchot, G., 294

Index

Pirie, N. W., 84
Pitts, W., 290
Plan, First Five Year, 57, 70–5, 84
Planck, Max, 65, 81, 282, 302f.
Playfair, John, 331
Pol, van der, B., 188
Polanyi, M., 63, 66, 123, 236f., 265
Pontecorvo, B. M., 185
Powell, C. F., 154, 281
Power, Eileen, 22
Priestley, R. E., 119
Princeton, 183
Progress of Science, 281
Protein structure, 109
Proton, negative, 106
Pugwash movement, 186

Quantum mechanics, new, 128

Rabi, I. I., 199, 292, 321
Radioastronomy, 118f.
Radioastronomy and Radar, 320, 327
Raman, C. V., 42, 140, 188
   effect, 42
Randers, G., 42, 253, 326
Rapkine, Louis, 93, 218f., 241, 252, 262, 265, 313
Ratcliffe, J. A., 320
Raushenbush, S., 294
Rayleigh, fourth Lord, 82, 200, 293
Relativity, general theory of, 108
Riley, D. P., 232, 234, 267, 271
Ritchie-Calder, Lord, 168, 191, 232, 275
Ribeiro, J. C., 321
Richardson, H. L., 271
Richmond, H. W., 9, 16f.
Rivera, Diego, 289f.
Robinson, Sir Robert, 231, 271
Rockefeller, David, 176
Rocket missiles, 116
Romilly, Esmond, 183
Rompe, 303
Roosevelt, F. D., 169, 197, 294
Roosevelt, F. D. Junior, 176
Rosbaud, Paul, 60–63, 116

Rosebury, T., 299
Rosenblueth, A. S., 290
Rosenblum, S., 93
Rosenfeld, L., 136
Rothschild, Lord, 241
Rothstein, Andrew, 53
Rous, Peyton, 171f.
Rowse, A. L., 19
Royal Society, 211, 235, 250f., 264
   rebellion in, 155
Rubenstein, M., 77
Ruhemann, Martin, 62, 113
Russell, Bertrand, 41
Russell, Sir John, 144, 239
Rutherford, Lady, 284, 297
Rutherford, Lord, 9, 12, 34, 38, 44, 49, 68, 83, 89f., 93, 95f., 100–3, 116f., 120f., 126, 130, 135f., 147–50, 155ff., 161f., 168, 175, 189f., 319, 325
   death of, 189
   his methods, 285
   Memorial Meeting, 269, 283f.
   on racialism, 158f.
Ruzicka, L., 188
Ryle, Sir Martin, 320

Saha, M. N., 284f.,
Samuel, Lord, 252
Sanders, A. G., 249
Sarton, G., 173, 176f.
Sarton, May, 177
Sarton, Mrs., 177
Schofield, H., 28, 58, 72
Science, a Ministry of, 273
Science and World Order, 232, 234
Science at the Cross Roads, 78
Science at War, 256
Science, British failure in utilization of, 209
Science Committee of the British Council, 250
Science for You, 43, 50
Science and the American Constitution, 173
   and social relations of, 197, 201, 206f., 232, 265ff., 270

Index

Science—*cont.*
  history of, 76–80, 178, 197
  writers, 84, 191, 274
  Writers, American National Association of, 274
  Writers, Association of British, 274
Science in Modern Society, 335f.
Science in Soviet Russia, 57
Science in War, 213
Scientific Advisory Committee, 220, 229f.
  equipment, 242
  journalism, 35, 41–51, 204
  Unions, International Council of, 248, 250, 270, 307
  Workers, Association of, 271
Scientific Correspondent of the *Manchester Guardian*, 49, 260
Scientific Types, 328, 336
Scientists of the Industrial Revolution, 335
Scott, C. P., 49, 67
Semenov, N. N., 131, 140, 274
Seward, Sir Albert, 69, 207
Shapley, H., 176f., 274
Sharp, Clifford, 35
Shaw, G. Bernard, 20
Sherrington, C., 132, 156
Short Stories in Science, 43, 50
Sidgwick, N. V., 33f., 38
Sigerist, H., 195
Simon, Sir Francis, 61, 81, 122, 164
Simpson, Esther, 243
Simpson, J. A., 271
Singer, Charles, 77, 178
Six Great Doctors, 326
Six Great Engineers, 327
Six Great Inventors, 320
Skobeltzyn, D. V., 105, 321
Slater, Sir William, 212
Soviet Science, 144
Soviet science, 76
  effect of integrating with the State, 57
Soviet theoretical physics, 139

Soviet Union of Scientific and Cultural Workers, 315
Smuts, General, 157
Snow, C. P. (Lord), 111, 202, 255, 272
Sobolev, 232
Social Function of Science, 204, 236
Social Relations of Sciences, 173, 210, 236f.
Society for Freedom in Science, 237
  for the Protection of Science and Learning, 313
  for Visiting Scientists, 240, 261, 267f., 270, 307, 313
Sokhey, General, 320
Sokolnikov, 142f.
Soltan, 322
Sommerfeld, A. J. W., 128
Stackpole, S. H., 174
Stallybrass, W. S., 50
Stalin, J. V., 73, 253, 330f.
Stamp, Lord, 167f., 202
Stapledon, Olaf, 236
Stapledon, R. G., 200
Statesmen of Science, 335
Steinhaus, H., 280
Stephenson, Marjory, 256
Stern, C., 65
Stern, Bernard, J., 182
Stern, O., 193, 199
Stott, V., 272
Strassman, 203
Stratton, F. J. M., 248, 291, 307
Suffolk, Earl of, 216f.
Sun, eclipse of, 42
super-conductivity, 164
Svedberg, T., 110
Szilard, L., 65

Tambovtzev, S. P., 58f., 73ff., 187, 315
Tamm, I., 136, 142
Tawney, R. H., 22f.
Taylor, S., 272
Taylor, Sir Geoffrey, 262
Technical education, 28ff., 74
  Marxist view of, 71f.
  universities, 74f.

Teissier, G., 188, 284
Telschow, E., 301
Thomson, Sir George, 284
Thomson, Sir J. J., 81, 89, 117, 225
Thompson, Sir D'Arcy W., 33f.
Timofeeff-Ressovsky, N. W., 188
Tippett, L. H. C., 19
Tiptree Hall, 23
Tizard, Sir Henry T., 33, 119, 218, 268
  Mission, 104
Todd, A. R., (Lord), 200
Toledano, Lombardo, 289, 314
Tolstoy, Ilya, 293
'Tots and Quots', 94, 210, 222
Trevelyan, C. P., 20, 58
Tripp, Brenda M. H., 229, 231
Truman, Harry S., 258, 294
T'U, Chang-Wang, 271
Tyndall, A. M., 153f.

UNESCO, 240, 243–8, 271, 273, 275f., 283, 285, 294f.
Universe, An Outline of the, 50f., 62, 67ff.
Urey, H. C., 153, 193
Usher, A. P., 175, 178
U. S. S. R., as ally, effects of, 229

Valéry, P., 187
Vavilov, N. I., 55, 77, 85
Vavilov, S. I., 55, 144
Veblen, O., 16, 194f.
Veblen, Thorstein, 194
Veksler, V. I., 324
Vogt, W., 294, 298
Voigt, F. A., 66f.

Waddington, C. H., 65, 168, 188, 211, 218
Wadsworth, A. P., 253, 258ff.
Wagner, Mrs., 66
Walker, Miles, 102f.
Walton, E. T. S., 93, 102f., 120
Warsaw, 277f.
Watson, D. M. S., 246
Watson-Watt, Sir Robert, 238, 251, 271, 300, 320

Weber, Max, 182
Wei Hseuh, 320
Weinberg, A. M., 324
Weinstein, 140
Weisskopf, V., 160
Wells, H. G., 21, 46, 168, 201f., 222, 232ff.
  Outline of History, 29
Weston, R., 327
Weyl, H., 194
Weyssenhoff, 281
Whiddington, R., 255, 258
White, Jessie, 26
Whitehead, J. H. C., 13
Wiener, Norbert, 290
Wiersma, E., 164
Wigner, E. P., 298
Wild, 175
Wilkins, J., 333
Wilkinson, Ellen, 246f.
Williams, E. J., 136f.
Wilson, C. T. R., 41, 162
Wilson, E. B., 176
Wilson, J. Harold, 207
Winant, J., 232
Woolf, L., 35, 41
Wooster, W. A., 274
Wootton, Lady, 236
World Congress of Intellectuals for Peace, 300, 308
  Federation of Scientific Workers, 186, 237–6, 285, 298, 308f., 318
Wren, Sir Christopher, 333f.
Wren, T. L., 9
Wrinch, Dorothy M., 168
Wright, Sir Almroth, 156
Wurmser, 188, 267

Young, Arthur, 327

Zavadovsky, B., 77
Zeppelins, 17
Zinn, W. H., 324f.
Zlotovsky, 186
Zuckerman, Sir Solly, 94, 210, 213, 218, 220f., 248
Zworykin, V. K., 207